Pacific Coast Ship China

Jacques Marc

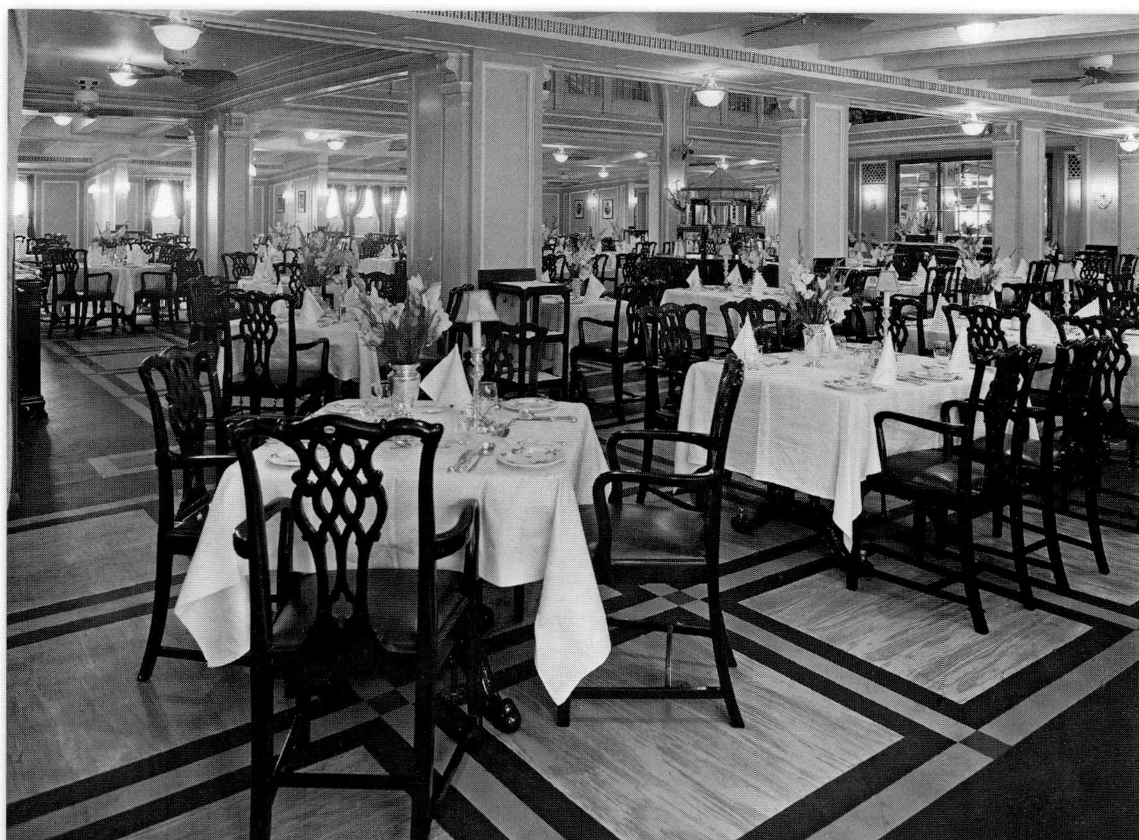

Pacific Coast Ship China

Jacques Marc

Royal BC Museum

Victoria, Canada

Published by the Royal BC Museum, 675 Belleville Street, Victoria, British Columbia, V8W 9W2, Canada.

Printed in Canada.

Edited by Andrea Scott and Gerry Truscott.
Layout by Reber Creative, in Bembo type.
Cover design by Chris Tyrrell.
Cover photographs by Kim Martin (© RBCM); china from the author's collection.
Printed in Canada by Friesens.

Library and Archives Canada Cataloguing in Publication

Marc, Jacques F., 1956–
Pacific Coast Ship China / Jacques Marc.

Includes bibliographical references and index.
ISBN 978-0-7726-5979-8

1. Commercial china – Collectors and collecting – Catalogs. 2. Porcelain – 19th century – Collectors and collecting – Catalogs. 3. Ceramic tableware – Pacific Coast (North America). 4. Naval porcelain – Pacific Coast (North America). 5. Steamboat lines – Collectibles – Pacific Coast (North America) – Catalogs. 6. Coastwise shipping – Pacific Coast (North America). 7. Shipping companies (Marine transportation) – Pacific Coast (North America). 8. Pacific Coast (North America) – Antiquities. I. Royal BC Museum II. Title.

NK8725.5.M37 2009 738.209'034075 C2009-900135-7

To my mother and in memory of my grandmother,
whose life-long fascination for all things literary inspired me.

Contents

Contents, *continued*

Preface

Gone are the days when the grand steamships plied the coastal waters of the Pacific coast of North America, crossed the Pacific Ocean to the Orient and travelled to New Zealand and Australia. The passenger ships of the steamship era – the late 1800s and early 1900s – have been replaced by aircraft and diesel-powered ships and ferries.

Early water transportation to and along the Pacific coast of North America was by sailing ship in which accommodation was marginal, food was poor and service was lacking. By the mid 1800s, steamships had become more reliable, and they rapidly replaced sailing ships as the preferred method for transporting passengers. Steamships were larger, faster and better equipped to provide amenities and comfort. As competition increased for business, steamship lines responded by offering improved service to win over patrons. A sit-down dinner, served on top of marked china on white linen, with silver flatware and hollowware was one of those benefits.

Cost has all but eliminated such china from most modern-day shipping companies. Today expediency has brought self-serve buffets with stainless steel cutlery and paper plates. The bygone era of the steamships is just a memory.

However, dining china from the steamers of the late 1800s and early 1900s survives today and provides a tangible link to the steamships of old. It is found at flea markets and antique shows, underwater at old dock sites and on shipwrecks. But just like the companies that operated those great ships, the identity of the china has also become lost with time. The purpose of this book is to provide collectors, museum technicians, divers, history buffs and others with the information they need to identify and date Pacific coast ship china.

Introduction

I found my first piece of top-marked ship china while scuba diving off the west coast of Vancouver Island in the early 1980s. It was a teacup clearly identified with the words "Canadian Pacific Railway Co." in a red garter and trimmed with red and green pinstripes. Upon trying to learn more about my discovery, I was surprised to find that no one could provide much information about it. Even a letter of inquiry to Canadian Pacific Railway was not much help. I was directed to Richard Luckin's book, *Dining on Rails*, as the most comprehensive reference on the subject.

Additional china finds collected underwater and at antique shows led to more questions. What shipping companies had the china made? When did these companies operate? When might the china have seen service? While looking for answers, I found several good books on railway dining-car china, but few dealing with steamship china. Those that do exist – Daniel Krummes' *Dining on Inland Seas*, Barbara Conroy's *Restaurant China Volume 2* and Karl Spence's *Ocean Liner Collectibles* – have focused on the china from the more luxurious Atlantic liners or on Great Lakes shipping.

The goal of this book is to provide you with an in-depth look at shipping china used along the Pacific coast of North America.

When conducting research for the book I tried to find accurate information on dating each china pattern. I visited numerous maritime museums and private collections to collect manufacturer and date-code information from as many pieces of china as possible. Through this process, I was able to develop time lines for each pattern. It is important to note that each time line relates to the period of manufacture, not use. Predicting the use span of a given pattern of china would be an impossible task. The use period for china depended on many factors, such as quantity produced, frequency of use and nature of use. Special-occasion china, like a VIP service, would last much longer than dinnerware used at every meal. At least one example exists where china was still in service ten years after production. The CPR was still using pieces of its Red Stripe china, last manufactured in 1908, aboard the *Princess Sophia*, which sank in 1918.

Identifying which company a pattern belonged to and when that company operated was equally challenging. Not all china is top marked or back stamped with the shipping company's name. Many just have a company house flag or initials on them. Others are stock patterns with no markings at all. To identify ownership I consulted books on the subject, shipping house flag-identification guides, corporate records, shipping company brochures and other sources.

This book deals exclusively with china used on board vessels or within shore establishments of shipping organizations registered on the Pacific coast of the United States and Canada. On these pages, the Pacific coast region includes the states of Alaska, Washington, Oregon, California, Idaho and Hawaii, the province of British Columbia and the Yukon Territory. The book covers china patterns used in coastwise, intercoastal and trans-Pacific services. Maritime china has a broader scope than just passenger vessels. Many freighter operations had their own china, as did oil companies, government services and yacht clubs. These have been documented where known.

Most of this book is written in the past tense, because few shipping companies today can afford the luxury of top-marked china. In some circumstances china was not used aboard ship but in shore facilities, such as naval messes, marine hospitals and shipping-firm cafeterias. These patterns are also documented. Shipping companies used china top marked with the company name, pretty floral china that was back stamped and stock patterns that were not marked at all. These have all been documented, including any verified stock patterns. The earliest recorded china pattern comes from the wreck of the USS *Saranac* from about 1875. The most recent pattern is from the BC Ferries *Coastal Renaissance*, delivered in 2007.

This book is for collectors who specialize in nautical china and for museum technicians and professionals who have to identify and date china as part of their job. It is also for researchers and anyone else interested in learning more about the china patterns used on the coastwise and trans-Pacific ships headquartered on the Pacific coast of Canada and the USA. I have not included a number of well-known shipping companies in this book (for example, Moore-McCormack, American-Hawaiian, Luckenbach, Sealand, Prudential and Grace Line). While these companies provided intercoastal service from the east coast to the west coast, they were not based here.

This is an identification and dating guide, not a value guide. That being said, there is often a strong correlation between rarity and value: generally the rarer a piece, the more valuable it is. As with value there are many factors that influence the rarity of particular patterns. This aspect of china is discussed in detail in Chapter 8.

Finally, this book summarizes 10 years of research and discoveries. It documents the history of 58 shipping companies, 15 government services and over 280 patterns of china.

1. The Rise and Fall of Steamship Service on the Pacific Coast

Captain James Cook was one of the first explorers to fully explore and map the Pacific Ocean. As in other parts of the world, he was followed first by traders and then colonizers. Early immigrants and settlers came to the west coast of North America aboard sailing ships, often crammed in dark, damp holds and subsisting on poor-quality food rations.

The development of the Pacific coast was shaped by conditions far different from those that prevailed in the Atlantic. The shores of the Pacific were sparsely settled by people in the early 1800s, the distances between communities were great and the Pacific was far from the centres of world power and trade.

The Hudson Bay Company's paddle steamer SS *Beaver*, built in 1835, was the first steam-powered vessel north of San Francisco. A multi-purpose vessel, it transported people and supplies between ports in British Columbia and Puget Sound for many decades. The first steamers dedicated to passenger service did not appear until many years later.

The discovery of gold in California in 1848 resulted in a rush of prospectors and fortune seekers from the eastern seaboard to the Pacific coast. This resulted in numerous small shipping lines being created to transport people and supplies from Panama to San Francisco, the gateway to the gold fields. These early steamers were generally small wooden vessels providing only the bare essentials of food, water and accommodation.

In 1849, Pacific Mail began monthly mail service between New York, San Francisco and Puget Sound. Passengers would travel by ship from New York to Colon, Panama, make their way across the isthmus on horseback or by wagon, then catch a Pacific Mail steamer from Panama City to Los Angeles, San Francisco or Puget Sound. The completion of the railroad between Colon and Panama City in 1855 significantly reduced travel time across the isthmus.

Ten years later, in 1858, a new gold rush – this time in British Columbia's lower Fraser River and Cariboo – led to another proliferation of steamship companies, as miners sought to get from San Francisco to Victoria and then to the gold fields. Just as San Francisco grew and prospered during the California gold rush, so did Victoria in 1858.

Initially, marine transportation companies concentrated on providing reliable coastwise service between the emerging communities along the Pacific coast. The first regular trans-Pacific steamer service did not occur until 1867, when the Pacific Mail Company inaugurated service to the Orient with the wooden side-wheel steamship, *Colorado.*

The completion of the first United States transcontinental railway in 1869 opened up the possibility of new travel routes from San Francisco to Australia and the Orient. Soon, many railroad companies became involved in the shipping business to connect their railroad ventures to the Orient. One of these, the Occidental and Oriental Line, was established by the Central Pacific and Union Pacific railroads in 1874, to ensure that passengers and freight from the Orient would land at west-coast ports.

The Canadian Pacific Transcontinental Railway was completed in 1886. Three years later, the company signed a mail contract with the British and Canadian governments to provide a through service between London and Yokohama via North America. In 1891, the *Empress of India*, the first of three new palatial Canadian Pacific liners, arrived in Vancouver on her maiden voyage from Hong Kong.

The Klondike gold rush of 1898 spurred the development of new communities in the north. This led to the proliferation of small steamship companies – such as North American Transportation and Trading Company, Alaska Commercial Company and Canadian Pacific Navigation Company –

delivering people and supplies from Vancouver and Seattle to ports in Alaska.

Numerous events from 1900 onward directly influenced the survival of many Pacific coast shipping companies. The outbreak of World War I in 1914 saw many ships commandeered by governments for the war effort. This resulted in the dissolution of Great Northern Pacific and Pacific Navigation, which only had two ships each. In February 1915, the US Congress passed the Seaman's Act, which saw the elimination of non-English speaking Chinese crews from all US ships. This made it difficult for US shipping lines to compete with Canadian and Japanese lines. As a result, many US lines, such as Pacific Mail, were forced to withdraw from trans-Pacific routes.

But the early 1920s saw new opportunities and optimism in the shipping business. There was a proliferation of new steamship companies that operated war surplus or Shipping Board vessels. Admiral Oriental Line, Los Angeles Steamship and American Mail were but a few that got their start in the 1920s. However, just as quick as it all started, the stock market crash of 1929 led to the decline of freight and passenger volumes. This placed companies like the Pacific Steamship Company (Admiral Line) and Los Angeles Steamship into serious financial difficulty.

The US seamen's strikes of 1934 (lasting 84 days) and 1936 (98 days) forced several more steamship companies into bankruptcy. The Merchant Marine Act of 1936 required passenger ships older than 20 years to significantly upgrade safety for passengers or restrict the number of passengers to 12. This act also cancelled all ocean-mail contracts and subsidies effective June 30, 1937. Without their mail subsidies many lines foundered. The financially overextended Dollar Line was one of them.

A summary compiled by the San Francisco Shipping Register in March 1938 illustrates how five years of depression, increased taxes and operating costs and continuing labour strife took their toll on the Pacific coast maritime industry. Between 1933 and 1938 the Pacific coast had lost 14 major steamship services – 7 in coastwise routes, 4 in intercoastal and 3 in offshore trade – with 89 vessels being withdrawn from service.

During World War II, governments again commandeered ships for the war effort. Many ships went down in the war. Canadian Pacific lost three of its four trans-Pacific *Empress* liners, and never resurrected its trans-Pacific service with the return of peace. The Inter-Island Steamship Company of Hawaii was also a casualty of WWII.

Finally, increasing costs and competition from the airline industry during the 1950s forced more shipping companies into receivership. Alaska Steamship dropped its passenger service to Alaskan ports in 1954, the final passenger trip being made September 24, 1954, by the SS *Denali*. Union Steamships of BC was not far behind, declaring bankruptcy in 1958. The last of the CPR coastal liners, SS *Princess Patricia,* was retired on October 12, 1981. The Pacific Far East Lines, one of the last US companies to offer trans-Pacific service, went into receivership in 1978.

The steamships of the 20th century have been replaced on short runs by modern car ferries of the Washington State Ferry system, Alaska Ferries and BC Ferries, while the ferry fleet of San Francisco Bay has been replaced by bridges.

2. Dining on Board

It took days and often weeks for ships to travel between coastal ports and across the Pacific, so it was necessary to feed people. Early steamship companies tried hard to win over passengers by providing cheaper rates, better service and good food on their routes. The dining room meant top-marked china, silver serving pieces, crisp table linens, etched crystal, full place settings of flatware and fresh flowers on the table. Dining aboard ship became a luxury that the average citizen might never experience on land.

During the late 19th century and early 20th century, most ships in coastwise and trans-Pacific service offered several classes of service. These were first class (saloon), second class (cabin/tourist) and third class, often referred to as steerage. While the class system primarily related to accommodation quality, it often extended to dining arrangements as well. On the trans-Pacific ships, passengers often ate in different dining rooms depending on the class in which they were travelling. On the smaller coastal ships, there was often only one dining room and class distinction was less prevalent.

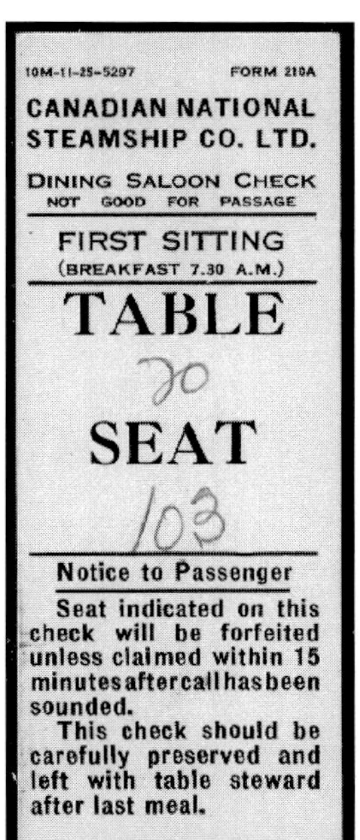
10M-11-25-5297 FORM 210A
CANADIAN NATIONAL STEAMSHIP CO. LTD.
DINING SALOON CHECK
NOT GOOD FOR PASSAGE
FIRST SITTING
(BREAKFAST 7.30 A.M.)
TABLE
20
SEAT
103
Notice to Passenger
Seat indicated on this check will be forfeited unless claimed within 15 minutes after call has been sounded.
This check should be carefully preserved and left with table steward after last meal.

Dining ticket.

Most coastal ship dining rooms could only accommodate about a third to half of the passengers at a sitting, so would have at least two sittings per meal. It is not uncommon to find memorabilia like shipping company meal tickets identifying the first, second or third sitting. On smaller ships, dining room chimes announced when meals were about to be served. Consisting of four or five metal strips of different lengths suspended over a sounding box, the chimes made pleasing tones when struck with a mallet.

The Northland Transportation Company ship SS *North Coast*, operating from Seattle to Alaska in 1938, is one example of a coastwise ship that had separate dining rooms based on class. It had a 58-seat first-class dining room forward and a 28-seat second-class dining saloon aft. The food for both dining rooms was prepared in one galley.

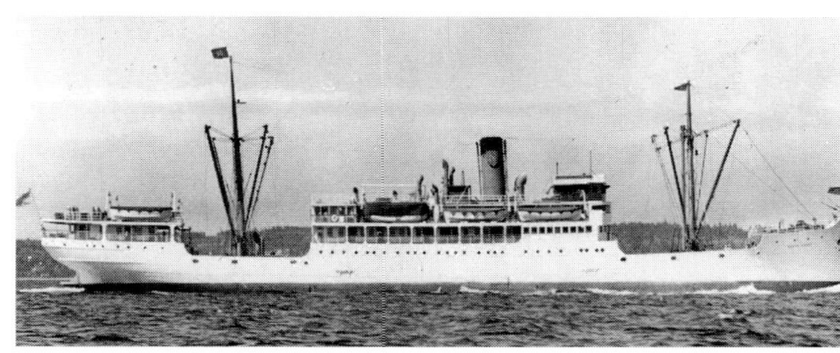

SS *North Coast*.

The quality of china and the decorations on it varied significantly between the different classes of dining rooms. On the large Canadian Pacific trans-Pacific liners, fine bone china made by famous makers like Limoges, Mintons and Copeland-Spode was used in first class. This china was often decorated with beautiful floral transfer prints and back-stamped with the line's name. Bone china, while extremely hard, chipped or cracked very easily. For this reason much ship china was normally of restaurant grade, generally referred to as ironstone or vitrified china. This grade of china was heavy, durable and made to withstand hard use. Much of this heavier china lasted for five or six years before having to be replaced.

On other lines, first-class and second-class china was decorated with pretty transfer prints, like the Matson Bombay pattern, or top marked with the line's name or house flag. Third-class and crew china was typically plain with few if any decorations or identifying features.

On some lines, the ship's captain had a table reserved for VIPs. The china used at this table was either a customized version of a standard pattern or entirely different. Canadian Pacific often added a gold band to its Tremblant pattern to distinguish first class or VIP service from the rest.

The Dining Experience

The best way to imagine the dining experience on the old steamships is to read some company promotional materials and passenger accounts.

> *If you would ask any of the tens of thousands of passengers who annually enjoy Admiral Line Service between Pacific coast cities, what more than anything else, pleased them, the answer most likely would be "the meals". In fact, many find it difficult to realize how such meals and services can be rendered, especially since the fares have been so drastically reduced. Above are three menus which are typical of the famously-good Admiral Line meals, not three selected carefully for advertising purposes, but menus that are actually typical of every-day meals. Note the fullness of the courses, the splendid variety of the items, and particularly the absence of prices! You eat as much or as little as you desire and your meals are included in the nominal fares. In addition to the regular meals, which are served convenient hours, bouillon and crackers are served on deck each morning, and tea and cookies each afternoon. In the evening those accustomed to a bed-time snack, may have sandwiches and coffee served in their staterooms.*
> – Pacific Steamship brochure, 1920s.

> *Dining on a Canadian Pacific liner is a rite. Few hotels present a menu so vast, so varied, so elaborately served. Each liner carries some six hundred tons of the world's different items of food; three hundred tons of the world's choicest delicacies. All yours: not a price in sight. And should you pause bewildered by such choice your solicitous steward inquires if perhaps you would like something else especially prepared!*
>
> *Each noon a truly wonderful cold table appears, attended by high-hatted white gloved chefs with gleaming knives. There will be famous Melton Mowbray Pie, the traditional Boar's Head, glazed and frilled with medieval magnificence, tremendous pink jelly fish, handsomely decorated cold chicken, hams unbelievably savoury to the nostrils and salads dazzling to eye and appetite.*
> – Canadian Pacific "Facts and Figures" booklet, 1937.

> *I was always given a little briefing by my father or mother before we went to the dining room about using the various knives, forks and spoons, as the dining room provided full silver service. Dining aboard the* Kathleen *was truly an elegant experience. My father was a "meat and potatoes" type of person, but he would usually order poached salmon, new potatoes with parsley and green peas for his entrée, if it was available. We always started with soup. My mother always enjoyed either crab or oyster cocktail after soup. Dessert for me was orange sherbet and my father usually opted for cheese. I think the CPR ships accounted for my first introduction to finger bowls.*
> – Ken Knox, a passenger aboard the SS *Princess Kathleen* (quoted in *Those Beautiful Coastal Liners* by Robert D. Turner, 2001).

> *One of the highlights of the nine-day Alaska cruise is the traditional Captain's Dinner. This gala, informal affair is usually held the evening before arrival at Vancouver. After dinner, passengers get together for a frolicking masquerade dance. The Ship takes on a festive atmosphere with a variety of fun and entertainment.*
> – Canadian National Steamships brochure, 1959, describing the SS *Prince George*.

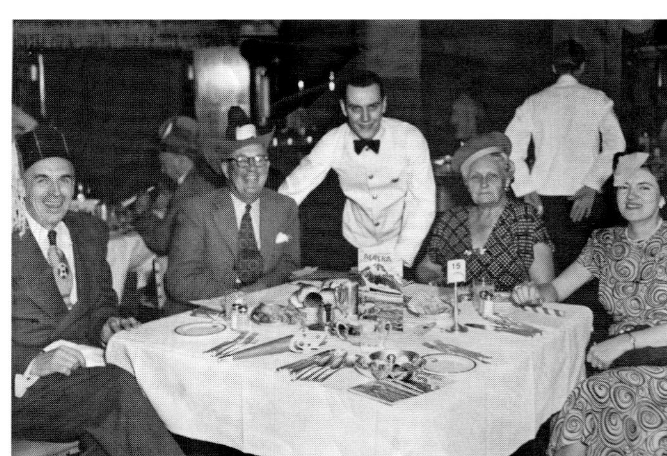

Captain's Dinner aboard the SS *Prince George II*. J. Marc collection.

Meals were announced with a xylophone. The steward would go around sounding it and call, "Rise and shine for the glory of the Union Line, first sitting please." The dining room on the Cardena *was roomy and clean with good light. The walls were decorated with nice watercolours and sketches done by various artists and passengers. The stewards would all be standing at the ready with a white towel over their arm. You didn't want to be late for your sitting or you would have to wait for the second or third sitting, which would not be as good. The food was good and wholesome and all you could eat. The baking was homemade and the soup was endless. There were often several choices on the menu. It was important to know your steward. They would tell you what was good and would steer you away from the poor choices. The* Cardena *was known as the friendly boat.*
– Bernard Bing, 2007, interviewed in 2007 about his experiences in the 1940s and 1950s.

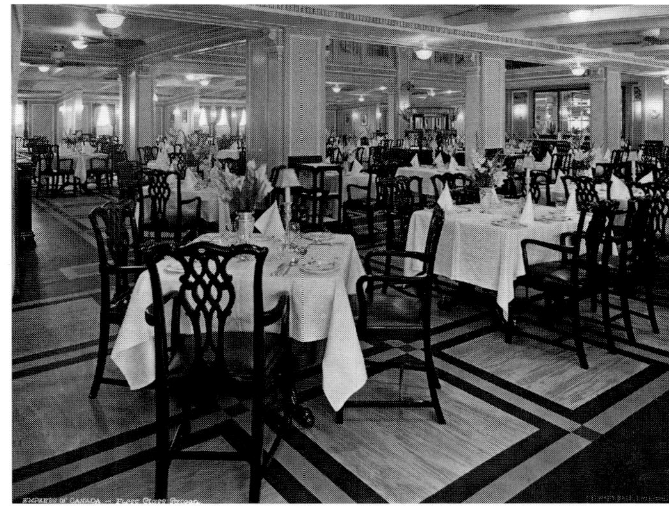

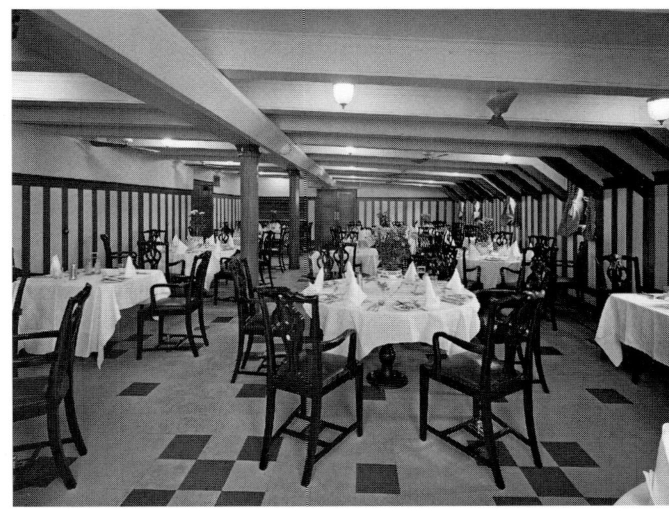

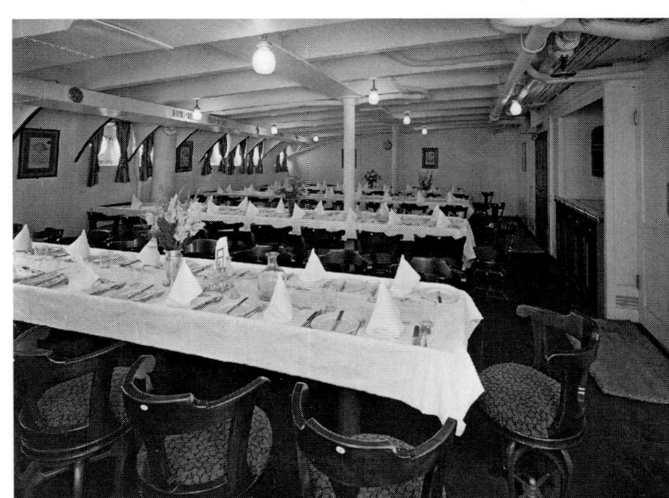

Empress of Canada first-class, second-class and third-class dining rooms. Maritime Museum of BC.

The Dining Saloon

The dining room varied in layout, decorations and capacity for each ship. Here are some examples.

Trans-Pacific Ships

The Canadian Pacific liner *Empress of Canada,* 653 feet (199 m) long, had three dining areas, for cabin- (first), tourist- (second-) and third-class passengers. The cabin-class dining room was on the main deck forward and decorated with paintings and lush plants as if you were in a high-end restaurant. The tables were set with the finest china and silverware. The tourist-class dining room was on the main deck amidships. The third-class dining room, located below the waterline, was utilitarian, but clean and bright. Tourist cuisine came out of the same kitchen as cabin class, but it was served less elaborately. Third-class had its own kitchen, which produced a surprising variety of food, wonderfully cooked and seasoned, and all you could eat.

The American Mail liner SS *President Madison,* 535 feet (163 m) long, could accommodate 199 first-class and 542 steerage passengers. The first-class dining room accommodated 182 passengers sitting at tables for 5 or 6 people. The steerage area was an area of large compartments around the ship's cargo hatches at the rear of the ship. A steerage class galley was on the main deck at the stern. Meals were prepared there and served on the square of the hatch in one of the compartments below the galley.

Coastwise Service

The Pacific Steamship Company vessel *HF Alexander* was 525 feet (160 m) long with accommodation for 585 passengers. The Colonial Dining Room on the *Alexander* was large, airy and inviting with seating for 270 persons. Table arrangement permitted groups of 2, 3, 4 or 6 persons.

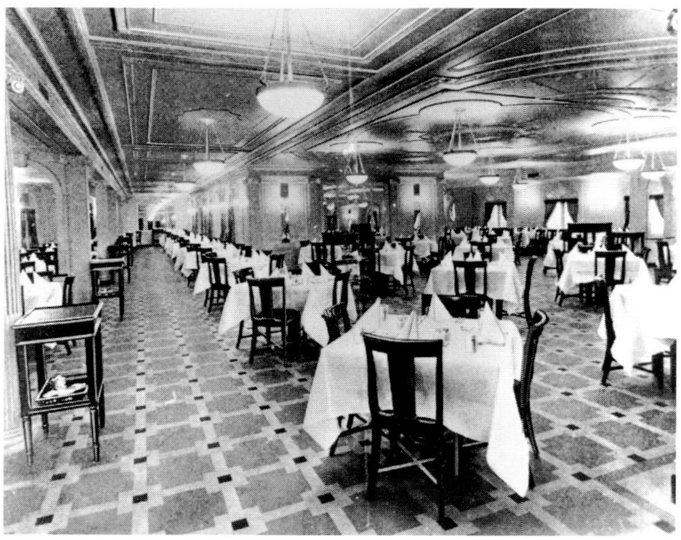

HF Alexander Colonial Dining Room. Puget Sound Maritime Historical Society neg. 11236.

> *The Canadian Pacific's popular coastal liner* Princess Kathleen *was 368 feet (112 m) long, could carry 1500 day passengers and had 136 staterooms accommodating up to 290. Her dining room could seat 166 guests depending on its configuration. The dining room was located at the rear of the ship at the waterline. It extended the full width of the ship and was decorated in a Louis XVI style, and was panelled in tones of ivory, with doors and other fittings in mahogany. Paired oval windows in the ships side gave light and were curtained in rich red silk. The flooring was green and white tile. Tables were arranged for parties of 6, 8 and 14.* – Robert D. Turner (*Those Beautiful Coastal Liners*, 2001).

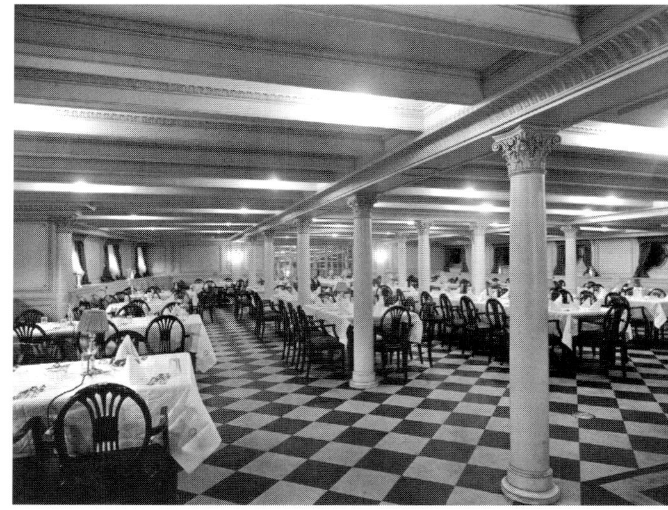

Princess Kathleen dining room. BC Archives A-06273.

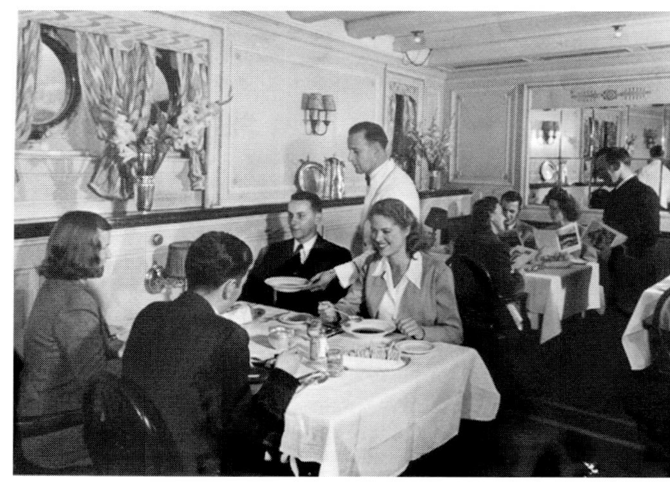

Dining aboard the *Princess Norah*. The Tremblant pattern is visible on the tables. J. Marc collection.

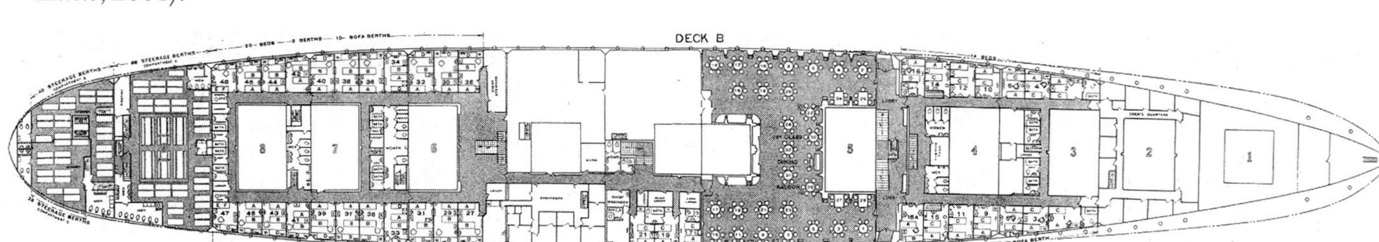

Deck plan of *President Madison*, showing the dining rooms.

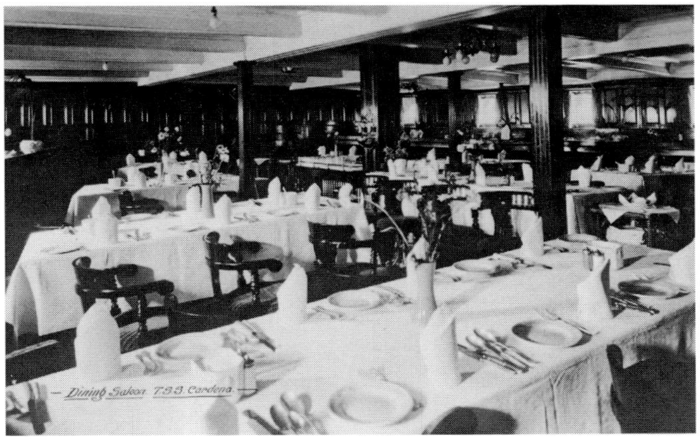

SS *Cardena* dining room. Lorrie Belveal collection.

The Union Steamship *Cardena* was a 235-foot (72-metre) coaster that served canneries and logging camps along British Columbia's coast. A combination freighter-passenger vessel, she was licensed to carry 250 day passengers. Her accommodation consisted of 132 cabin berths and 60 deck settees. Her club-style oak-panelled dining room was located amidships on the main deck, and extended across the entire deck. It had a red-and-cream-coloured tile floor. It could seat 68 people at tables of 6 or 8, serving basic, good-quality meals. When the boat was full, it took three sittings to feed everyone on board.

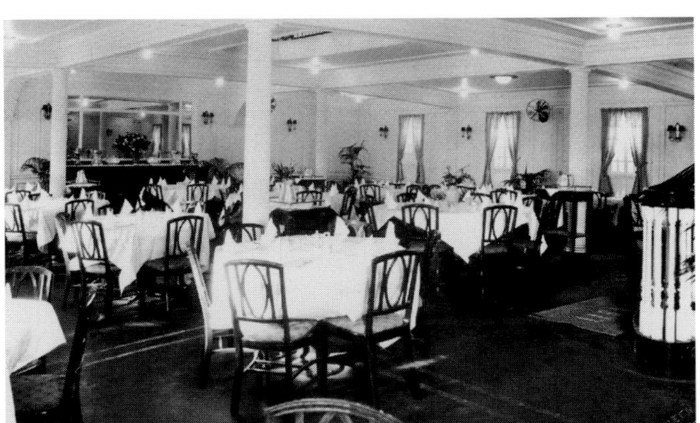

SS *Haleakala* dining room. Puget Sound Maritime Historical Society collection, neg. 3317-3.

The inter-island SS *Haleakala* was 360 feet (110 m) long, with accommodation for 334 first-class and 100 second-class passengers. Linen for staterooms and the dining room was imported from Ireland, the crockery from Germany. The dining room, located forward amidships, extended the full width of the ship. It could seat 134 people, 6 to each small table.

Provisioning a Voyage

The amount of food consumed on board an ocean-going ship is impressive. On the *Empress of Britain*, about 900 passengers and a crew of 428 consumed the following quantity of food on a round-the-world cruise in 1934: 60,000 pounds (27,000 kg) of beef; 45,000 pounds (20,000 kg) of poultry; 41,000 pounds (18,500 kg) of fish; 40,000 pounds (18,000 kg) of lamb, veal and pork; 25,000 pounds (11,000 kg) of ham and bacon; 2000 game birds; 1000 pounds (455 kg) of caviar; 100 tons of potatoes; 50 tons of flour; 17.5 tons of sugar; 12 tons of butter; 7500 pounds (3400 kg) of coffee; 5000 pounds (2300 kg) of tea; 5500 pounds (2500 kg) of cheese; 150,000 eggs; 2200 cases of fruit and vegetables. The cost to provision the ship for this trip totalled approximately $135,000 or slightly more than $1000 per day.

The Admiral Oriental Line ship *President Jackson* carried 300 first-class and tourist-class passengers. In 1921, a return voyage to Japan took 55 days and called for the serving of 100,000 meals. The provisions and stores for the trip cost $20,000.

Menus

The selection of food found on the liners of the 19th and 20th centuries was nowhere near as extensive as today's modern cruise ships. This is evident by reading the early menus.

However, the food was wholesome and plentiful. As the size of ships increased and the ability to keep food fresh evolved, the selection of items on the menu increased. To a certain extent the menus reflect this evolution.

Menus at the turn of the century were typed onto card stock pre-printed with a shipping line header or photograph. Pre-1900 menus were often written out in longhand on this card stock. By the early 20th century, the menu for each meal was added by typewriter and copied by mimeograph machine.

As ships got bigger and more luxurious the menus became fancier, featuring a cover photograph or drawing. Inside, the contents also reflected the increasing diversity of food.

In time, menus became promotional items, not only documenting the food served but advertising the destinations visited by the shipping company. It was not uncommon for shipping lines serving Alaska to feature mountain fjords, ice fields or totem poles on the cover page of their menus. The lines serving the Orient used images of the people and cities visited. The lines serving Hawaii and the South Pacific also highlighted special features of the countries they visited. The Matson Macouillard watercolour series is an excellent example of this: 10 menu covers each featuring a different scene.

Because food was such a promotional aspect of any ocean trip, several companies began printing their menus on folding postcard stock so that passengers could mail them to friends and relatives, perhaps to make them envious of the offerings. This was particularly true of Pacific Steamships (Admiral Line), Inter-Island and Los Angeles Steamships. The Alaska Line, BC Coast Steamships and Northland Navigation Company also intermittently produced menu postcards.

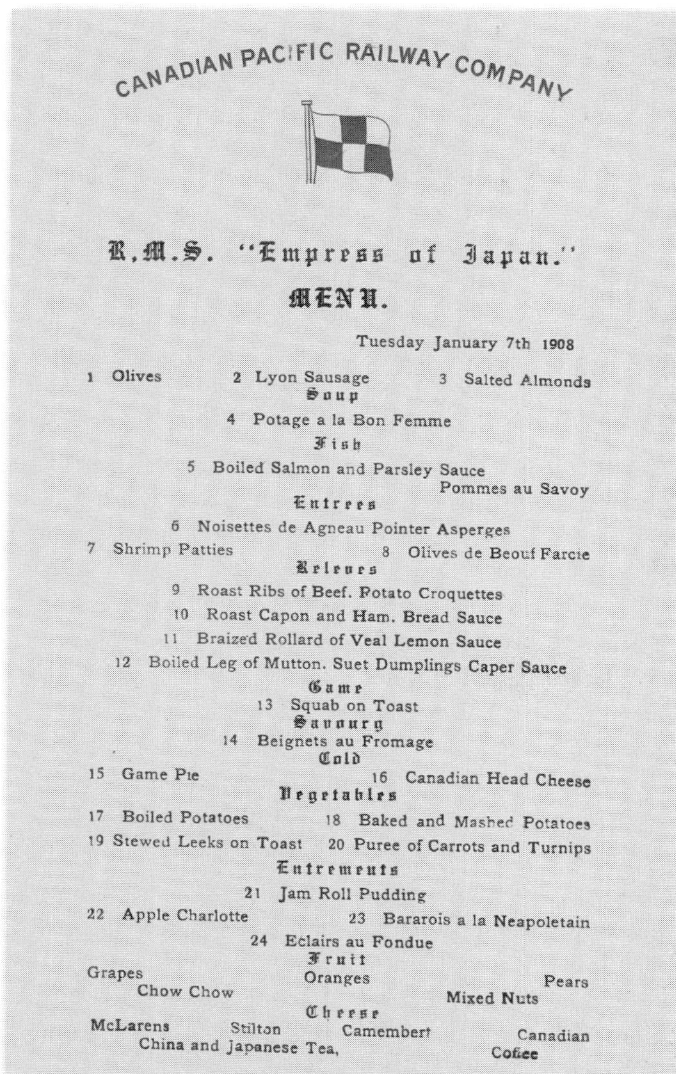
CANADIAN PACIFIC RAILWAY COMPANY

R.M.S. "Empress of Japan."

MENU.

Tuesday January 7th 1908

1 Olives 2 Lyon Sausage 3 Salted Almonds

Soup

4 Potage a la Bon Femme

Fish

5 Boiled Salmon and Parsley Sauce Pommes au Savoy

Entrees

6 Noisettes de Agneau Pointer Asperges

7 Shrimp Patties 8 Olives de Beouf Farcie

Releves

9 Roast Ribs of Beef. Potato Croquettes

10 Roast Capon and Ham. Bread Sauce

11 Braized Rollard of Veal Lemon Sauce

12 Boiled Leg of Mutton. Suet Dumplings Caper Sauce

Game

13 Squab on Toast

Savoury

14 Beignets au Fromage

Cold

15 Game Pie 16 Canadian Head Cheese

Vegetables

17 Boiled Potatoes 18 Baked and Mashed Potatoes

19 Stewed Leeks on Toast 20 Puree of Carrots and Turnips

Entremets

21 Jam Roll Pudding

22 Apple Charlotte 23 Bararois a la Neapoletain

24 Eclairs au Fondue

Fruit

Grapes Chow Chow Oranges Mixed Nuts Pears

Cheese

McLarens Stilton Camembert Canadian

China and Japanese Tea, Coffee

1908 *Empress of Japan* menu. J. Marc collection.

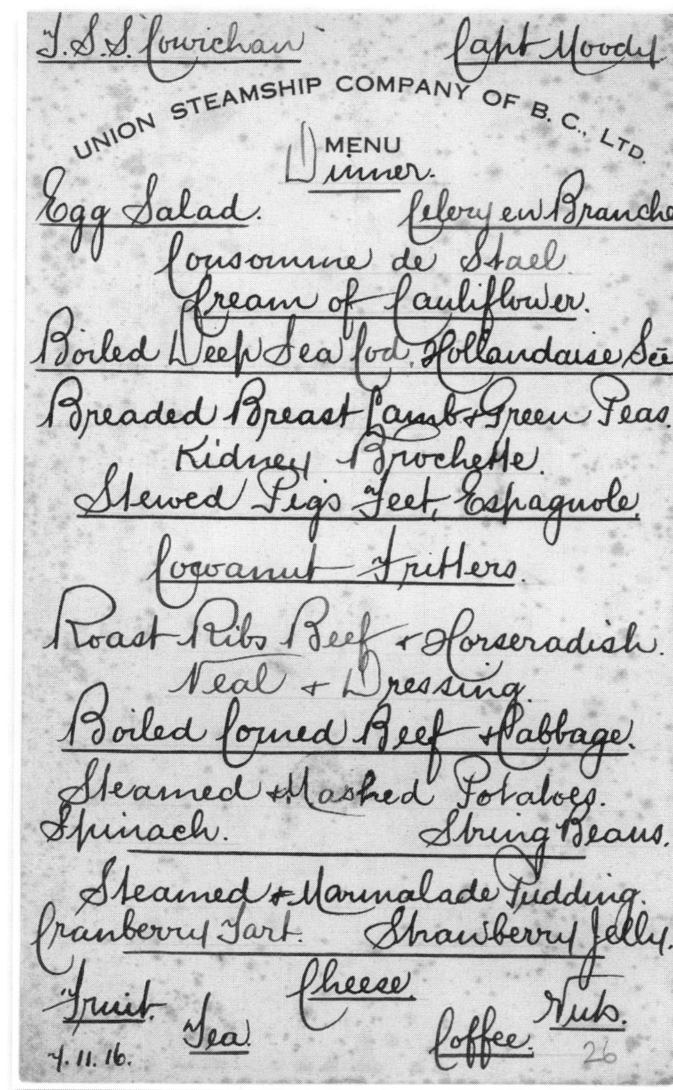
S.S. Cowichan Capt Moody

UNION STEAMSHIP COMPANY OF B.C., LTD.

MENU

Dinner.

Egg Salad. Celery en Branche

Consomme de Stael

Cream of Cauliflower.

Boiled Deep Sea Cod, Hollandaise Sce.

Breaded Breast Lamb + Green Peas

Kidney Brochette.

Stewed Pigs Feet, Espagnole

Cocoanut Fritters.

Roast Ribs Beef + Horseradish.

Veal + Dressing.

Boiled Corned Beef + Cabbage.

Steamed + Mashed Potatoes.

Spinach. String Beans.

Steamed + Marmalade Pudding.

Cranberry Tart. Strawberry Jelly.

Fruit. Cheese Nuts.

Tea Coffee.

4.11.16. 26

1916 SS *Cowichan* menu. Lorrie Belveal collection.

3. Ship China

Ship china is the crockery used aboard ships. It includes dinnerware, oven-to-table serving pieces and sanitary ware. Ship china was made from vitrified china, true porcelain, bone china or early ironstone. The most popular ship china was vitrified, a medium-to-heavy-gauge china made of clay, feldspar and quartz and fired at a very high temperature. During the firing process the material would fuse together, creating non-porous china, which was very resistant to breakage from normal use. This heavy-duty china was then decorated with colourful floral transfer prints or with the company's logo or house flag. A guide to the shapes, uses, sizes and characteristics follows.

Most ship china made for Pacific coast Canadian and US lines was made in England or the United States. Before 1900, almost all of it came from England; the two most prolific manufacturers were Grindley Hotelware and John Maddock and Sons. Early US manufacturers included Greenwood China, Union Porcelain and Onondaga Pottery. Later, some china was also produced in France by Limoges or in Germany by Bauscher. Appendix 2 contains a guide to manufacturers.

Stock vs Custom Patterns

Stock

Stock patterns are off-the-shelf china patterns, purchased without any identifying marks or logos to indicate steamship usage. Many of the china patterns used aboard ship were stock patterns. These patterns were designed and produced "on spec" by the manufacturers and made available to any customer who wanted to buy them. This book includes only those stock patterns known through china company records, steamship company files and photographic evidence, or recovered from wrecks. An example is the BC Ferries Malaspina pattern: its name and usage were determined from BC Ferries requisition information.

Semi Custom

In some instances, shipping companies wanted to have their name appear on their china but did not want to go to the time and expense of having a proprietary pattern designed. China manufacturers often produced stock china on speculation with a gap left in decorative bands for this purpose. Companies could add their name, image or company logo to the china for a small additional charge. In other cases, a shipping company would take a stock pattern and have it customized by simply adding a back stamp to the chinaware. The Canadian National Steamship Jasper pattern is a good example of a semi-custom pattern: a stock Royal Doulton pattern (known as Bristol, CN-3079) customized with a back stamp identifying ownership and usage. The Matson Bombay pattern is another good example.

Custom

Custom or formal china was designed especially for a particular shipping line. In some circumstances, these patterns were proprietary or registered. The Canadian Pacific Empress pattern is a registered pattern. Canadian Pacific registered it with the number 774878 in 1932 so that it could switch manufacturers if necessary.

The Los Angeles Steamship Company Poppy pattern was also registered with the US patent office. Custom china flourished from the early 1900s through the 1950s when shipping lines wanted to impress the public and influence their choice of ship or company serving a particular route. Custom china all but disappeared by the 1960s, when most shipping lines were fighting just to survive.

Shapes and Uses of Ship China

Every manufacturer produced a variety of products, and the shapes and sizes of the pieces often dictated their use. These diagrams will help identify what the different pieces and shapes of china were used for. Commercial china manufacturers have no standard terminology, but most use the item names listed here. Except for airline or oriental pieces, a bowl is seldom called a bowl, but referred to by the food it is most often used to serve.

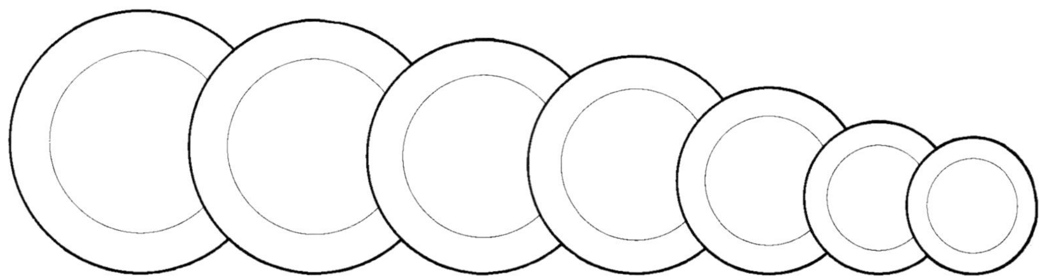

(1)

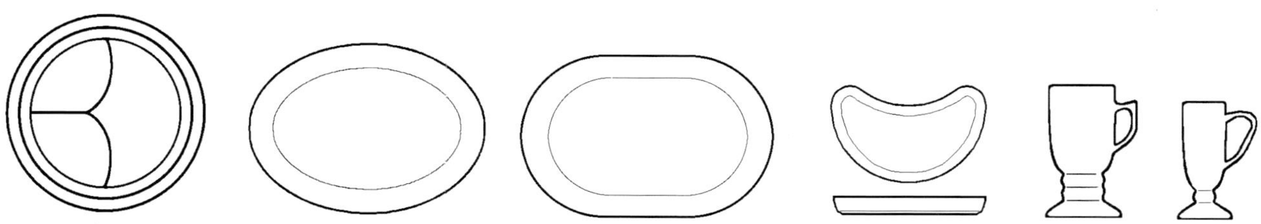

(2) (3) (4) (5) (6) (7)

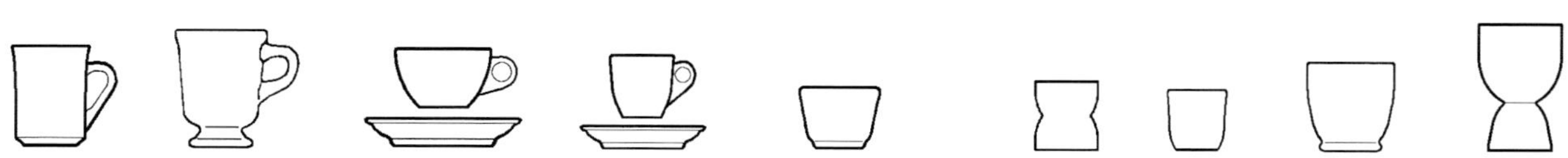

(8) (9) (10) (11) (12) (13) (14) (15) (16)

1. Plates. Manufacturers identify plates by size rather than type, except for chop, service, compartment and bone plates. A chop plate is a serving plate that is 28 cm (11 in.) or more in diameter. A service plate is 25–29 cm (10–11.5 inches) in diameter and almost always custom decorated; it's on the table when the patron arrives and removed just before the meal is served, but occasionally the appetizer, soup or salad plate is placed on top of the service plate, which is then removed before the entree is served.
2. Compartment (grille or grill) plate, popular during the Great Depression of the 1930s.
3. Platter – any oval or oblong plate, regardless of size.
4. Platter.
5. Crescent salad (or bone) plate.
6. Irish coffee.
7. Café au lait.
8. Mug – placed directly on the table, whereas a cup is placed on a saucer.
9. Mug, footed.
10. Cup and saucer.
11. A.D. (after dinner) cup and saucer (demi tasse).
12. Oriental (Chinese tea) cup.
13. Egg stand.
14. Egg stand.
15. Egg cup, footed.
16. Double-egg cup (combination egg cup and stand).

Illustrations on pages 14–16 from *Restaurant China*, vol. 1, by Barbara J. Conroy.

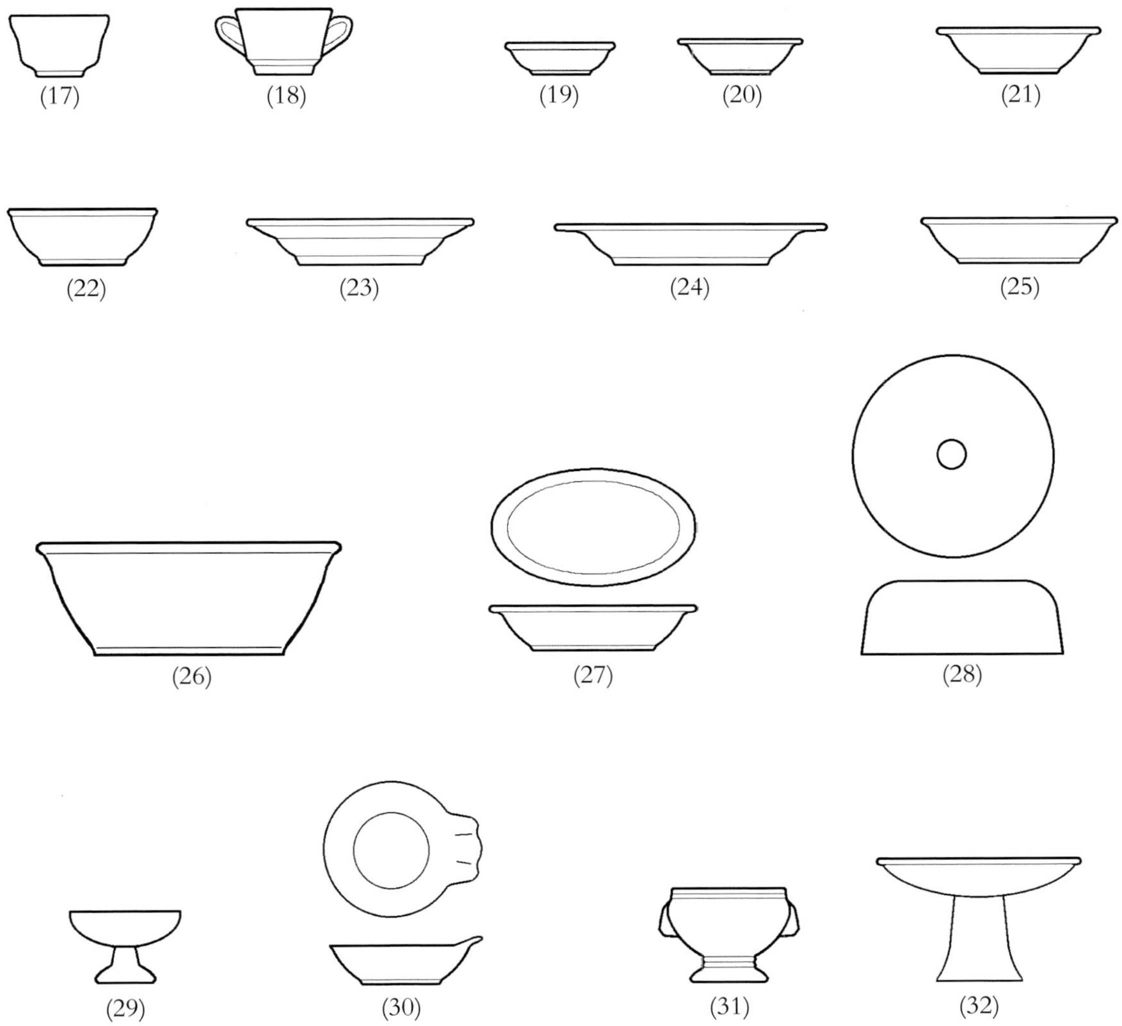

17. Bouillon.
18. Bouillon, handled.
19. Fruit, coupe (rimless), rolled-edge.
20. Fruit, narrow rim (or nappy).
21. Grapefruit, narrow rim.
22. Oatmeal, rolled-edge.
23. Soup, wide rim (or soup plate).
24. Soup, wide rim (or soup plate).
25. Soup, coupe (rimless), rolled-edge.
26. Salad – also in smaller sizes, but proportionally deeper than a fruit, grapefruit or soup bowls.
27. Baker, narrow rim – all oval bowls are bakers regardless of size.
28. Cake (or toast) cover.
29. Sherbet (or footed egg cup).
30. Ice-cream shell.
31. Compote (or comport), lion head.
32. Compote (or comport), rolled-edge.

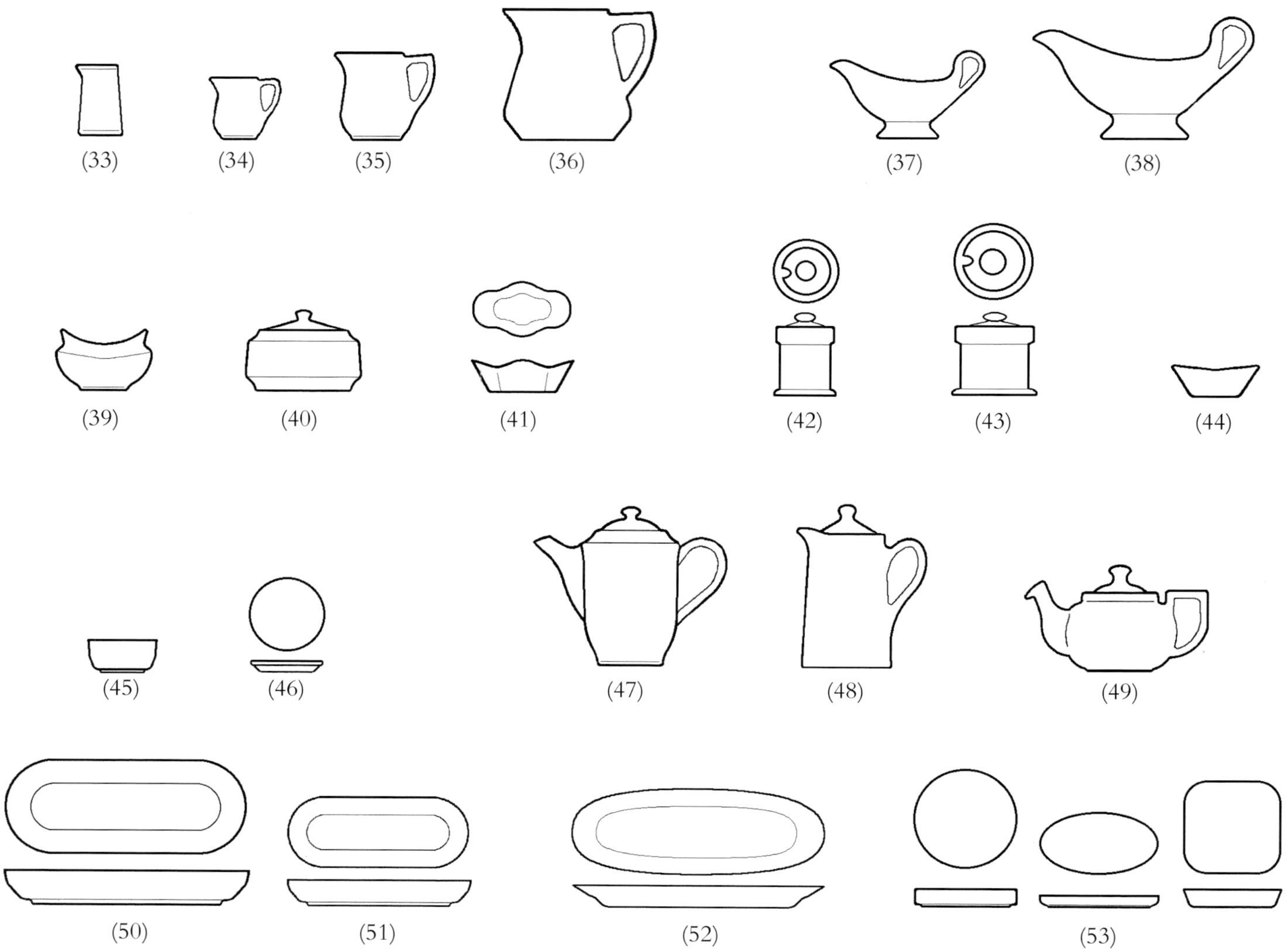

33. Creamer (or cream).
34. Cream pitcher, individual (or jug).
35. Cream pitcher (or jug).
36. Pitcher (or jug) – up to 3.2 litres (109 oz.) capacity, sometimes perhaps more.
37. Sauceboat, individual.
38. Sauceboat.
39. Sugar, open (or sugar, room service).
40. Covered sugar – sugars and covers are individual items; many sugars were supplied without covers.
41. Sugar, open (or sugar, room service).
42. Mustard.
43. Mustard (or jam).
44. Jelly.
45. Butter pot, sauce or nut.
46. Butter (or butter chip, butter pad or butter pat).
47. Coffee pot – produced in numerous shapes, but taller than a tea pot.
48. Cocoa pot.
49. Tea pot – produced in numerous shapes, but shorter than a coffee pot (Boston teapot illustrated).
50. Celery.
51. Pickle (or small celery); a pickle is a shallow oval, oblong, elongated shell or rounded diamond-shaped piece, approximately 20 cm (8 in.) long.
52. Corn tray.
53. Trays – small custom-decorated trays used to hold match books, mints, chocolates or other small amenities.

4. Histories of Shipping Companies and Government Services

On the Pacific coast, 58 steamship companies and government services used marked china. The federal government services of Canada and the United States operated on both coasts. These short histories of each will help the collector determine when and where these companies and services operated. All are listed alphabetically by the registered name, followed by the recognized line names.

Company Name	Abbreviation
Admiral Oriental Line	AOL
Alaska Commercial Co.	ACC
Alaska Cruise Lines	ACL
Alaska Steamship Co.	ASC
American Mail Line	AML
American President Lines	APL
American Yukon Navigation Co.	AYN
Black Ball Lines – Canada	BBL
Bonners Ferry & Kaslo Transportation Co.	BFK
Boscowitz Steamship Co.	BSC
British Columbia Ferries	BCF
British Columbia Salvage Co.	BCS
British Columbia Steamship Co.	BSS
California Navigation & Improvement Co.	CNI
California Transportation Co.	CTC
Canadian National Steamship Co.	CNS
Canadian Pacific Navigation Co.	CPN
Canadian Pacific Railway Co.	CPR
China Mail Steamship Co.	CMS
Coeur d'Alene & St Joe Transportation Co.	CDA
Dalles, Portland & Astoria Navigation Co.	DPA
Dollar Steamship Co.	DSC
Grand Trunk Pacific Coast Steamship Co.	GTP
Great Northern Pacific Steamship Co.	GNP
Great Northern Steamship Co.	GNS
Imperial Oil Co.	IOC
Inland Navigation Co.	INC
Inter-Island Steam Navigation Co.	ISN
Los Angeles Steamship Co.	LAS

Company Name	Abbreviation
Matson Navigation Co.	MNC
McCormick Steamship Co.	MCS
Monticello Steamship Co.	MSC
North American Transportation & Trading Co.	NAT
Northland Navigation Co.	NLN
North Pacific Steamship Co.	NPS
Northwestern Steamship Co.	NWS
Oceanic Steamship Co.	OSC
Oregon Railway & Navigation Co.	ORN
Pacific Alaska Navigation Co.	PAN
Pacific Coast Steamship Co.	PCS
Pacific Far East Line	PFE
Pacific Mail Steamship Co.	PMS
Pacific Navigation Co.	PNC
Pacific Steamship Co.	PSC
Panama Mail Steamship Co.	PMC
Pope & Talbot Lines	P&T
Puget Sound & Alaska Steamship Co.	PSA
Puget Sound Navigation Co.	PSN
San Francisco & Portland Steamship Co.	SFP
Seaboard Shipping Co.	SSC
Southern Pacific Co.	SPC
Standard Oil Co. of California / BC	SOC
States Steamship Co.	SSS
Terminal Steamship Co.	TSS
Union Oil Co. of California	UOC
Union Steamship Co. of BC	USC
Western Canada Steamship Co.	WCS
Western Steam Navigation Co.	WSN

Government Agency	Abbreviation
Canadian Coast Guard	CCG
Canadian Government Merchant Marine	CMM
Canadian Department of Marine & Fisheries	DMF
Royal Canadian Navy	RCN
US Army Transport Service	USAT
US Bureau of Fisheries	USBF
US Coast & Geodetic Survey	USCS
US Coast Guard	USCG

Government Agency	Abbreviation
US Lighthouse Service	USLS
US Marine Hospital Service	USMH
US Maritime Commission	USMC
US National Oceanic & Atmospheric Admin.	NOAA
US Navy	USN
US Revenue Cutter Service	USRC
US Shipping Board	USSB

Shipping Companies

Admiral Oriental Line (AOL), 1922–26

The Pacific Steamship Company (PSC; Admiral Line) first began sending cargo ships across the Pacific Ocean, between Seattle and Singapore, in 1917. In late 1918 it chartered a number of cargo steamers from the United States Shipping Board (USSB) to expand this service. In 1921 the USSB assigned five passenger/cargo vessels, called 535s (in reference to their length), to the PSC to manage in overseas service. Passenger service between Seattle and the Orient was inaugurated with the sailing of the *Wenatchee* on April 9, 1921. Shortly after this, the USSB decided that the administration of its 535s should be separate from PSC. The Admiral Oriental Line Inc. was formed on October 17, 1922. By the end of the year, AOL offered passenger and freight service to Yokohama, Shanghai, Hong Kong and Manila, and freight-only service to Dairew and Singapore. AOL vessels were named after US presidents, as was the practice with all Pacific Coast Shipping Board vessels. The *Wenatchee* became the *President Jackson*. The other four were the *President McKinley*, *President Grant*, *President Madison* and *President Jefferson*. In 1926, the USSB sold all five ships to the Dollar Line. With no ships left in its fleet, AOL ceased to exist. The Dollar Company announced that the new name for AOL would be the American Mail Line. The *President Grant* sailed under its new house flag on June 15, 1926.

Alaska Commercial Company (ACC), 1868–1901

The Alaska Commercial Company was incorporated in San Francisco on September 17, 1868, by a consortium of businessmen led by Hayward Hutchinson and Louis Sloss. It was the intention of the new company to develop mercantile trade in the new territory of Alaska and to obtain a sealing lease on the Pribiloff Islands. To carry out its business, the new company purchased the assets of the Russian-American company from Hutchinson, Kohl & Co. who had acquired them at auction. The assets included the steamers *Alexander*, *Constantine* and *Fideliter*, and trading stations at Sitka, Kodiak, Kaluk, Nushagak, St Paul Island, St George Island, Unalaska, St Michael and 41 other sites.

In August 1870, the new company realized its second goal by signing a 20-year lease agreement with the US government to harvest 100,000 seals annually from the Pribiloff Islands. From its inception, ACC operated a number of ocean steamers on its route from San Francisco to Alaska, delivering supplies to the northern outposts and returning with cargos of seal skins and furs. ACC added river operations to its repertoire in 1869 when it placed the 50-foot (15-metre) sternwheeler SS *Yukon* in service on the Yukon River. The small *Yukon* was followed in 1889 by the 125-foot (38-metre) *Arctic*. Many larger, far more luxurious riverboats were added as the years progressed.

Between 1880 and 1890, ACC established six salmon canneries in the Kodiak region. In 1896 the ACC transported the first gold from the Yukon to San Francisco. The peak for river operations occurred during the height of Klondike gold rush in 1898. The first of the large sternwheelers, *Alice* and *Bella,* were built at St Michael for operation on the Yukon River in 1895. They were followed in the spring of 1898 with the deluxe sternwheelers *Susie*, *Sarah* and *Hannah*. Built in the style of Mississippi riverboats, their dining rooms were palatial with mahogany woodwork. Dishes, bed linen and blankets were monogrammed with the company initials. Between 1899 and 1901, ACC operated 12 riverboats (*Sarah*, *Susie*, *Hannah*, *Louise*, *Margaret*, *Leah*, *Alice*, *Bella*, *City of Paris*, *Victoria*, *Luella* and *Sadie*), 2 tugs and 19 barges.

In the spring of 1899, ACC landed more freight and passengers at St Michael than all the other transportation companies combined. In 1900, ACC operated the steamer *St Paul* on the Nome run, while the *Bertha*, *Portland* and *Dora* provided feeder service to southwestern ports. But by early 1901, ACC officials could see that the company could no longer make a profits. The fur trade had dwindled to a trickle and the gold rush was over. In order to survive into the future, ACC joined with the Alaska Exploration Company and the Empire Transportation Company in 1901 to form the Northern Navigation Company for transportation operations and the Northern Commercial Company for mercantile trade. Alaska Commercial Company removed its signs from the Yukon River trading stations and its pennant from the steamers on April 6, 1901. They were replaced with those of the new Northern Commercial Company.

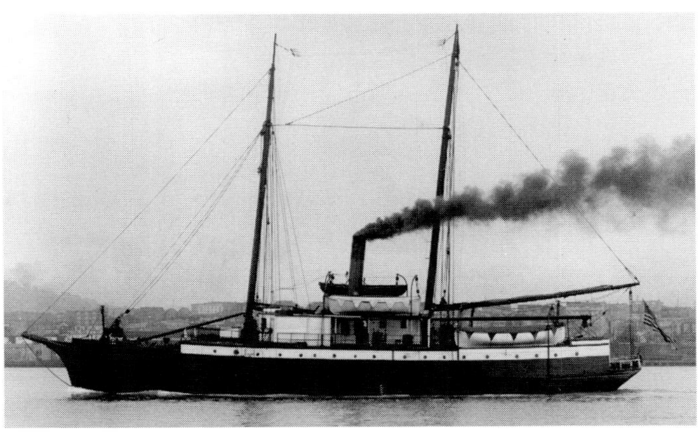

SS *Dora*. UASBC Archives.

Alaska Cruise Lines Ltd (ACL), 1957–71

In 1957, Charles B. West of Seattle formed Alaska Cruise Lines, a subsidiary to his travel company, Westours. He first leased and later purchased the vessels *Coquitlam II* and *Camosun III* from the bankrupt Union Steamship Company of BC in 1958, renaming them *Glacier Queen* and *Yukon Sta*r, respectively. Both vessels remained under Canadian registry and were operated in the summer cruise service to southeast Alaska ports. By the late 1960s, the *Yukon Star* and *Glacier Queen* were too small to handle the growing tourist trade. In 1967 West added a third Canadian-registered vessel, *Polar Star* (*ex Delos*), to his fleet. Then, in 1970, he purchased *Cabo Izarra* from Spanish owners Naviera Cruceros, renaming it *West Star*, and also chartered the Greek liner, *Orpheus*. The *Yukon Star* and *Glacier Queen*

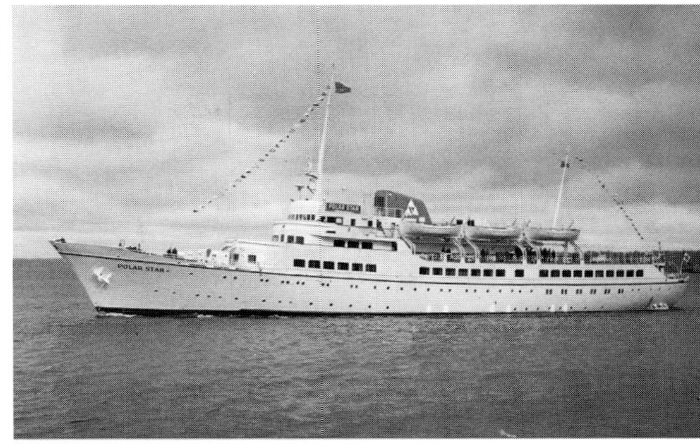

Polar Star. C.P. Johnston Co.

were subsequently withdrawn from service. Late in the 1970 season, ACL changed the departure location for its southeastern Alaska cruise ships to Victoria. Westours, including Alaska Cruise Lines, was sold to Holland America Line in 1971. Holland America still operates Westours today.

Alaska Steamship Company (ASC), 1894–1971

The Alaska Steamship Company was incorporated in December 1894 in Port Townsend, Washington. The new company intended to compete with the venerable old Pacific Coast Steamship Company and thereby reduce passenger fares and freight rates to Alaska. ASC purchased the SS *Willapa* and advertised her first trip to Alaska on March 3, 1895. As business increased, the company added the steamer *Alliance* to its fleet. Then, in 1899, it purchased the *Dirigo* and *Farallon*, and these four ships constituted the line's modest fleet through the gold-rush era.

Timing is everything. The new company prospered during the gold-rush era as the demand for passage to Alaska often outweighed the availability of ships. ASC merged with the Northwestern Steamship Company on January 1, 1908, keeping the name, Alaska Steamship Company. With this merger the company controlled 15 vessels.

After World War II, there was hardly enough business between Puget Sound and Alaska to support one steamship company. In March 1949, ASC bought the assets of competitor Northland Transportation. A series of strikes in the early 1950s tied up operations and cost the company needed revenue. Increasing costs and competition from the airline industry forced ASC to announce on July 6, 1954, that it was getting out of the passenger business. The last passenger trip was made by the SS *Denali* on September 24, 1954. The company maintained a freight service until 1971, when it also folded. Over its lifetime, ASC owned 66 ships and leased or chartered numerous others.

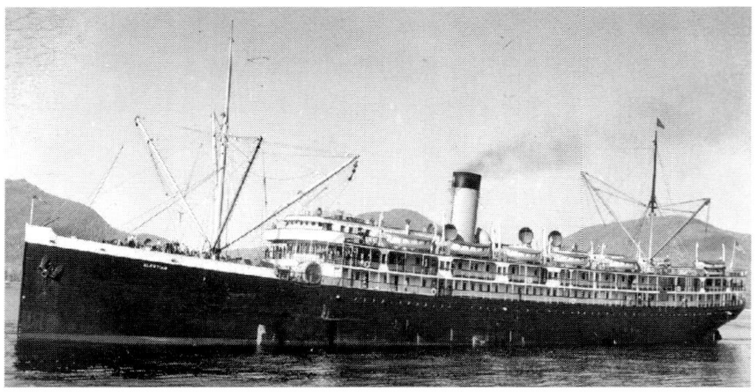

SS *Aleutian* in Alaska.

American Mail Line (AML), 1926–73

In November 1926, the Dollar Line acquired the ships of the Admiral Oriental Line and announced that the name of its new subsidiary would be the American Mail Line. This name received legal status on October 17, 1930. AML ships offered passenger and freight service between Seattle and the Orient. Ships departed from Seattle and sailed to Yokohama, Kobe, Shanghai, Hong Kong and Manila. From the Orient they returned to Los Angeles. From here they went back to the Orient and then returned to Seattle. This service, known as the horseshoe route, was provided by five liners: *President Jefferson*, *President McKinley*, *President Jackson*, *President Grant* and *President Madison*.

In 1936, the US Merchant Marine Act was passed, requiring ships older than 20 years to significantly upgrade their safety features or carry no more than 12 passengers. AML chose to restrict the number of passengers to 12 and operate as a freight only service. On June 22, 1938, a government ocean-mail contract expired and was not renewed. Heavily in debt, AML ceased operations shortly thereafter.

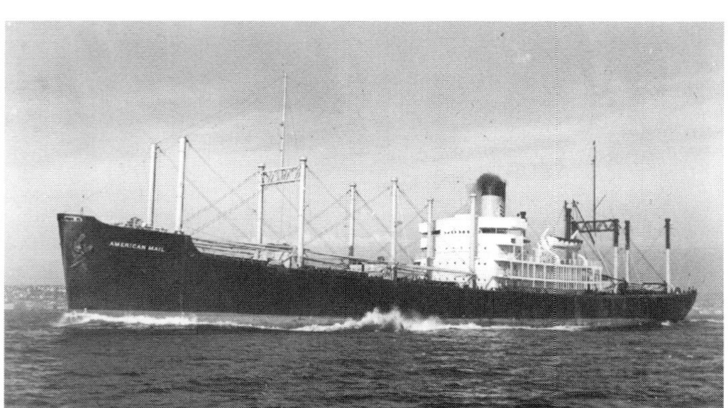

SS *American Mail*. J. Boyd Ellis Post Card Co.

In 1939, the US Maritime Commission proposed to replace AML with a government-owned service. A consortium of businessmen from Puget Sound formed the Pacific Northwest Oriental Line (PNWOL) and lobbied to act as managing agents for the Maritime Commission ships. Service to the Orient resumed on September 2, 1939, using four former Shipping Board freighters: *Satartia*, *Coldbrook*, *Capillo* and *Collingsworth*.

In February 1940, the Maritime Commission solicited bids for the purchase or charter of the four freighters operated by PNWOL. AML, having overcome its financial difficulties (by selling off the *President Madison* and transferring ownership of its other four ships to the government), tendered the successful bid and re-entered the Orient service. From April to July 1941, AML took delivery of 3 new Maritime Commission ships, *Cape Alava*, *Cape Fairweather* and *Island Mail*. During World War II, AML operated 58 vessels for the War Shipping Administration. At the end of the war, it purchased 7 war-surplus vessels from the Maritime Commission (*Ocean Mail*, *Canada Mail*, *Washington Mail*, *Oregon Mail*, *India Mail*, *American Mail* and *Java Mail*). All were fitted with refrigeration systems and limited passenger accommodations.

In the summer of 1954, controlling stock of AML was sold to American Presidents Line (APL) of San Francisco. APL continued to operate AML as a separate entity and in 1961 AML began a ship construction project to replace its aging fleet with new Mariner-type (C-4) vessels. The first vessel, *Washington Mail*, was launched in May 1961. *Philippine Mail* and *Japan Mail* followed, then in 1965, the *Oregon Mail* and *Canada Mail*. The new ships had deluxe air-conditioned accommodation for 12 passengers, a spacious lounge, library and service bar.

AML continued to replace its older vessels with larger, faster and more sophisticated ships. *Alaska Mail*, the first of five C-5 Mail-liner class ships was launched on April 1, 1968, followed by *Indian Mail* on July 2, 1968. The last of the C-5s, *American Mail*, was delivered in 1969. The Seattle-based AML name and flag disappeared from trans-Pacific shipping in 1973 when American President Lines absorbed its subsidiary.

American President Lines (APL), 1938–present

On November 5, 1938, the US government assumed control of the debt-ridden Dollar Line and renamed it American President Lines. It transferred 15 ships built in the 1920s from Dollar to APL and immediately ordered 7 new ships to replace the aging fleet. *President Jackson* and *President Monroe* joined the round-the-world service in 1940, and the *President Hayes* in 1941. The *President Garfield*, *President Van Buren*, *President Adams* and *President Polk* were taken over by the US Navy upon completion.

After the attack on Pearl Harbor, all APL ships were taken over for war duty. During World War II, APL acted as an agent for the War Shipping Administration, overseeing the manning, equipping, overhaul and repair of vessels, and the handling of cargo and passengers. At the end of the war only two ships (*President Polk* and *President Monroe)* returned to APL's round-the-world service. Two additional ships (*President Cleveland* and *President Wilson*) were built in 1947–48 for trans-Pacific service.

In 1952, American oil man Ralph K. Davies and a group of investors purchased APL from the US government for $18 million. The passenger ship *President Hoover* was added to trans-Pacific service in 1957 and the *President Roosevelt* was placed in round-the-world service in 1962. By the mid-1960s, air travel had resulted in a sharp decline in passenger service. APL suspended trans-Pacific passenger service in 1973 with the retirement of its last ocean liner *President Wilson* and turned its attention to cargo. In 1974 the company's fleet comprised 15 container ships and 5 bulk carriers. In 1977 the company suspended round-the-world cargo service and focused on its trans-Pacific market. In 1993 the company operated 5 freighters, each capable of carrying 12 passengers to the Orient.

On April 13, 1997, APL was bought by Neptune Orient Lines of Singapore, but retained its separate corporate name and Oakland, California, headquarters. Since then, all new ships have been built in Germany and Korea and none have been named after US presidents. Today, after several mergers, APL serves Europe, Latin America and the Pacific Rim.

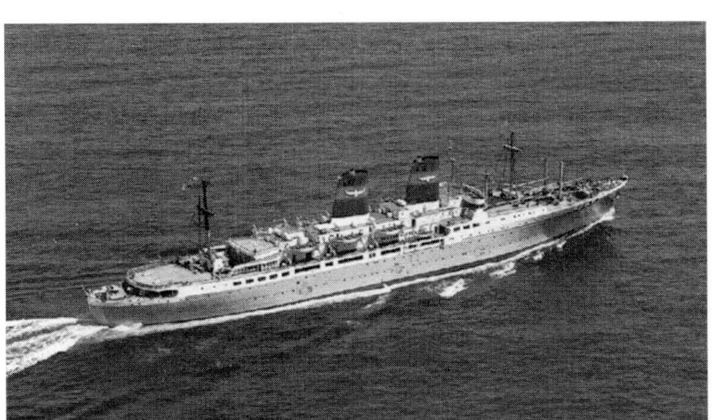

SS *President Roosevelt* at sea. APL postcard.

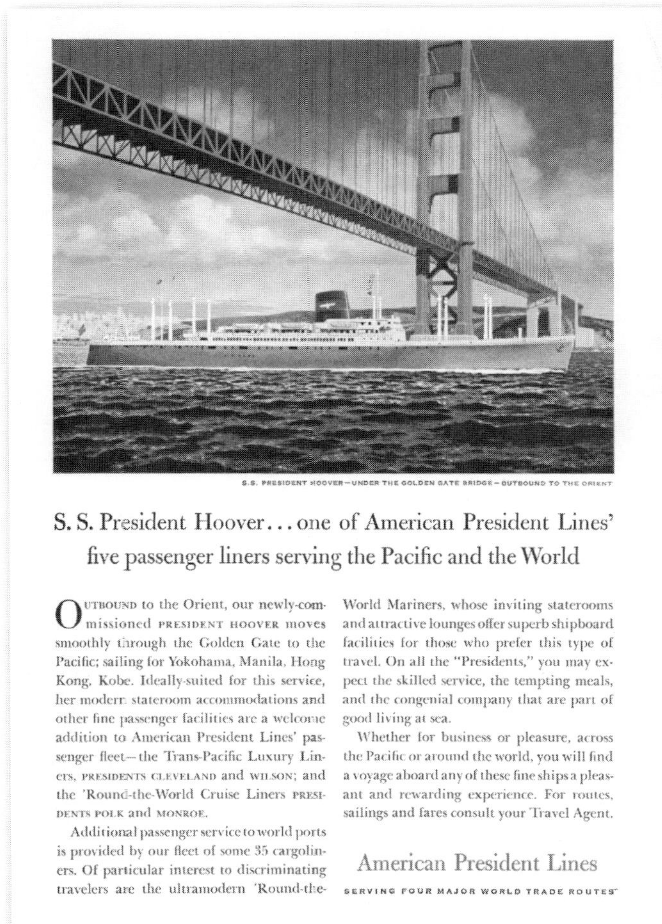

1950s *National Geographic* magazine advertisement for American President Lines.

American Yukon Navigation Company (AYN), 1913–43

The American and Yukon Navigation Company was formed in 1913 as a subsidiary to the White Pass and Yukon Route to operate sternwheelers on the US portion of the Yukon River in competition with the Northern Navigation Company. AYN started with two new sternwheelers, the SS *Alaska* and the SS *Yukon*, both built in Whitehorse in 1913, plus several others that were borrowed from the British Yukon Navigation Company, another White Pass subsidiary. Times were lean after the gold-rush era. The shipments of gold, freight and passengers had been declining for some time on the Yukon River, as had the population of the region, making river competition cut-throat. A ruinous rate war ensued between AYN and Northern Navigation during the 1913 operating season, and both firms lost money.

As there was only enough business for one company, White Pass and Yukon Route bought out Northern Navigation Company on April 10, 1914. The sale included 42 steamboats, 54 barges and all marine facilities on the Yukon River. Since many of the vessels were surplus, most were pulled out of the water and left to rot at St Michael or Dawson. When the Alaska Railroad was completed between Seward and Fairbanks in 1922, AYN closed its operation on the Yukon River below Tanana, sold the fixed property at St Michael to the Northern Commercial Company and sold movable stores to White Pass. After 1924, AYN operated only the SS *Yukon* on the run to Nenana. The SS *Alaska* was returned to the British Yukon Navigation Company to operate in Canada. In 1927 AYN downsized further by selling four of its boats (*Alice*, *White Seal*, *Minneapolis* and *Reliance*) to the Alaska Railroad.

AYN continued to operate its steamer, *Yukon*, on a tourist run between Dawson and Nenana until 1940. In 1942 it leased the SS *Yukon* to the Alaska Railroad to augment its fleet moving military supplies from Nenana to points on the lower Yukon River. In 1943 AYN sold the *Yukon* and its four other remaining sternwheelers (*Julia B*, *Suzie*, *Schwatka* and *Seattle No. 3*) along with all of its routes in Alaska to the Alaska Railroad for $101,125. Only the *Yukon* was fit for service; the rest became derelicts at St Michael and Dawson. All Canadian operations on the American side of the Yukon River now ceased.

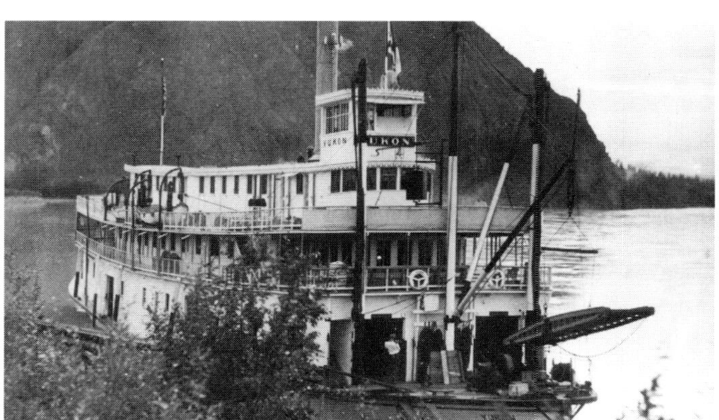

SS *Yukon* on the Yukon River above the Arctic Circle. Gowen Sutton Co. photo.

Black Ball Lines – Canada Ltd (BBL), 1951–61

When the State of Washington bought out the Puget Sound Navigation Company in 1951 it moved some of PSN's vessels north across the border to British Columbia. Black Ball Lines – Canada Ltd was incorporated September 27, 1951. The company's president was George Paulin of Victoria; the chairman of the board of the parent, Black Ball Line Ltd, was Captain Alexander M. Peabody. BBL began a ferry service between Gibsons Landing and Horseshoe Bay on August 11, 1951, so linking the Sunshine Coast to Vancouver; the MV *Bainbridge* carried most of the load, supplemented by the MV *Smookwa*. The MV *Quillayute* connected Earls Cove and Saltery Bay. In 1953, the MV *Kahloke* began operation between Horseshoe Bay and Nanaimo, and the MV *Chinook II* joined her in 1954.

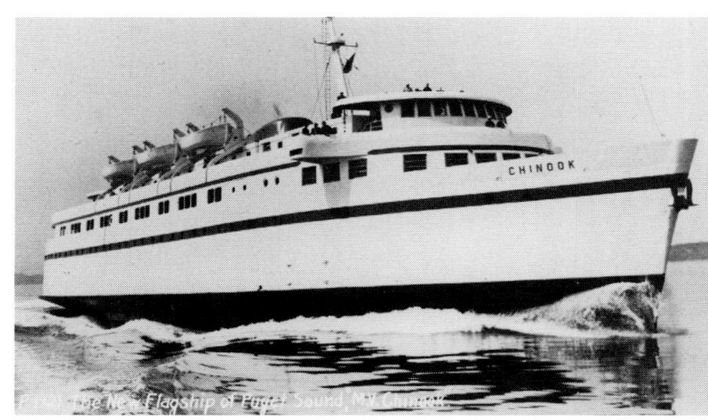

Black Ball Lines MV *Chinook*.

On May 17, 1958, employees of Canadian Pacific Steamships went on strike, leaving Vancouver Island dependent entirely on the BBL. But then BBL was served with 72-hours strike notice less than a month later. The provincial government invoked the Civil Defence Act to take possession of and use the property of the Black Ball Line to maintain service to Vancouver Island. As a result of this labour strife, the British Columbia government decided to establish its own ferry system, and bought the BBL on November 30, 1961.

Bonners Ferry and Kaslo Transportation Company (BFK), 1893–94

The Bonners Ferry and Kaslo Transportation Company was formed in 1893 to place a large first-class sternwheeler in service on Kootenay Lake. The 140-foot (43-metre) *State of Idaho* was built by Depuy, Lannen & Rutter at Bonners Ferry, Idaho, and entered service on Kootenay Lake in competition with the Columbia and Kootenay Steam Navigation Company in May 1893. Competition from the *State of Idaho* did not last long. The BFK overextended itself and ran into financial difficulties. Her owners had to keep the *Idaho* north of the border to avoid seizure by creditors. The company's luck didn't improve. On November 10, 1893, the sternwheeler ran aground and sank near Ainsworth Hotsprings. A survey confirmed the vessel to be a constructive loss. It was sold at auction for $350 and the BFK ceased to exist.

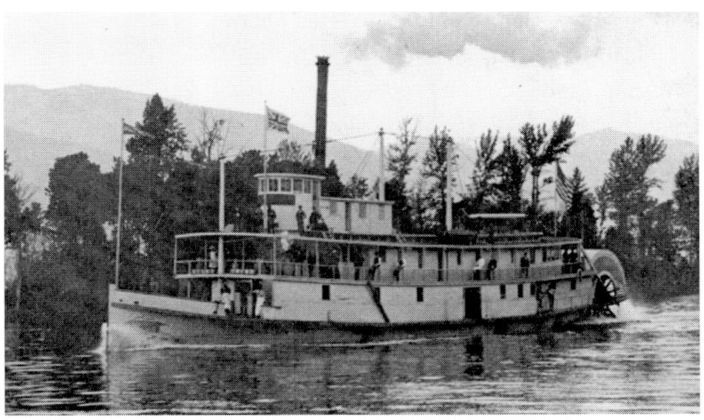

SS *State of Idaho* on the Kootenay River, 1894. BC Archives B-07155.

Boscowitz Steamship Company Ltd (BSC), 1899–1911

The Boscowitz Steamship Company was incorporated March 18, 1899, at Victoria, BC. The principals were Captain James D. Warren, William McDowell, John Barnsley, L. Dickinson, George McGregor and Walter Morris. Initially, the new line operated the small wooden passenger steamer *Barbara Boscowitz* from Victoria to Stewart and Skagway, servicing canneries and small outposts en route. The *Barbara Boscowitz* was wrecked on a reef in Johnstone Strait on October 2, 1904. BSC replaced her with the wooden passenger-cargo steamer SS *Venture* in November 1904. It enlarged its fleet in 1907, buying a 191-foot (58-metre) steel passenger-cargo steamer, the *Vadso*, from Gotenbury, Sweden.

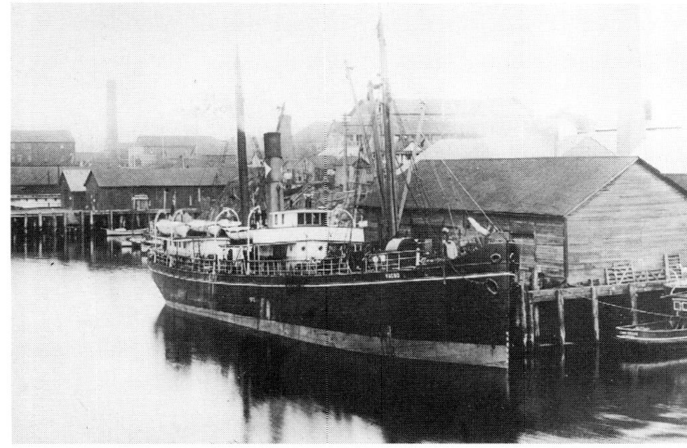

SS *Vadso*. Lorrie Belveal collection.

The *Vadso* could only carry 50 passengers, so BSC operated her chiefly as a cannery freighter.

On March 25, 1909, the *Venture* was destroyed by fire at the Inverness Cannery on the Skeena River. Captain Warren quickly chartered a small replacement steamer, the *St Dennis*, from San Diego for the 1909 season. A new *Venture*, launched on February 16, 1910, was 180 feet (55 metres) long, made of steel and licensed for 186 passengers. It served on the cannery route to the Skeena and Nass rivers via Prince Rupert. The Boscowitz Steamship Company was bought out by the Union Steamship Company of BC (J.H. Welsford & Company of Liverpool) on September 23, 1911.

British Columbia Ferries (BCF), Dogwood Fleet, 1960–present

The British Columbia Ferry system was born out of labour strife in the late 1950s (see Black Ball Lines, page 22). On July 18, 1958, the government of British Columbia announced that it would build two new vessels to operate between Tsawwassen, south of

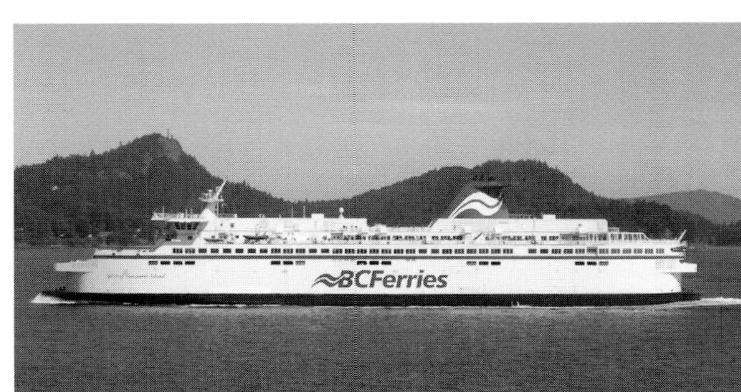

Spirit of British Columbia near Active Pass. J. Marc collection.

Vancouver, and Swartz Bay on Vancouver Island. The MV *Sidney* was launched October 6, 1959, and the MV *Tsawwassen* followed on November 28. The two ships entered service in June 9, 1960.

By 1961, the government decided that it was in the ferry business to stay, so purchased Black Ball Lines – Canada from its parent company, the Puget Sound Navigation Company. With this purchase, BCF acquired terminals at Horseshoe Bay, north of Vancouver, and at Departure Bay in Nanaimo, along with the motor vessels serving that run, *Chinook II* and *Kahloke*; it also acquired docking facilities at Langdale, Earls Cove and Saltery Bay, and the vessels serving Howe Sound (MV *Quillayute*, MV *Bainbridge* and SS *Smokwa*). In 1962, it added two new ferries to the Tsawwassen-Swartz Bay route: *Queen of Victoria* and *Queen of Vancouver*.

By 1966, the fleet had expanded to 14 vessels and operated regularly scheduled service to the Gulf Islands. That year it built the *Queen of Prince Rupert* to provide service from northern Vancouver Island to Prince Rupert through the Inside Passage. In 1976, it added two jumbo C-class ferries, the *Queen of Cowichan* and *Queen of Coquitlam*, each capable of carrying 362 cars. Next came the *Queen of Alberni*, built specifically to carry over-height commercial and recreational vehicles, then two more C-class ferries in 1979. In May 1980, the *Queen of the North* (formerly *Stena Danica*) started service on the Inside Passage.

Two S-class ferries joined the fleet in the early 1990s, each able to carry 470 cars: the *Spirit of British Columbia* (1993) and the *Spirit of Vancouver Island* (1994). In 1999 the *Pacificat Explorer,* the first of three catamaran ferries was commissioned. The *Pacificats* were in service intermittently until March 2003 when all three were sold due to high operating costs. Today BCF operates 38 vessels serving 48 destinations in British Columbia.

British Columbia Salvage Company Ltd (BCS), 1901–18

The BC Salvage Company was organized by a group of investors headed by H.F. Bullen of Victoria as early as 1896. Early on the company, headquartered in Esquimalt, relied on chartering vessels from the Canadian Pacific Navigation Company to carry out its business. BC Salvage Company was officially incorporated September 9, 1901.

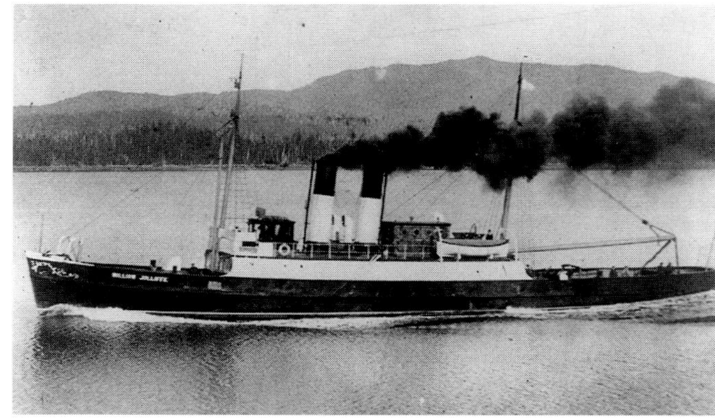

SS *Salvage Chief*. J. Marc collection.

BCS purchased its first salvage vessel, the *Maude*, in 1901, then added the CPR iron steamer *Danube*, renaming it *Salvor*, in 1905, and the *William Joliffe* (purchased from Joliffe Tug Co. of England) in 1907. BCS participated in many wreck rescue and recovery operations during its 22-year history (1896–1918), notably the *Pass of Melfort* in 1905, the *Valencia* disaster of 1906, the *Princess May* in 1910, the SS *Spokane* in 1911, and the SS *Congress* in 1916. It sold the *William Joliffe* to the Canadian Pacific Railway in 1914. On March 28, 1918, BCS was succeeded by the Pacific Salvage Company, which sold the SS *Salvor* and replaced it with the former CPR ship, *Tees*.

British Columbia Steamship Company (1975) Ltd (BSS), 1975–88

The British Columbia Steamship Company was formed by the government of British Columbia in 1975 in response to the vacuum left when Canadian Pacific Railway Co. (CPR) suspended its Victoria-Seattle passenger service and retired the *Princess Marguerite* on September 30, 1974. BSS bought the

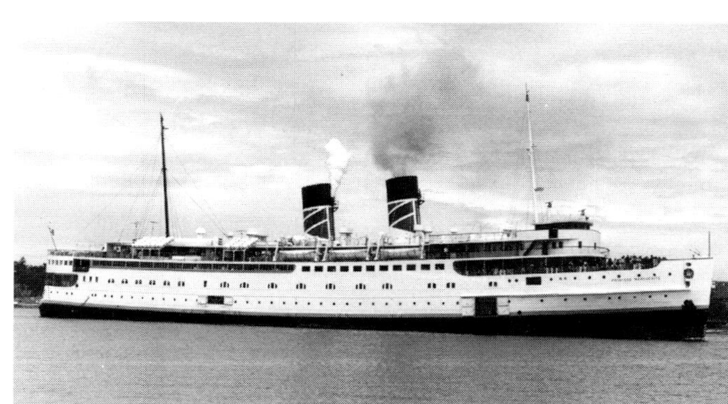

Princess Marguerite. Maritime Museum of BC.

Princess Marguerite from the CPR on April 1, 1975, had her refurbished and placed her back in service between Victoria and Seattle in the spring of 1975. In the fall of 1975 BSS bought a recently retired passenger ship, the SS *Prince George*, from the Canadian National Steamship Co. for possible cruises and to augment the Victoria–Seattle service. But the *Prince George* never saw active service and was re-sold in 1976. In the fall of 1979, the *Princess Marguerite* was found to be unsafe for passenger service. BSS quickly acquired the *Queen of Prince Rupert* from the BC Ferries, renamed it *Victoria Princess* and operated it for the 1980 season. The *Princess Marguerite* was upgraded and returned to active service in the spring of 1981.

In 1987 BSS added the *Vancouver Island Princess* (formerly the *Princess of Vancouver*) to the Victoria-Seattle run. This permitted morning and afternoon sailings from each city. Both ships were equipped with casinos and slot machines in 1987. BSS also tried operating in partnership with the Boeing Corporation and other companies to offer jet-foil service between Victoria and Seattle in 1976 and 1980. Neither venture proved successful. In November 1988 the BC government decided that it wanted out of the shipping business and sold BSS to the Stena Line.

California Navigation and Improvement Company (CNI), 1889–1927

The California Navigation & Improvement Company was formed June 5, 1889, to offer passenger and freight service between Stockton and San Francisco. It also served landings on the upper San Joaquin River and operated a large boat yard in Stockton. The line's more important steamers and their years of construction were the *Mary Garrett* (1878), *T.C. Walker* (1885), *J.D. Peters* (1889) and *J.R. McDonald* (1899). In addition to transporting passengers, the CNI also operated freight steamers that towed barges carrying farming materials and grain shipments. The *J.D. Peters* and *T.C. Walker* were well-known in the towns along the river.

Competition on the Stockton-San Francisco route intensified with the formation of the Union Transportation Company in 1891 – both companies offered overnight steamer service between the two cities. On November 27, 1898, the *T.C. Walker* blew up near Stockton, killing nine men; it was rebuilt and ran for many more years. Two older steamers, the *A.C. Freese* and the *Leader*, were still active in 1907.

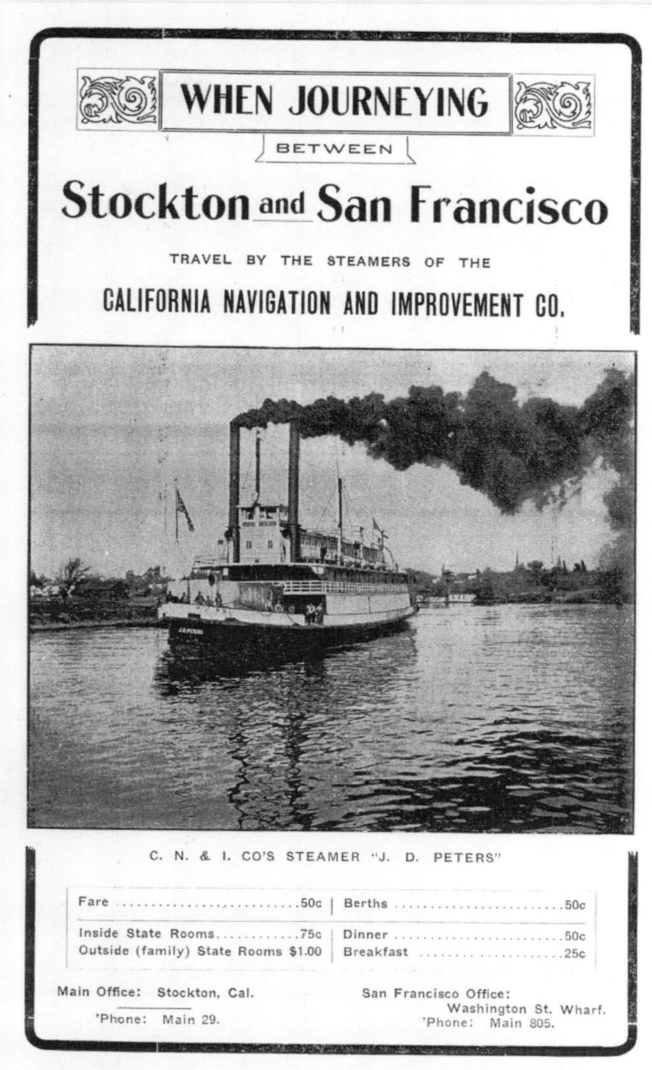

California Navigation & Improvement Company ad, ca 1905. Bill Stritzel collection.

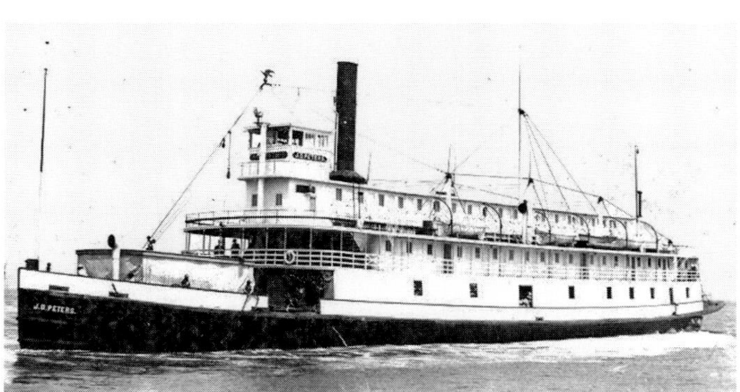

The *J.D. Peters* later in her career; originally she had two stacks.

In the early 1920s the California Transportation Company bought controlling interest in CNI. Both firms continued operating independently until 1927, when they merged to become the California Transportation Company.

California Transportation Company (CTC), 1875–1948

The California Transportation Company was formed by Captain Nels E. Anderson and Andrew Nelson of San Francisco on July 31, 1875. Anderson and Nelson had been in partnership providing freight services on the Sacramento River since 1856. The CTC was noted for its service between San Francisco, Sacramento and the San Joaquin River delta. Early on it was engaged heavily in transporting fruits and vegetables from the delta farmlands to San Francisco. The first CTC boats were the *Reform*, *Pioneer*, *Relief* and *Washington*. As its operations grew, the company added more steamers, including the *Constance*, *Onward*, *Onisbo*, *Pride of the River*, *Aurora*, *Isleton* and *Fort Bragg*.

In 1905, CTC bought out the Union Transportation Company, which operated steamers between Stockton and San Francisco. In October 1907, it established a freight-and-passenger service between Sacramento and San Francisco. It launched the deluxe Sacramento steamers *Capital City* in 1910 and *Fort Sutter* in 1913 to meet ever-increasing passenger and freight traffic. These floating palaces had three passenger decks and introduced plumbing to individual staterooms. These sternwheelers became very popular on the San Francisco-Sacramento route.

By the early 1920s the CTC had acquired controlling interest in its competitor, the California Navigation & Improvement Company. The two firms merged on March 31, 1927, to become the California Transportation Company. In June of that year, the new company replaced the *Capital City* and *Fort Sutter* with palatial steamers, the *Delta King* and the *Delta Queen*. In 1932, the worst year of the Great Depression, the CTC and two river-freight operators formed a unified service called the River Lines. By the late 1930s, the River Lines (CTC) was operating only the *Delta King* and *Delta Queen* on the Sacramento run. The River Lines suspended passenger service September 29, 1940. In fall of 1941 the CTC sold the *Delta King* and *Delta Queen* to the Isbrandsten Steamship Co. of New York. The formal corporate existence of the California Transportation Company ended on January 1, 1948.

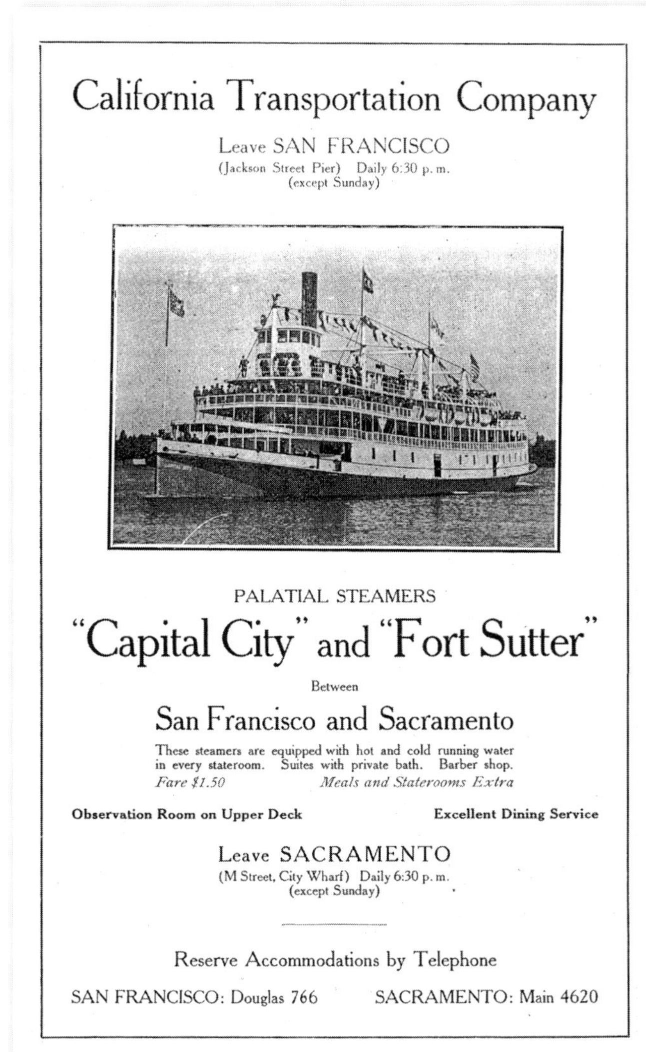

California Transportation Company ad, from about 1914. Bill Stritzel collection.

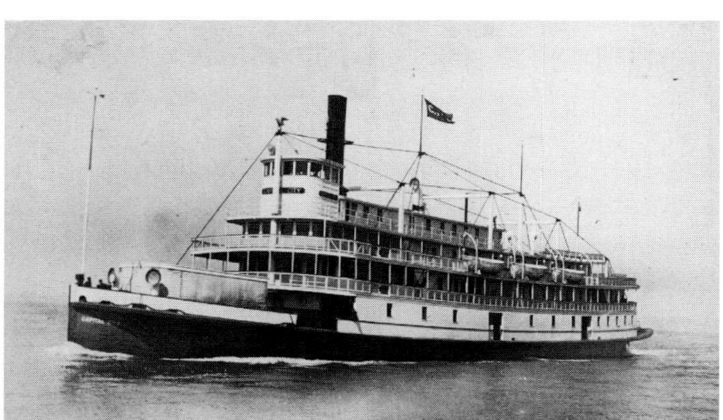

SS *Capital City* postcard. Bill Stritzel collection.

Canadian National Steamship Company Ltd (CNS), Prince Line, 1925–75

The Canadian National Steamship Company was formed on February 25, 1925, inheriting a fleet of four *Prince* ships from the bankrupt Grand Trunk Pacific Railway (GTP). CNS ran an express service from Vancouver to Prince Rupert and a triangle route between Vancouver, Victoria and Seattle, in competition with CPR liners. CNS ships also continued the service provided by GTP to the Queen Charlotte Islands and southeastern Alaska. The large, fast ships, *Prince Rupert* and *Prince George*, operated on the Vancouver-Prince Rupert run, and the smaller *Prince John* and *Prince Albert* served the Queen Charlotte Islands. In its first year, CNS added the *Prince Charles* to the service from Prince Rupert to the Queen Charlotte Islands, and retired the *Prince Albert*. Then in 1930 it added the *Prince William* to supplement the Queen Charlotte Islands service, but found this ship to be woefully unsuited for the exposed Hecate Strait crossing and removed her from the run in less than a year. That year saw the arrival of two new large ships: the *Prince Henry* entered the Alaska cruise service on July 23, and the *Prince David* re-established the triangle run between Vancouver, Victoria and Seattle on August 12. The *Prince Robert* arrived in 1931 and joined the tri-city run. The new ships were not successful. CNS sent the *Prince Henry* and *Prince David* to the Atlantic coast to serve as cruise liners. The *Prince Robert* continued on the west coast as a relief ship until 1939 when she was purchased by the navy and converted to an armed merchant cruiser.

In June 1940, CNS sold the two small ships, *Prince John* and *Prince Charles*, to the Union Steamship Company of BC. In September 1945, the *Prince George* burned at Ketchikan, leaving only the *Prince Rupert* in service. CNS replaced the *Prince George* with a new ship of the same name in June 1948. The *Prince Rupert* was retired in 1956, after 46 years of service, leaving only the new *Prince George* to carry on the Pacific coast service. By the 1970s, the *Prince George* was operating in limited tourist cruises to Alaska. She was retired in the spring of 1975, bringing an end to CNS operations on the west coast.

In addition to its Pacific coastal operations, CNS ran a fleet of five *Lady* ships between east-coast ports (Montreal and Halifax) and the West Indies from 1928 to 1958.

Canadian Pacific Navigation Company Ltd (CPN), 1883–1901

The Canadian Pacific Navigation Company Ltd was incorporated January 17, 1883, at Victoria, BC. It was created when Captain John Irving merged his Pioneer Line of Fraser River paddlewheelers with the fleet of vessels owned by the Hudson's Bay Company. The fleet of the new company comprised the steamers *Princess Louise*, *Enterprise*, *Otter*, *R.P. Rithet*, *William Irving* and *Reliance*. In its first year of operation, it added the sternwheelers *Western Slope* and *Gertrude* and the side-wheeler *Yosemite*.

The new line operated daily passenger service between Victoria, Vancouver and New Westminster. CPN also offered two trips per week from Vancouver to Tacoma, Port Townsend and Seattle. Semi-monthly trips followed to the north and central coast of BC, with stops at Alert Bay, Bella Bella, the Skeena River, Metlakatla, Fort Simpson and intermediate ports. On June 7, 1886, with the aid of a government subsidy, CPN initiated service to the west coast of Vancouver Island. The steamer *Maude* made monthly trips to Port Alberni, calling at way points along the coast,

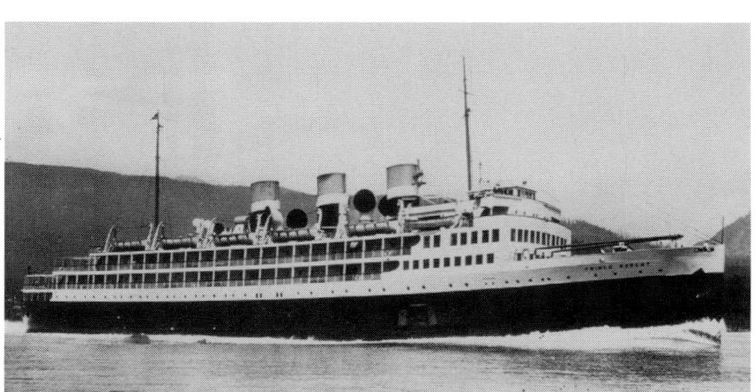

SS *Prince Robert* entering Vancouver Harbour. Gowen Sutton Co.

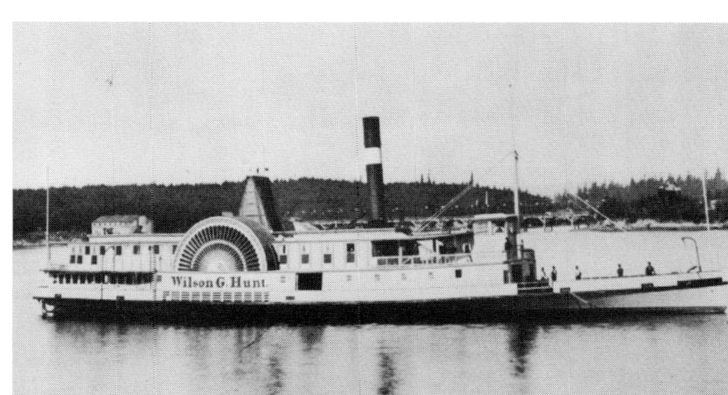

Wilson G. Hunt J. Marc collection.

at Barkley Sound and at Clayoquot. By 1892, CPN had increased its fleet by purchasing or building five more steamers and steamships: *Wilson G. Hunt*, *Premier*, *Sardonyx*, *Islander* and *Danube*.

In 1892, CPN advertised that the northern trips would be extended to Alaska during the tourist season. It expanded service to the north significantly in the late 1890s to accommodate the increased business from the Klondike gold rush. In 1901, Canadian Pacific Railway purchased CPN's fleet of 14 vessels. CPN continued to operate as a separate entity until May 1903, when it was absorbed into the CPR as its BC Coast Steamship Service.

Canadian Pacific Railway Company (CPR), Princess Line, 1881–present

The Canadian Pacific Railway Company was incorporated on February 16, 1881, to construct a railway across Canada. CPR entered the water transportation business in 1883 by using tugs and barges on the Great Lakes to assist with the construction of the transcontinental railway. The first three CPR ships (*Alberta*, *Algoma* and *Athabasca*) were delivered for use on the Great Lakes in 1884. The railway was completed in June 1886.

In 1887 the CPR chartered three steamships (*Abyssinia*, *Batavia* and *Parthia*) to connect Vancouver, the railway's west-coast terminus, to the Orient. It replaced these vessels in 1891 with the first three *Empress* ships (*India*, *China* and *Japan*). A fleet of four *Empress* ships maintained the trans-Pacific service until January 11, 1941, when three of them were lost in World War II – the service was never resurrected.

In addition to its ocean-going *Empress* fleet on the west coast, the CPR operated the BC Lakes and River Service and the BC Coast Service. The Lakes and River Service ran paddlewheelers and tugs on the interior lakes from 1897 until the mid 1930s. Passenger service on Okanagan Lake ended with the retirement of the *Sicamous* in January 1935. Passenger service on Kootenay Lake ended on December 31, 1930. In 1901 the CPR bought out the Canadian Pacific Navigation Company and started the BC Coast Steamship Service with its fleet of *Princess* ships. The *Princess* ships ran on the express service from Vancouver to Prince Rupert and on the triangle run between Vancouver, Victoria and Seattle, in opposition to the CNR liners. In addition to its BC service the CPR also operated ships, like the *Princess Sophia* and *Princess Norah*, to Alaska and the *Danube* fortnightly to Portland.

The last CPR BC Coast Service passenger vessel was the *Princess Patricia*, which operated in the Alaska cruise trade until October 12, 1981.

Chevron Shipping, 1984–present

Chevron Shipping is the tanker subsidiary of the Chevron Corporation. See Standard Oil of California on page 45.

China Mail Steamship Company Ltd (CMS), 1915–23

The China Mail Steamship Company was incorporated at San Francisco in 1915 to provide passenger and freight service between San Francisco and Hong Kong.

The CMS operated three ships in its short career. It inaugurated its service with the SS *China*, bought from

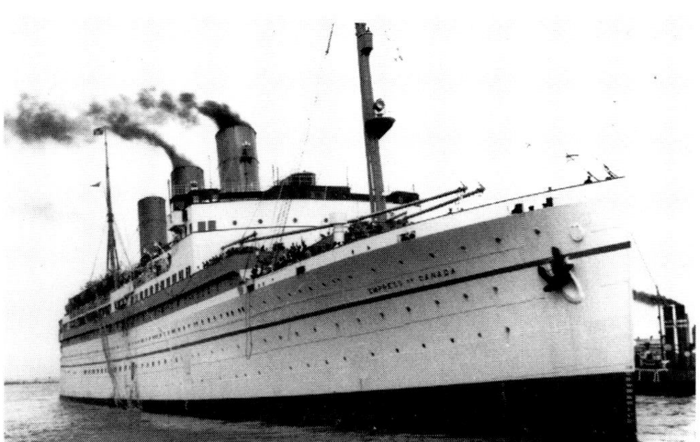

Empress of Canada. J. Marc Collection

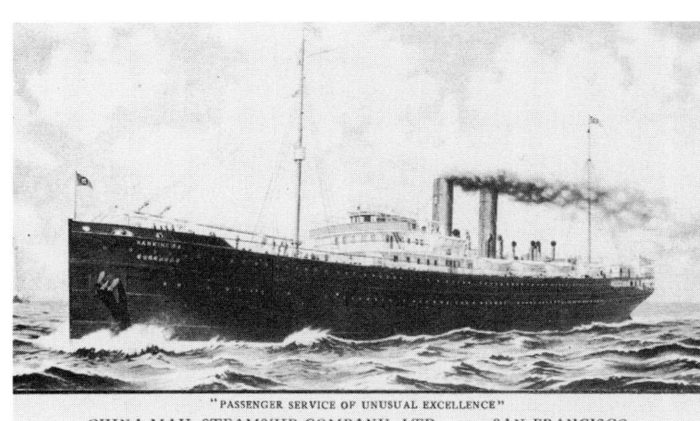

SS *Nanking* at sea.

Pacific Mail Steamships in October 1915. The *Nanking* followed in 1917 and the *Nile* in 1919. Though CMS was 51% US owned, the US government ruled that its ships were foreign and could not participate between Hawaii and San Francisco. This ruling, along with perennial smuggling problems, forced CMS to cease operations on March 13, 1923.

Coeur d'Alene and St Joe Transportation Company Ltd (CDA), Red Collar Line, 1903–22

In 1902 father and son James and Harry Spaulding began laying the groundwork for a new steamboat company to provide passenger and freight service on Lake Coeur d'Alene and the St Joe River. A year later they formed the Coeur d' Alene & St Joe Transportation Co. with J.C. White, William Dollar and others, incorporating it on March 21, 1903. The company launched its first vessel, the 146-foot (45-metre) side-wheeler *Idaho*, on June 13, 1903.

In February 1904 the company bought out its major competition, the propeller boats *Spokane* and *Colfax*, from J.D. McDonald and George Reynolds. CDA's unofficial name, Red Collar Line, came about soon after, when the company began using a red collar on the smokestacks of its steamers. Within a few months of being bought out, McDonald and Reynolds built a new sternwheeler, the *Boneta*, and were back in competition with Red Collar Line, doing business as White Star Navigation Company.

CDA bought out White Star Navigation in 1908 and became a monopoly on Lake Coeur d'Alene, operating 9 passenger steamers, 2 tugs and 14 barges. The *Georgie Oakes*, acquired with the purchase of White Star, was the largest and fastest steamboat on the lake; it had a fine interior, including staterooms, and could carry up to 1000 passengers. That year, the Milwaukee Railroad built their main line down the St Joe River, reducing the need for steamers to deliver passengers and cargo. This development, in combination with the increased use of automobiles and motorboats, led to fewer and fewer people using the steamboats. Dwindling business and a minor recession forced J.C. White and the CDA to declare bankruptcy in late 1922.

CDA's assets were purchased by lumberman Fred Herrick and others, who used them to start the Red Collar Steamboat Company in February 1923. But Herrick and the company went bankrupt in just six years. The Edward Rutledge Timber Company purchased the assets in 1929 and formed the Red Collar Line Inc. The *Flyer*, the last of the passenger vessels, operated until 1935. The name Red Collar Line Inc. was used by Rutledge Timber for its tug operations until 1961, when the company was dissolved.

Dalles, Portland and Astoria Navigation Company (DPA), Regulator Line, 1891–1915

The Dalles, Portland and Astoria Navigation Company, known as the Regulator Line, was formed on May 1, 1890, to compete with the Oregon Railway and Navigation Company on the Columbia River. It launched two fine sternwheelers in 1891: the *Regulator* to run on the middle Columbia between Portland and The Dalles, and the *Dalles City* to run on the lower Columbia between Portland and Astoria.

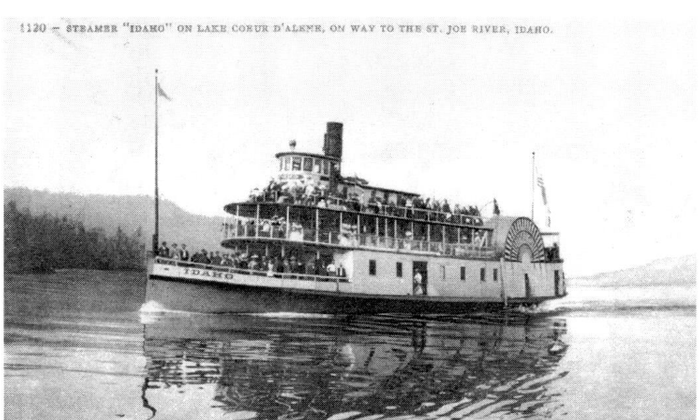

Idaho on Lake Coeur d'Alene, ca 1908.
Edward H. Mitchel photo.

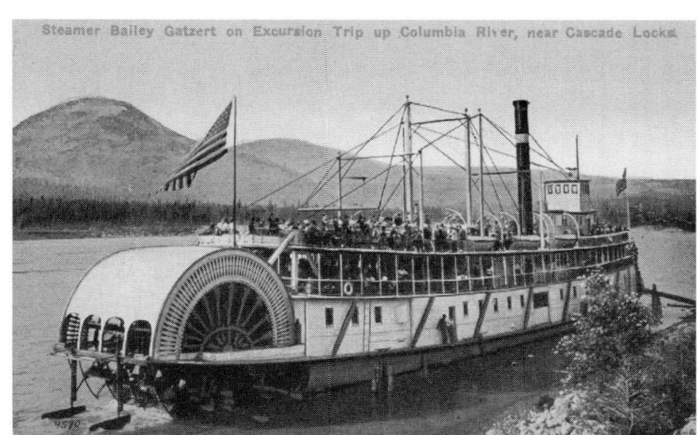

Paddle steamer *Bailey Gatzert*.

On May 30, 1902, the Columbia River & Northern Railway bought DPA but continued operating it as an independent subsidiary. DPA merged with the Columbia River and Puget Sound Navigation Co. (White Collar Line) in 1903, adding the *Telephone* and *Bailey Gatzert* to the Regulator Line fleet. The Columbia River & Northern Railway continued operating DPA boats until it was purchased by the Spokane, Portland and Seattle Railway on March 30, 1908. SPSR operated the steamboats until the line was sold April 26, 1915, to meet the requirements of the Panama Canal Act (i.e., a railroad may not maintain a boat service parallel to its own railway line). Over the course of it existence the DPA operated numerous sternwheelers, including those mentioned above, the *State of Washington* and the *Stranger*.

Dollar Steamship Company (DSC), Dollar Line, 1900–1938

The Dollar Steamship Company was formally incorporated August 15, 1900, by Captain Robert Dollar. After establishing a demand for his lumber in Asia, Captain Dollar began acquiring ships to transport it there. In 1923 the company bought seven war-surplus 502-type *President* liners from the United States Shipping Board and began a west-bound round-the-world service. The *President Harrison* inaugurated this service, departing from San Francisco on January 5, 1924.

In 1925 DSC purchased five more *President* liners (535 type) from the USSB and put them on the trans-Pacific service. In 1929 it purchased the Panama Pacific liners *Manchuria* and *Mongolia*, renamed them the *President Johnson* and the *President Fillmore*, and entered them in its round-the-world service. In 1931 DSC built two large, luxurious vessels (*President Coolidge* and *President Hoover*) to compete on trans-Pacific routes with the new NYK and Canadian Pacific liners.

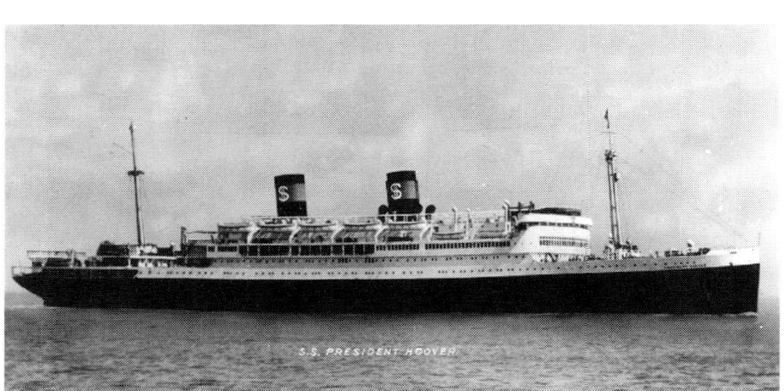

SS *President Hoover*.

In 1937 the *President Hoover* ran aground in Taiwan and became a total loss. The Great Depression of the 1930s resulted in the loss of trade and financial troubles for the DSC. In 1938 the US government took over its management and reorganized the company as the American Presidents Line.

Grand Trunk Pacific Coast Steamship Company (GTP), 1910–25

The Grand Trunk Pacific Coast Steamship Company was incorporated on May 26, 1910, to provide regular passenger and freight service between Prince Rupert and Vancouver, BC. GTP ships also provided service from Prince Rupert to the Queen Charlotte Islands and to ports in southeastern Alaska (Ketchikan, Wrangell, Juneau and Skagway). The names of all GTP ships began with *Prince*.

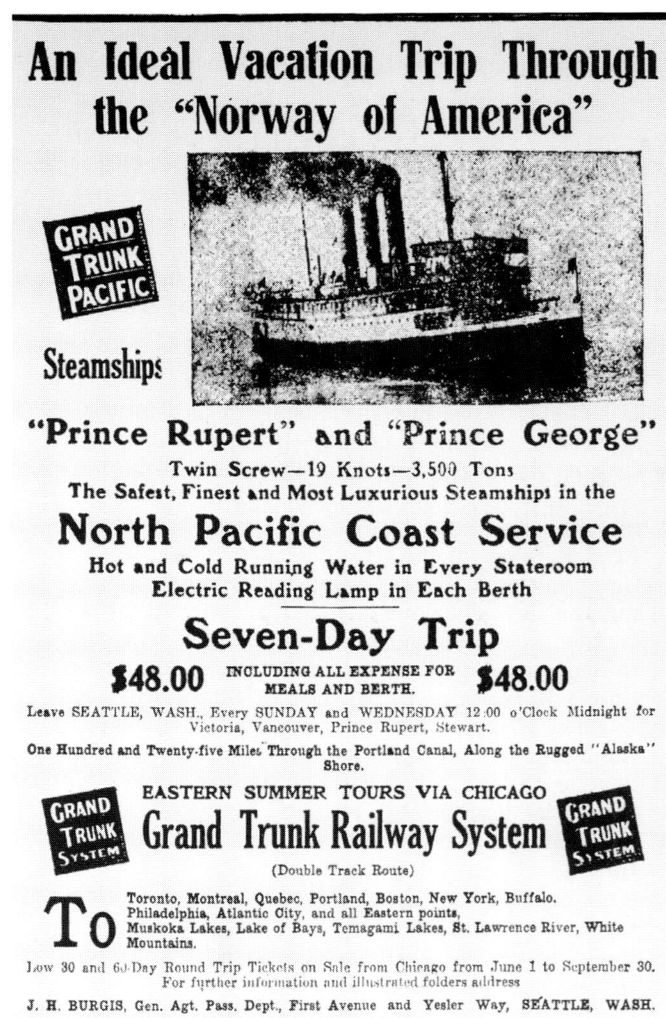

Grand Trunk Pacific advertisement.

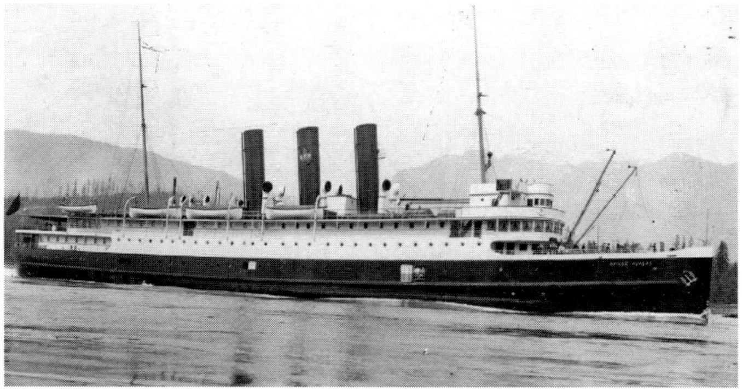

SS *Prince Rupert*. Gowen Sutton Co. photo.

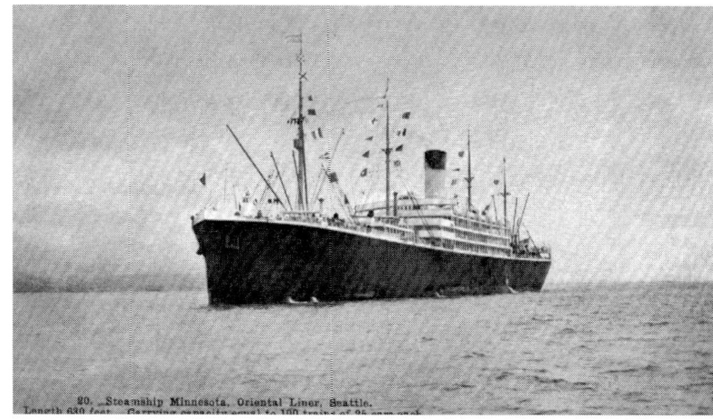

SS *Minnesota*. Rhodes Brothers photo.

GTP's end began on March 7, 1916, when the Canadian government forced it into receivership, because it could not pay its enormous debt. In 1924 the Canadian Government Merchant Marine briefly took over operation of two former GTP ships, *Prince Rupert* and *Prince George*, placing them on the Vancouver-Skagway run during the spring and summer. The BC Coast Steamship Services of Grand Trunk Pacific became part of the Canadian National Steamship Company on February 26, 1925.

Great Northern Pacific Steamship Company (GNP), 1914–17

The Great Northern Pacific Steamship Company was formed by James J. Hill on September 30, 1914. The steamship line was created to transport rail passengers from Flavel, near Astoria on the Columbia River, to San Francisco. The company's name was a combination of the names of Hill's two transcontinental railroads, Great Northern and Northern Pacific. The GNP operated two ships, the SS *Great Northern* and SS *Northern Pacific*. During the winter of 1915–16, the ships also ran to Hawaii. That service was not a success and Hill put the line up for sale when World War I began; but the US government commandeered the two ships on September 17, 1917, ending the line.

Great Northern Steamship Company (GNS), 1904–15

Rail baron James J. Hill organized the Great Northern Steamship Company in 1900 to provide trans-Pacific steamship service between Portland and the Orient. He ordered two large passenger vessels to provide this service. The *Minnesota* arrived in Seattle in December 1904 and the *Dakota* followed in June 1905. The *Minnesota* made its first trip to the Orient on January 21, 1905. The *Dakota* was wrecked off the coast of Japan on March 5,1907, during its seventh voyage. Hill did not replace the *Dakota*, but maintained the *Minnesota* in service to the Orient until October 1915, when he sold her to the Atlantic Transportation Company, so ending Great Northern Steamship Company.

Imperial Oil Company (IOC), 1880–present

The Imperial Oil Company was formed in London, Ontario, in 1880 when 16 oil refiners pooled their resources to find, produce, refine and distribute petroleum and its products in Canada. In 1896 IOC sold a majority interest to Standard Oil of New Jersey to gain expansion capital, and took over all of Standard's Canadian assets in 1899. IOC expanded its operations to British Columbia on November 23, 1899. In 1907 it opened its first gasoline service station in Vancouver, and in 1914 it built a refinery on Burrard Inlet near Port Moody. To supply its new refinery Imperial began importing oil from South America.

IOC operated a fleet of tankers to move its oil. The *Albertolite* (1929–46), among others, brought crude from South American and Californian ports to IOC's BC refinery. The *Marvolite* (1926), *Beeceelite* (1937) and *Argo* (1944) moved the finished product to coastal communities. In 1959 the corporate name of the company became Imperial Oil Limited. The coastal tanker *Imperial Skeena* was launched in 1969 to replace *Imperial Vancouver* (1938), carrying petroleum products to BC coastal ports. The *Imperial Tofino* followed in 1973 to replace *Imperial Nanaimo* (formerly *Beeceelite*). In 1978 a subsidiary, Esso Resources Canada Limited, was formed to manage all the company's natural-resource activities. Three years later Esso Petroleum Canada was formed to manage the company's petroleum-product operations. Imperial Oil Limited is still in operation today doing business as Esso Canada.

Inland Navigation Company (INC), 1935–59

The Inland Navigation Company was formed at Seattle in 1935 by R.S. Whaley and H.A. Dent as the principle stockholders. From 1937 to December 1938 INC's operations on the Columbia River were limited because the Bonneville Dam was being constructed and vessels could not go above it. INC's first purpose-built vessel, the *Inland Chief*, was launched in January 1937. This vessel was used to transport petroleum products and other goods up river and wheat down river. In 1941 INC and its two subsidiaries, the River Terminals Company and Columbia Snake River Towing Company, merged with Captain Leppaluoto's Upper Columbia River Towing Company. Captain Leppaluoto became manager of all marine equipment. In the late 1940s INC instituted service above The Dalles. It used small gasoline-powered vessels to provide freight, passenger and mail service from Lewiston, Idaho, up the Snake River to Johnston's Bar. By 1950 INC was operating 11 towboats and small river craft and 25 barges on the Columbia River, in coastwise service and on the Snake River.

In 1958 Whaley and Dent purchased Captain Leppaluoto's interest in INC, which consisted of nine firms. In 1959 Whaley and Dent sold Upper Columbia River Towing and most of the affiliated water carriers to Pacific Inland Navigation headed by John L. Bullock.

Inter-Island Steam Navigation Company (ISN), 1883–1950

Captain Thomas R. Foster and associates incorporated the Inter-Island Steam Navigation Company on February 7, 1883, to haul people and cargo between the Hawaiian Islands. Captain Foster had three wooden steamers built to start his business: the *James Makee* (1879), the *C.R. Bishop* (1880) and the *Iwalani* (1881). ISN provided service to ports on Kauai and Oahu, as well as the ports of Kona, Kukuihaele, Honokaa and Kukaiau on Hawaii, and Lahaina, Maalaea Bay and Makena on Maui's leeward coast. By 1898 ISN's fleet consisted of 10 vessels. In July 1905 it purchased the Wilder's Steamship Company.

INC was not substantially affected by World War I, as none of its steamers were taken over by the US government. In 1923 it had a new passenger vessel built for the Volcano Route (Honolulu to Hilo): the SS *Haleakala* was 360 feet (110 metres) long and could accommodate 334 first-class and 100 second-class passengers – she was the largest ISN ship in service.

On the eve of the Great Depression, ISN brought out three new steamships: the *Waialeale* (1928) and the *Hualalai* and *Humuula* (1929). By 1934 ISN needed only five vessels in its active fleet. The *Waialeale* and *Hualalai* each made three round trips a week to Hilo (via Lahaina), to Kahului (via Kaunakakai) and Nawiliwili. The *Humuula* and *Hawaii* served west Hawaii and island landings where cattle and sugar were available. The *Naia* ran bi-weekly between Kaumala-pau, Lanai and Lahaina. After Pearl Harbor, the US government took over all ISN ships for military service. When it returned the ships at the end of the war, ISN sold or chartered all but one of them to other companies. It kept only one passenger ship, the *Hualalai*. Increased operating costs and the loss of passengers due to air travel resulted in ISN incurring heavy financial losses. It ceased business in 1950.

Los Angeles Steamship Company (LAS), 1920–37

The Los Angeles Steamship Company was formed on June 10, 1920, to provide passenger service between Los Angeles and San Francisco. It purchased two ships famous on the west coast, the *Yale* and the *Harvard*, from the US Navy. On May 2, 1921, the *Yale* made a festive and gala departure from Los Angeles for her inaugural run to San Francisco, and the *Harvard* followed on August 5. Each vessel made four scheduled sailings a week and carried 565 passengers at an average speed of 24.6 knots.

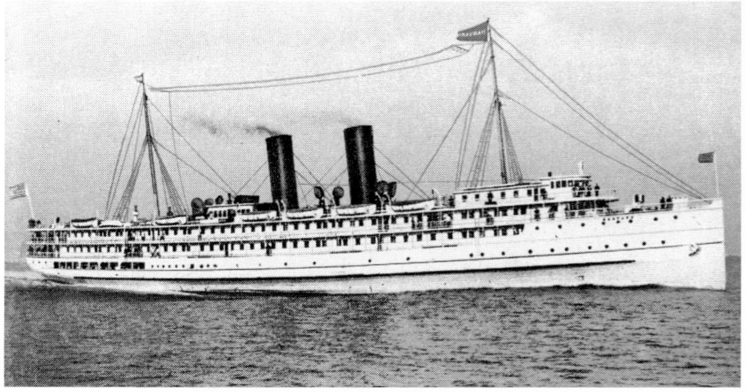

SS *Harvard*.

LAS expanded its service to Hawaii in December 1921, using two German-built war-prize liners chartered from the US Shipping Board. The *City of Los Angeles* departed on her maiden voyage to Hawaii on September 11, 1922, with a gala send-off. Through the 1920s LAS grew and expanded its service to San Diego. The stock market crash of 1929 resulted in a decline of freight and passenger volume. Then in May 1930, a fire gutted the *City of Honolulu (II)* while she was berthed at Honolulu. Now financially overextended, LAS was absorbed by its competitor, Matson, in October 1930. At the time of transfer, on January 1, 1931, LAS owned 12 ships. For a while, Matson operated the LAS ships from the Los Angeles office as a subsidiary. The venerable old *Yale*, last of the LAS ships, was retired in July 1936. On December 30, 1937, the Los Angeles Steamship Company was formally dissolved.

Matson Navigation Company (MNC), 1901–present (passenger service 1908–71)

Matson Navigation was founded in 1882 by Captain William Matson, who began a freighting service between California and Hawaii with the schooner *Emma Claudia*.

The company was officially incorporated at San Francisco in 1901. The first passenger sailing was made by the steamer *Lurline* on June 6, 1908. The 126-passenger vessel *Wilhelmina* followed in 1910. During World War I, the US Navy took over Matson's three passenger ships, *Wilhelmina, Maui* and *Matsonia*, and did not return them until 1920.

In 1926 MNC purchased the Oceanic Steamship Company and its three ships, the *Sierra, Sonoma* and *Ventura*, which had been providing regular freight and passenger service to Australia. With increasing passenger traffic to Hawaii, it added the *Malolo* in 1927. In October 1930 MNC absorbed the financially strapped Los Angeles Steamship Company. Between 1930 and 1932 it built three new liners: *Mariposa, Monterey* and *Lurline*. MNC entered the Australia trade with the sailing of the *Mariposa* on May 7, 1932, and the *Monterey* followed. The *Lurline* joined the *Malolo*, servicing Hawaii. The New Zealand/Australia service ceased in January 1942 when the *Monterey* and *Mariposa* were converted to armed transports for war service. All four passenger vessels served as troop ships during World War II.

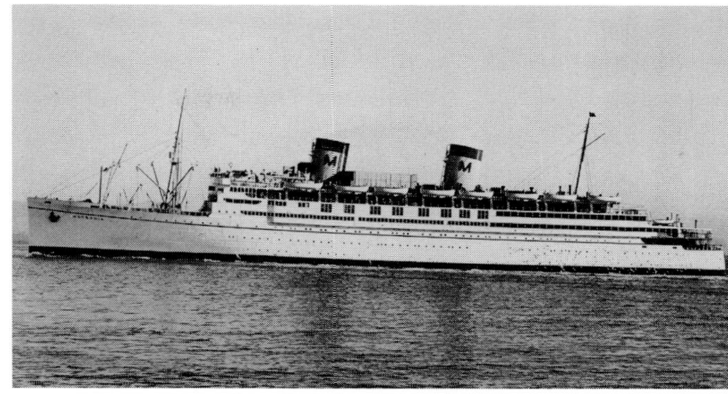

SS *Matsonia*. Matson Navigation Co. postcard.

Only the *Matsonia* returned to passenger service after the war. The *Lurline* replaced her in 1948. In 1955, MNC undertook a major shipbuilding program that produced the South Pacific liners *Mariposa* and *Monterey*; it also rebuilt the wartime *Monterey* and renamed her *Matsonia*. The *Lurline* entered the Pacific coast-Hawaii service until she was sold in 1963. The *Matsonia* was renamed *Lurline* and continued in the Hawaii service until 1970 when she too was sold. In 1971 MNC decided to concentrate on its Pacific coast-Hawaii freight service. It sold its two remaining passenger ships, *Mariposa* and *Monterey*, to the Pacific Far East Line. Matson continues to operate as a freight-only line with a large fleet of container ships.

McCormick Steamship Company (MCS), 1923–40

The McCormick Steamship Company was incorporated in California on April 27, 1923, to consolidate and develop the various steamship businesses of the Charles R. McCormick Lumber Company of Delaware. The new company started with five wooden steam schooners (*Celilio*, *Everett*, *Multnomah*, *Wahkeena* and *Wapama*) purchased from the McCormick Lumber Company. In 1924 MCS purchased the SS *Rose City* from the San Francisco & Portland Steamship Company, formed the Rose City Steamship Company and began regular passenger service between Portland, San Francisco and Los Angeles. The SS *Newport* joined the *Rose City* on this run between 1925 and 1928. Every effort was made to attract a first-class passenger trade, by providing on-board activities and meals of variety and quality.

In 1925 MCS's parent company, Charles R. McCormick Lumber Company of Delaware, purchased all the holdings of the Pope & Talbot Company,

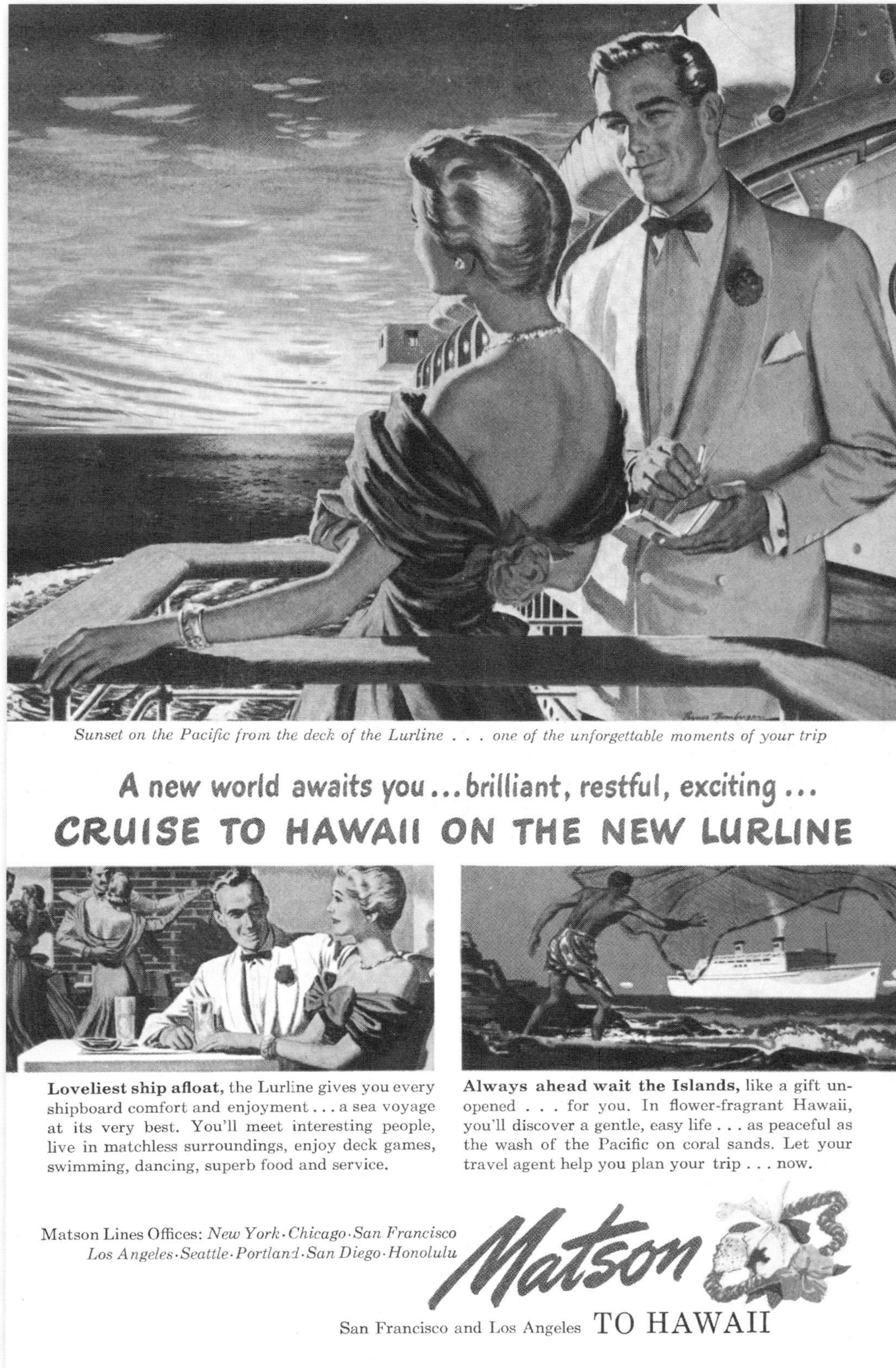

Matson Navigation Co. advertisement in *National Geographic* magazine.

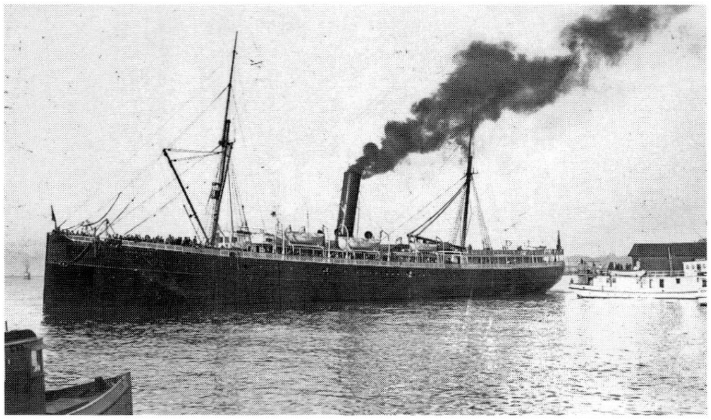

SS *Rose City*. Pacific Photo Co.

including its ships used in coastwise and intercoastal trade and their docks at Seattle, Portland and San Francisco. By the end of 1925 MCS had 25 vessels in coastwise and intercoastal trade. In January 1926 it acquired from the US Shipping Board 6 vessels and the route of the Pacific Argentine Brazil Line. The PABL served Pacific coast ports, then passed through the Panama Canal to east coast ports in South America.

By 1928 MCS was one of the largest operators in the USA, owning some 20 vessels and having another 25 under charter. That year, it expanded its passenger service, offering trips from Tacoma to San Francisco aboard the steam schooner *Wapama*. In 1930 MCS sold the *Rose City*, its only all-passenger vessel. By the mid 1930s MCS's parent company had woefully overextended itself financially. A foreclosure suit was filed against the Charles R. McCormick Lumber Company on February 15, 1938, when it failed to pay off its debt to Pope & Talbot. McCormick's lumber mills and various steamship enterprises passed into the hands of the Pope & Talbot Company, and was reorganized as Pope & Talbot Inc. in 1940. The McCormick Steamship Company was formally dissolved on August 31, 1940, but P&T continued to operate its ships under the McCormick name until 1947.

Monticello Steamship Company (MSC), 1904–27

The Monticello Steamship Company was formally incorporated in 1904 by brothers Zepheniah and Charles Hatch. The Hatches had been operating small steamers in the San Francisco Bay area, under the name Hatch Brothers Steamship Company. With the Monticello Steamship Company, they provided ferry service between San Francisco, Vallejo and Mare Island. Initially, MSC operated the small steamers *Monticello* (126 ft; 38 m) and *Arrow* (157 ft; 48 m). As business improved, it acquired two larger steamers, the *General Frisbi* (187 ft; 57 m) and the *Sehome*. In 1905 MSC co-ordinated its ferry service with the electric rail service of the Napa Valley Route. The trip from Vallejo to San Francisco was approximately 48 kilometres and it took MSC boats an hour and 45 minutes to complete.

As its business grew the company expanded to meet the need. The steel steamer *Napa Valley* (230 ft; 70 m) was added to the fleet in 1910, the *Asbury Park* (*City of Sacramento*) in 1918 and the *Calistoga* in 1924. The *Asbury Park* and *Napa Valley* were converted to carry automobiles in 1922. To offer passenger amenities the MSC opened concessions aboard its boats, such as restaurants, bars, shoeshine stations and barber shops. In 1925 the company transported 96,000 vehicles on its ships. In February 1927 Monticello Steamship Company was sold to the Golden Gate Ferry Company.

North American Transportation and Trading Company (NAT), 1892–1906

The North American Transportation and Trading Company was formed at Seattle in 1892 to compete to the Alaska Commercial Company. The investors and founding members included, John J. Healy, John Cudahy, Michael Cudahy, Portus B. Weare and the Corn Exchange Bank of Chicago. In April 1892, Healy and Weare chartered the steamer *Alice Blanchard* in Seattle and sailed for St Michael with machinery and material for a river vessel. At St Michael they built a log house and assembled the *Portus B. Weare*. In the 1893 season they began establishing mercantile outlets along the Yukon River.

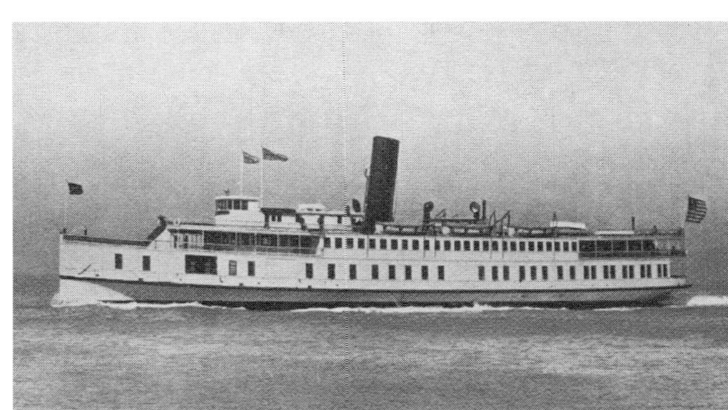

SS *Napa Valley*, ca 1920. Edward H. Mitchell photo.

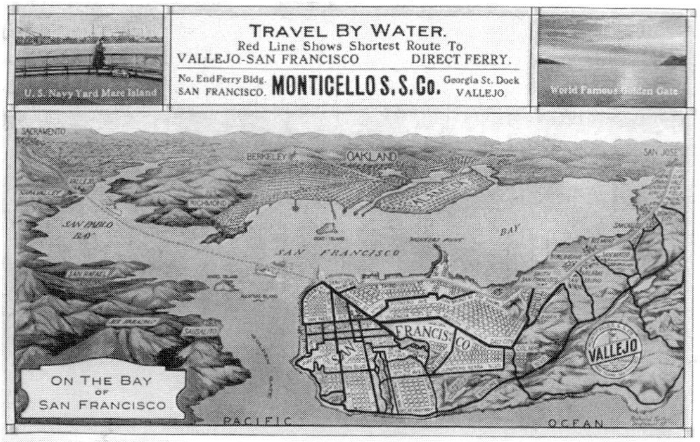

Monticello Steamship Co. advertising card.

By the 1895 navigation season NAT was able to throw its marine and mercantile competition at the veteran Alaska Commercial Company, trading station for station, boat for boat, and price for price. In 1896 NAT built the 175-foot (53-metre) sternwheeler *John J. Healy* at St Michael for service on the Yukon River. It also chartered the SS *Portland* to provide connecting service from Puget Sound with its river steamers at St Michael. The *Portland* was returned to her owners in the fall of 1897, after one season. In August 1897 NAT ordered a fleet of vessels from the Moran Shipbuilding Company of Seattle. It replaced the SS *Portland* with the SS *Roanoke* May 29, 1998, on its service between Puget Sound, St Michael and Nome.

At the height of the gold rush in 1998–99 NAT operated eight sternwheelers (*T.C. Power*, *John Cudahy*, *Portus B. Weare*, *Charles H. Hamilton*, *Klondyke*, *John J. Healy*, *John C. Barr* and *Will H. Isom*) and eight barges on the Yukon River. In 1903 it chartered three small Great Lakes steel freighters from Globe Navigation Co. to move cargo from Seattle to St Michael. NAT chartered its fleet of river steamers to Capt. Omar J. Humphrey during the 1904 season but returned to operating its own fleet in 1905. It sold the steamer *Roanoke* to Charles Doe & Co. of San Francisco in 1905, who had just started the North Pacific Steamship Company. NAT withdrew from the marine field entirely at the end of 1906, selling its fleet of steamboats and St Michael terminals to the Merchant's Yukon Transportation Company.

Northland Navigation Company (NLN), 1951–80

Northland Navigation Company was incorporated by Captain H.J. Terry on June 8, 1951. At its peak Northland owned and operated 11 ships, providing freight and passenger service along the west coast of Vancouver Island and weekly from Vancouver to Prince Rupert, the Queen Charlotte Islands, Stewart and way points. As it grew, NLN purchased the *Queen of the North* (formerly *Princess Norah*) and the *Princess of Alberni* from the Canadian Pacific BC Coast Service, and in January 1959 it acquired the assets and vessels of the bankrupt Union Steamship Company.

In 1972 the Canadian Transport Commission refused to renew the $130,000 annual subsidy for NLN's Vancouver Island west-coast route. Consequently, the *Skeena Prince* made her final run on July 4, 1972. Four years later (October 31, 1976) NLN's north-coast route subsidy expired and was not renewed by the federal government. NLN withdrew its three remaining passenger/freight ships from service and sold them, so ending passenger service from Vancouver to up-coast fishing, logging and mining towns as far north as Stewart. It continued tug and barge service from Vancouver to Kitimat until January 1, 1980, when heavy losses forced the company to put its tugs and barges up for sale, bringing an end to the company.

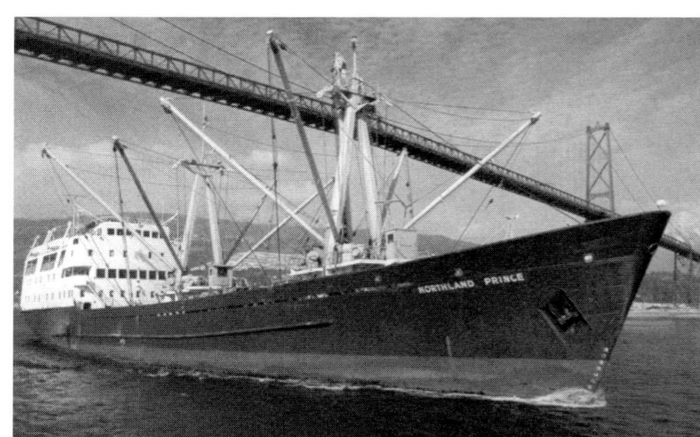

MV *Northland Prince* passing under Lions Gate Bridge. Johnny Clarke photo, Natural Colour Reproductions.

North Pacific Steamship Company (NPS), Thru Line, 1905–17

Two North Pacific Steamships operated on the Pacific coast. The first company was organized by G.E. Dodwell and William Pearce in the early 1890s to operate between Tacoma and the Far East. That line was short-lived, lasting until 1898. The second North Pacific Steamship Company was organized by Charles P. Doe and associates in San Francisco on April 1, 1905. The first ship purchased by the new NPS was the veteran Alaska Coast passenger steamer *Roanoke,* which was placed in service between San Francisco and Portland. San Francisco became the focal point of company operations. The steamer *Alliance* joined NPS in 1908 and provided service to smaller ports between San Francisco and Portland. The *George W. Elder* was added in 1909 and began running in the opposite direction to the *Roanoke* on the Portland-San Francisco route.

The North Pacific Steamship Company reached its zenith during the 1912–17 period, operating seven passenger ships on five routes with over 14 ports of call from Portland to San Diego. As NPS was the only steamship company operating a through service from Portland to San Diego, it adopted the name "Thru Line". Its fortunes began to change with the wreck of the steamer *Santa Clara* at Coos Bay on November 1, 1915, which saw 15 lives lost. Just six months later, on May 9, 1916, the *Roanoke* foundered off Point San Luis with the loss of 47 lives. These two disasters resulted in bad press and an erosion of public confidence. Patronage fell off sharply and the company began losing money. A year later, on May 8, 1917, NPS sold its last three ships, *F.A. Kilburn*, *Breakwater* and *George W. Elder* and the directors voted the company out of existence.

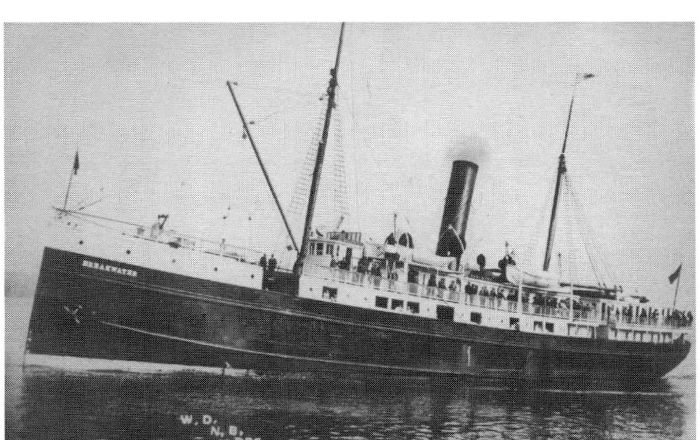

SS *Breakwater.*

Northwestern Steamship Company (NWS), 1904–08

In 1900 John Rosene and associates of Seattle incorporated the Northwestern Commercial Company to operate a mercantile establishment and lighterage service at Nome. The Northwestern Steamship Company was incorporated as a subsidiary of the Northwestern Commercial Company in 1904. NWS provided passenger and freight service from Seattle to northern ports in Alaska. To inaugurate its northern service NWS purchased three large steamers, the *Tacoma*, *Olympia* and *Victoria*. Having close affiliations with Russia, John Rosene dispatched all three vessels to Vladivostok with supplies for the Russians who were at war with the Japanese. On January 1, 1905, NWS purchased the vessels of the Alaska Pacific Navigation Company and took over the Valdez run in addition to its existing Nome run. The company continued to expand and in 1906 added four more vessels (*Pennsylvania*, *Saratoga*, *Yucatan* and *Orizaba*) to its fleet. The Guggenheim-Morgan copper interests who owned Kennecott Copper bought NWS in 1907. In December 1907 the new owners merged NWS with the Alaska Steamship Company. All company assets and vessels were transferred to the new Alaska Steamship Company on January 1, 1908.

Oceanic Steamship Company (OSC), Spreckels Line, 1880–1926

John, Adolph and Rudolph Spreckels, sons of sugar baron Claus Spreckels, formed the John D. Spreckels & Brothers Company in 1880 to provide freight and passenger service between Honolulu and San Francisco. The name of the company was changed a short time later to the Oceanic Steamship Company. OSC ordered two modern ships for this service: the *Mariposa* was delivered July 15, 1883, and the *Alameda* followed a short time later. In 1885 OSC received a contract worth $150,000 per year to provide mail service to Australia. The *Mariposa* and *Alameda*, along with the *Mararoa*, provided that service while two other ships, the *Australia* and *Zealandia*, maintained the San Francisco to Honolulu service.

OSC was economically dependent on trans-Pacific mail contracts from the US government. It thrived when it had contracts but fell into economic problems when it didn't. In 1907 the US government cancelled its $29,000 portion of the Australia mail contract.

As a result OSC dropped its service to Australia but maintained service to Honolulu. Matson Navigation Company purchased OSC in May 1926 and operated it as a Matson subsidiary thereafter. MNC maintained the Oceanic name through the 1930s, sometimes coupling it with Matson, as Matson-Oceanic.

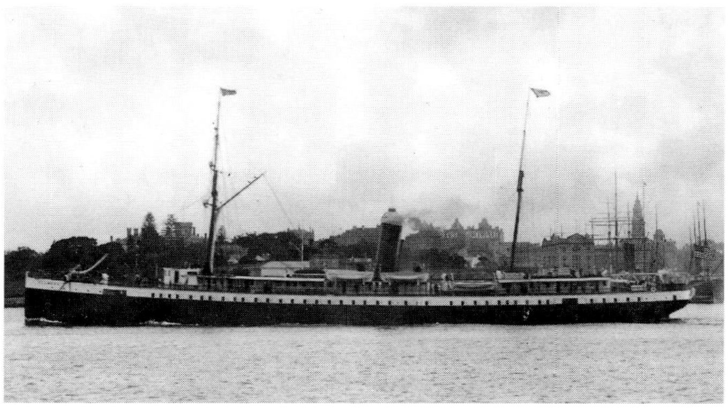

SS *Alameda*.

Oregon Railway and Navigation Company (ORN), 1879–1910

The Oregon Railway and Navigation Company was incorporated by Henry Villard on June 14, 1879. It took over the floating assets of the Oregon Steam Navigation Company on July 1, 1879. One year after incorporation ORN was operating four modern steamers (*George W. Elder*, *City of Chester*, *Oregon* and *Columbia*) in coastwise service, as well as thirty river steamers and 350 kilometres of track. Villard had the SS *Columbia* built especially for the Portland-San Francisco service in 1880 and it was one of the first vessels to have electric lights.

ORN operated paddlewheelers on the Snake, Willamette and Columbia rivers. In May 1881 it purchased the Starr Line's steamers of Puget Sound and began providing steamboat service on the sound and to Victoria and New Westminster, BC. By 1882 ORN had 10 passenger steamers and 5 freight steamers in its ocean service, plying from San Francisco to Puget Sound, San Diego and intermediate ports. The *State of California* ran to Portland, the *Victoria* to New Westminster and the *Constantine* to Mendocino, all iron-propellered steamers. The wooden-propellered *Idaho* ran to British Columbia; the wooden-propellered *Los Angeles* and *Alexander Duncan* and the wooden side-wheelers *Dakota*, *Orizaba*, *Ancon* and *Senator* operated passenger service on the southern coast of California.

In 1896 severe financial problems forced ORN into receivership. It was reincarnated later in the same year as the Oregon Railroad and Navigation Company, incorporated in Washington State on November 21, 1898. By early 1900 ORN had passed into the control of E.H. Harriman's Union Pacific Railroad. Union Pacific organized the San Francisco and Portland Steamship Company in 1904 to operate the ORN's ocean services, while ORN continued to operate river-boat service on the Snake, Willamette and Columbia rivers. The company abandoned all of its navigation on the Willamette River south of Oregon City around 1907. In late 1910 Union Pacific created the Oregon-Washington Railroad & Navigation Company. The ORN was merged into the new company on December 23, 1910. By 1912 all river navigation on the Willamette River south of Portland had been abandoned.

Pacific Alaska Navigation Company (PAN), Admiral Line, 1912–16

In 1907 Hubbard Foster Alexander bought the financially ailing Alaska Pacific Navigation Co., which operated passenger service between Seattle, San Francisco and San Pedro. A few months later he also acquired the Alaska Coast Co., which operated freight and passenger service to Alaska. In 1912 Alexander organized the Pacific Alaska Navigation Company as a holding company for his two shipping lines. On January 1, 1914, Pacific Alaska Navigation Company, Admiral Line, became the operating company, taking over the operation of all vessels owned by Alaska Coast and Alaska Pacific Steamship companies. A 1914 brochure lists PAN as operating six *Admiral* vessels: *Farragut*, *Schley* and *Dewey* maintained

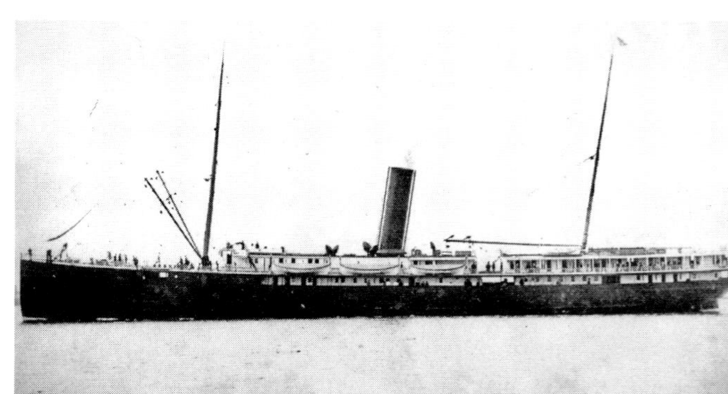

SS *State of California*. Bruce Lanham collection.

service between Seattle, Tacoma and San Francisco, while *Sampson*, *Evans* and *Watson* ran from Seattle to Alaskan ports as far north as Kodiak. PAN provided connecting service from San Francisco to the southern ports of Los Angeles and San Diego aboard the Pacific Navigation Company vessels *Yale* and *Harvard*.

The Admiral Line incurred its first loss on August 26, 1914, when the Canadian Pacific liner *Princess Victoria* rammed and sank the *Admiral Sampson* off Point No Point in Puget Sound. On July 31, 1916, PAN announced that it has leased the *Harvard* and *Yale* from the Metropolitan Steamship Company to maintain service from San Francisco to points south. October 1, 1916, the Pacific Alaska Navigation Company merged fleets with the Pacific Coast Steamship Company of San Francisco to form the Pacific Steamship Company, operating under the trade-name Admiral Line.

Pacific Coast Steamship Company (PCS), 1877–1916

Charles Goodall and Chris Nelson of Seattle formed a steamship company in 1867 to operate steamers on the west coast of the United States and Canada. In 1877, the Goodall, Nelson and Perkins Company reorganized to become the Pacific Coast Steamship Company. At the time of its inception, PCS inherited a fleet of four side-wheelers and ten screw steamers from its predecessor. In 1881 it acquired the government mail contract to Alaska and began running the *Ancon*, *Idaho* and *Eureka* to Alaska once a month to fulfil this commitment. PCS also provided regularly scheduled service from Seattle to San Francisco and points south.

In 1882 the company's steamers *Idaho* and *Geo. W. Elder* sailed three times a month from San Francisco to Olympia, stopping at Victoria, Port Townsend, Seattle and Tacoma along the way. The *Orizaba* and *Ancon* serviced the southern route, stopping at Port Harford, Santa Barbara, San Pedro and ending at San Diego. In 1882 PCS had three new iron-screw steamers, *State of California*, *Queen of the Pacific* and *Santa Rosa* built in eastern yards. It also purchased the *City of Topeka*, *Pomona*, *Curacao* and *Carona* to supplement its expanding coastal service. The PCS fleet grew again in 1887–88 with the purchase of *Walla Walla*, *Umatilla* and *City of Puebla*.

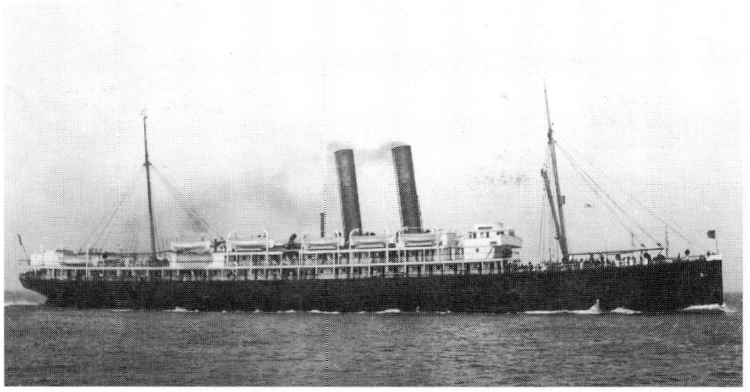

Pacific Coast Steamship Company's SS *Governor* in San Francisco Bay, 1913.

During the 1890s PCS became the leader in both US coastal shipping and the Alaska service. When news of the discovery of gold reached Seattle in 1897, PCS was in the perfect situation as it already maintained regularly scheduled steamship service to Alaskan ports. During the late 1890s the company bought the *Cottage City* and *Senator* for service in the gold-rush trade.

A series of disasters befell the company in the early 1900s. On January 2, 1902, the *Walla Walla* collided with the sailing bark *Max* and sank, drowning 30 people. On October 27, 2004, the SS *Queen* caught fire, killing 14 people. On January 22, 1906, the *Valencia* ran ashore on the west coast of Vancouver Island and 136 people drowned. In 1911 the company's dominance in the San Francisco to Los Angeles service was challenged by the Pacific Navigation Company steamers *Yale* and *Harvard*. Ships of the Pacific Alaska Navigation Company (Admiral Line) and Great Northern Pacific Line were challenging PCS on other routes as well. The loss of the *Cottage City* and *Santa Rosa* in 1911 and the *State of California* in 1913 didn't help to bolster company fortunes. In 1915 PCS began negotiations to sell the company's assets to H.F. Alexander (Pacific Alaska Navigation). On October 1, 1916, the fleets of the Pacific Alaska Navigation Company and Pacific Coast Steamship Company merged to form the Pacific Steamship Company (Admiral Line).

Pacific Far East Line (PFE), 1946–78

The Pacific Far East Line was founded by Thomas E. Cuffe at San Francisco in July 1946 to take advantage of the availability of surplus wartime cargo ships. Initially the company operated five cargo ships under bare-boat charter between California and the Far East.

By 1949 PFE was operating 31 ships across the Pacific and was in a strong position shipping US military cargoes. By the mid 1960s the PFE fleet had been reduced to 13 cargo carriers, 9 of which were the fast, modern Mariner type.

All PFE ship names ended with the word *Bear*. In 1971 PFE purchased the passenger ships *Mariposa* and *Monterey* from the Matson Navigation Company, which was getting out of the passenger-transportation business. PFE offered cruises to Alaska, Hawaii, New Zealand, Australia and Europe. The company filed for bankruptcy in 1978.

Pacific Mail Steamship Company (PMS), 1848–1925

The Pacific Mail Steamship Company was founded in New York on April 18, 1848, by a group of merchants led by William Henry Aspinall. The year before, the company had won a government contract to carry mail from Panama to California. PMS ordered three new steamers early in 1848 to meet the conditions of the contract. The first, *California*, departed from New York on October 6, 1848, and arrived at San Francisco February 28, 1849. It and several other vessels maintained PMS's coastwise service for nearly 20 years.

In 1865 the US Congress approved a subsidy for a new line of steamers from the west coast to the Orient and selected PMS to provide that service. The first scheduled trans-Pacific steamer, *Colorado*, departed from San Francisco for Yokohama and Hong Kong on January 1, 1867. In the next three years, PMS added four new large ships (*Great Republic*, *China*, *Japan* and *America*) to the Orient run. In October 1875, the company began carrying British mail from San Francisco to New Zealand and Australia, but dropped the route in 1885 when the contract was not renewed.

PMS took delivery of the liners *Korea* and *Siberia* in 1902 for service from San Francisco to the Orient, and *Manchuria* and *Mongolia* in 1904 for the Seattle-Orient route. In February 1915 Congress passed the Seaman's Act, which forced the elimination of non-English-speaking Chinese crews. Unable to compete with Canadian and Japanese lines, PMS withdrew from the Pacific trade on August 25, 1915. It sold the *Manchuria*, *Mongolia*, *Korea*, *Siberia* and *China* to the Atlantic Transportation Company. American International Corporation and W.R. Grace & Company purchased what was left of the PMS coastal fleet and continued to operate Pacific Mail as a subsidiary of Grace. PMS continued to operate between San Francisco and Panama, including stops at Mexican ports.

In 1916 PMS purchased three new ships (*Ecuador*, *Venezuela* and *Columbia*) to resume trans-Pacific service. From 1918 to 1922 the company managed 40 vessels for the United States Shipping Board and 12 of its own. In 1921 PMS chartered 5 large new 535-class passenger ships from the United States Shipping Board as it saw its future in the trans-Pacific trade. But on March 25, 1925, under pressure from commercial interests, the Shipping Board put up for auction all of the 535s chartered to PMS. With the loss of its trans-Pacific ships PMS was left with only its older vessels on the Central America run, together with three intercoastal passenger steamers. W.R. Grace created a new subsidiary, called the Panama Mail Steamship Company, to operate the coastal ships. The Dollar Line acquired the 535s and eventually

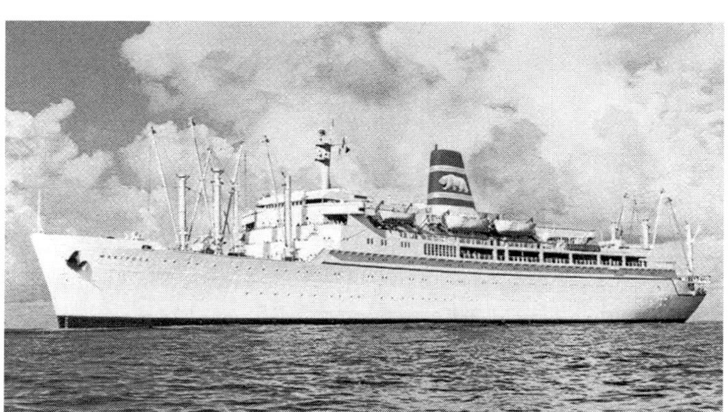

SS *Mariposa*.

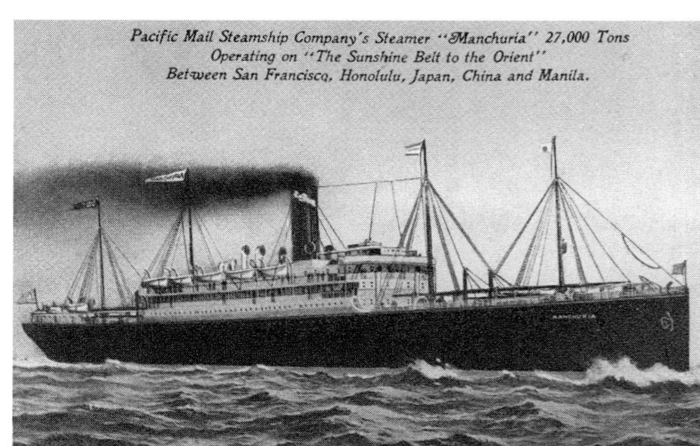

SS *Manchuria* on Sunshine Belt to the Orient. Edward Mitchell photo.

the line name and house flag. The last of the PMS ships, *President Lincoln*, sailed from San Francisco on May 16, 1925, under Dollar Line flag.

Pacific Navigation Company (PNC), 1910–16

In 1910, Bennett and Goodall (formerly Goodall, Perkins and Co.) formed the Pacific Navigation Company to provide passenger and freight service from San Francisco to points south. It chartered the triple-screw turbine sister ships *Harvard* and *Yale* from the Metropolitan Steamship Company of New York. The two ships arrived at Los Angeles on December 16, 1910, and began an overnight service between San Francisco and San Pedro (Los Angeles). PNC extended the service to San Diego in 1911.

H.F. Alexander's Alaska Pacific Navigation Company, which had been operating a through service from Puget Sound to San Francisco and Los Angeles in competition with Pacific Coast Steamship Company, decided to end its service at San Francisco and work in conjunction with PNC for its through service. The two companies coordinated their sailing schedules, tickets and bills of lading. PNC reciprocated in its business between San Francisco and Puget Sound. In the spring of 1916 the Pacific Navigation Company's charter expired for the white flyers *Yale* and *Harvard*. On July 31, the Pacific Alaska Navigation Company (Admiral Line) announced that it had leased the *Harvard* and *Yale* from the Metropolitan Steamship Company. The Pacific Navigation Company was dissolved and the PAN house flag was hoisted on the two ships on September 1, 1916. This arrangement was short-lived as both ships were taken over by US government in 1917 for the war effort.

Pacific Steamship Company (PSC), Admiral Line, 1916–35

On October 1, 1916, shipping baron H.F. Alexander merged the fleets of the Pacific Alaska Navigation Company and Pacific Coast Steamship Company to form a new company known as the Pacific Steamship Company (Admiral Line). With a combined fleet of 20 vessels, PSC operated freight and passenger ships to all major US and Canadian west-coast ports. Less than a year after the Admiral Line had been formed, the United States was at war. In October 1917 the ships *Umatilla* and *Senator* were assigned to the Puget Sound-Orient cargo run, and the *President* and *Governor* were chartered to Matson Lines for wartime service between Seattle and Hawaii. In the spring of 1918 the flyers *Yale* and *Harvard* were taken over by the US Navy for service as transports. After the war PSC began rebuilding its fleet and routes.

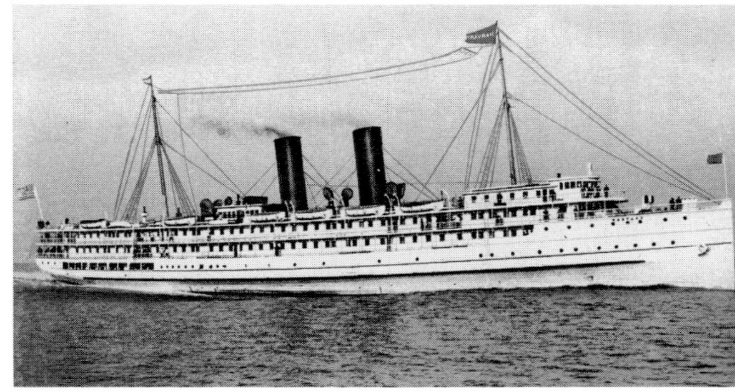

SS *Harvard*. Detroit Publishing Company photo.

In March 1922 PCS chartered the former German liner *Callao*, renamed her *Ruth Alexander*, and placed her on the Seattle-San Francisco run. It also purchased the SS *Great Northern* from the navy, renamed her *H.F. Alexander* and operated her on the Seattle-San Francisco route opposite the *Ruth Alexander*. Later that year the company rebuilt the SS *President*, renamed her *Dorothy Alexander* and began running her between Seattle and San Diego. In November 1923 the Admiral Line bought the China Mail ship *Nanking*, renamed her *Emma Alexander*, and placed her on the Portland-San Francisco route. This gave PSC a fleet of four fast, luxurious ships for southern coastal service.
A number of smaller ships provided service to Alaskan ports. During this same period the US Shipping Board assigned five 535-class passenger-cargo vessels to PCS to manage in overseas service. The company inaugurated passenger service between Seattle and the Orient with the sailing of the *Wenatchee* on April 9, 1921. Early in 1922 the Shipping Board decided that the administration of its 535s should be separate from PSC, so Admiral Oriental Line was formed on October 17, 1922 to operate them.

In 1925 PCS built a great pier, warehouse and office building complex on the Seattle waterfront, providing a modern freight and passenger terminal and administrative headquarters for the company. In the fall of 1927 the Admiral Line added two more passenger vessels, *Admiral Bensen* and *Admiral Peoples*, to its fleet.

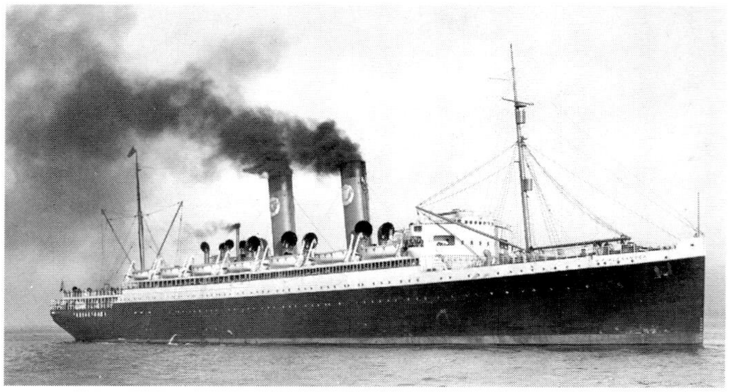

The Pacific Steamship Company's flagship, *H.F. Alexander.*

By 1929 the big steamship company was in trouble, because the automobile was making big inroads in the coastal shipping business. The Great Depression of the early 1930s was the final blow. H.F. Alexander resigned as president of PSC early in 1931 and was succeeded by Robert Stanley Dollar. On April 15, 1932, the Dollar interests (majority shareholders of PSC) consolidated the traffic and operating departments of PSC and the American Mail Line. The coastwise business of the consolidated lines was to be carried out by a new firm, Pacific Steamship Lines Ltd

On February 21, 1933, the fleet of Admiral Line was transferred to Pacific Steamship Lines and Harold Dollar became president of the new company. The 1934–35 period of economic depression, coupled with labour unrest, led to the eventual announcement in early September 1935 that Pacific Steamship Lines was bankrupt and closing down. Admiral Line creditors tried operating Pacific Steamship vessels on the Puget Sound/British Columbia/California route from May to September 1936. But work stoppages and crew walkouts that preceded the waterfront and shipping strike of 1936 made maintaining the service difficult. The last passenger ship, the *Ruth Alexander*, made the final sailing for the company, departing from Seattle for San Francisco on September 28, 1936. The names of ships in the Pacific Steamship Company fleet began with *Admiral* or ended with *Alexander.*

Panama Mail Steamship Company (PMC), 1925–31

In the spring of 1925 the historic Pacific Mail Steamship Company (see page 40) ceased to exist. Its five chartered US Shipping Board vessels that operated in trans-Pacific service were sold to the Dollar Line and its passenger and freight vessels that operated in intercoastal service and to Central American ports were purchased by W.R. Grace & Co.

In June 1925 Grace created a subsidiary called the Panama Mail Steamship Company. The new firm was staffed by former Pacific Mail employees to operate the intercoastal and Central American ships. The vessels *Columbia*, *Ecuador* and *Venezuela* ran from San Francisco to New York, while the *City of Panama* and *City of San Francisco* continued to operate between San Francisco and Panama. Two smaller steamers, the *San Juan* and *Corinto*, operated from San Francisco to Mexican and Central American ports. In 1928 Grace Line transferred two of its ships, the *Santa Ana* and *Santa Luisa*, to Panama Mail for intercoastal service, renaming them *Guatemala* and *El Salvador.* The Panama Mail name and service were short-lived. In January 1931, W.R. Grace absorbed PMC, and its ships adopted the Grace Line's green, white and black funnel colours.

Pope & Talbot Lines (P&T), 1947–61

The Pope & Talbot dynasty on the west coast began on December 1, 1849, when Andrew Jackson Pope and Frederic Talbot from East Machias, Maine, stepped ashore at San Francisco. Captain William Talbot joined them on March 4, 1850, arriving aboard his own brig, *Oriental*, with a load of lumber and passengers. On December 20, 1852, Pope, Captain Talbot, Charles Foster and Captain J.P. Keller formed the Puget Mill Company to manufacture lumber at Puget Sound in the territory of Oregon. The company soon purchased vast timberlands in the region, built mills and operated a fleet of lumber schooners and tugs to move its wood to market. The Puget Sound Commercial Company was formed in 1877 to own and operate Pope & Talbot's lumber vessels.

In 1925 the Charles R. McCormick Lumber Company purchased the holdings of Pope & Talbot, including four large sawmills, its steamships for coastwise and intercoastal trade, and docks at Seattle, Portland and San Francisco. Later, on February 15, 1938, a foreclosure suit was filed against the Charles R. McCormick Lumber Company when McCormick failed to pay off its debt to Pope & Talbot management. On April 27, the modern Pope & Talbot Lumber Company took over management of the foreclosed mills and timberlands of the Charles R. McCormick Lumber Company.

On June 29, 1940, it was reorganized as Pope & Talbot, Inc. The McCormick Steamship Company (MCS) became a division of Pope & Talbot on August 31, who operated it through the war years until early 1947.

In September 1939 MCS operated a fleet of 26 vessels, but at the end of the war only 5 remained afloat. In the spring of 1946 Pope & Talbot re-entered all of MCS's former routes with the exception of the Pacific Argentine Brazil Line's run to the east coast of South America, which had been taken over by the Moore-McCormack Line. To supplement their diminished fleet, Pope & Talbot Lines (P&T) purchased 7 C-3 freighters from the Maritime Commission in 1947: the *P&T Seafarer*, *P&T Forester*, *P&T Pathfinder* and *P&T Trader* went into service between Pacific-coast ports and South America, and the *P&T Adventurer*, *P&T Explorer* and *P&T Navigator* were put in coastwise service. Each ship was equipped to carry 12 passengers. When P&T purchased the new vessels, it replaced the McCormick name with Pope & Talbot. (The "M" disappeared from the white funnel five-pointed star and was replaced with "P&T" in black.)

P&T terminated its intercoastal steamship service in November 1961. The last of its intercoastal ships (*P&T Explorer*, *P&T Forester*, *P&T Navigator* and *P&T Seafarer*) were sold to American Foreign Steamship Corporation of New York in 1963.

Puget Sound and Alaska Steamship Company (PSA), 1890–99

On September 17, 1889, the Northern Pacific Railway purchased the Washington Steamboat Company from Captain J.B. Jackson and associates and began operating steamers on Puget Sound to connect with its trans-Pacific liners. Captain Jackson became general manager of the new railroad-owned Puget Sound and Alaska Steamship Company.

PSA operated two palatial liners – the *City of Kingston* and *City of Seattle* – on Puget Sound and to Vancouver and Victoria in British Columbia. The *City of Kingston* was lost in a collision off the Tacoma waterfront April 24, 1899. Its loss prompted the Northern Pacific Railway to retire from the steamship business in the following May, turning over its routes to the Washington and Alaska Steamship Company of Dodwell and Company.

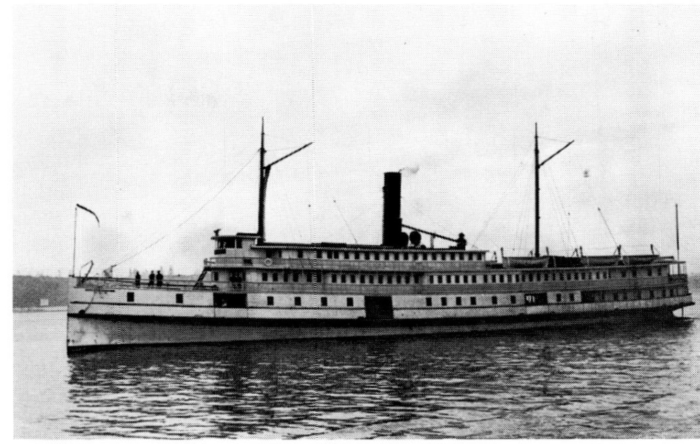

City of Kingston on Puget Sound. Puget Sound Maritime Historical Society photo 546-5.

Puget Sound Navigation Company (PSN), Black Ball Line, 1898–1951

In 1898 the founders of the Alaska Steamship Company formed the Puget Sound Navigation Company to provide transportation service to points around Puget Sound. PSN's smaller vessels served all the routes on the sound, while its larger passenger steamers operated between Seattle, Port Townsend, Port Angeles and Victoria. As highways increasingly linked communities around Puget Sound, the role of water transportation shifted from carrying passengers to carrying passengers and automobiles. By the mid 1920s the conversion of passenger steamers into car ferries was commonplace. For the first 25 years of its history, PSN's vessels flew a distinctive blue house flag with a red star within a white diamond. In 1927 the PSN house flag was changed to a black ball on an orange flag and the company began operating under the name Black Ball Line.

In 1935 the streamlined futuristic-looking diesel ferry *Kalakala* was placed in service on the Seattle-Bremerton run. In the same year, PSN bought out its rival, the Kitsap Transportation Company, making it the largest inland fleet in the world. Between 1937 and 1942, it continued to grow, buying up 17 used ferries from California (made obsolete there by the opening of the Golden Gate Bridge). In 1942 PSN operated 23 vessels and monopolized Puget Sound ferry transportation. But its monopoly would be its downfall. In February 1948 the company proposed a 30% rate increase. The Washington state government refused to allow the rate hike and made plans to purchase

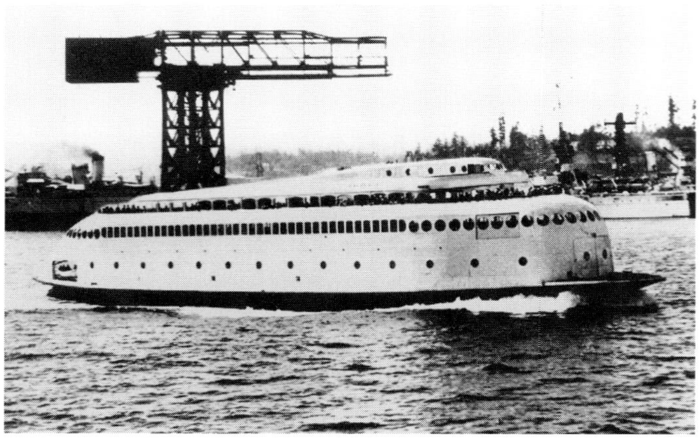

MV *Kalakala* leaving Bremerton. Lorne Campbell collection.

the fleet. On December 30, 1950, PSN's shareholders agreed to sell, and the State of Washington took control of 16 ferries and 20 terminals at midnight June 1, 1951.

San Francisco and Portland Steamship Company (SFP), 1904–24

The Union Pacific Railroad organized the San Francisco and Portland Steamship Company in 1904 to operate the Oregon Railway and Navigation Company's ocean division. SFP operated independently but remained one of E.H. Harriman's interests. Initially it ran the steamers *Columbia* and *George W. Elder* on the "big-three" route (Portland, San Francisco and Los Angeles). In 1910 SFP took delivery of two new steamers, the *Bear* and *Beaver*. The *Bear* was wrecked at Cape Mendocino on June 14, 1916, with the loss of five lives. The *Beaver* and running mate *Rose City* continued on the big-three route until 1917 when the *Beaver* was seconded to the war effort. In 1921 SFP chartered the steamship *Alaska* from the Alaska Steamship Company to assist the *Rose City*. The *Alaska* was lost off Blunts Reef on August 6, 1921, with the loss of 42 passengers and crew. The end came for SFP in 1924 when the McCormick Steamship Company purchased the *Rose City*, the last of SFP's vessels. Union Pacific agreed to the sale on the understanding that McCormick would provide connecting passenger service with Union Pacific trains.

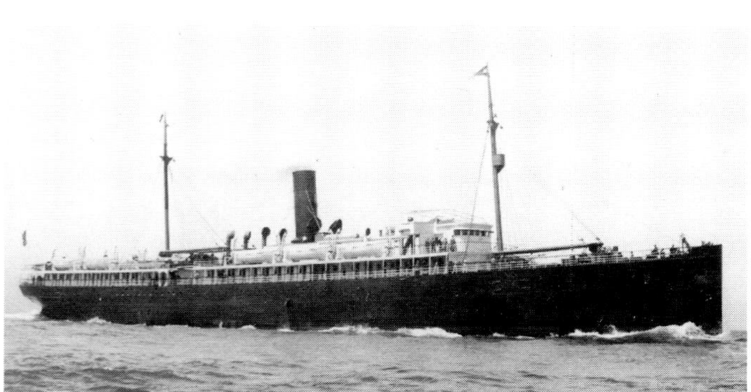

SS *Bear*, 1913.

Seaboard Shipping Company (SSC), 1936–present

The Seaboard Shipping Company was incorporated at Vancouver, BC, in June 1936 to charter ships to transport lumber for a consortium of lumber companies doing business as Seaboard Lumber Sales. By 1939 Seaboard had 143 vessels under charter, most of British registry. With the coming of World War II, SSC's activities were cut drastically because the British government requisitioned the country's merchant fleet for the war effort. Partway through the war, SSC was assigned 10 Canadian government ships to administer by the Park Steamship Company. After the war these ships were offered for sale and SSC purchased 6 of them. A subsidiary company, called Seaboard Owners Ltd, was created to own the ships while SSC acted as the managing operator. The ships' names, *Enterprise*, *Pioneer*, *Queen*, *Ranger*, *Star* and *Trader*, all began with *Seaboard*. These ships operated mainly between British Columbia and the United Kingdom, although they made a few voyages to the Orient in the late 1940s. By 1957 all the ships had been sold off or transferred to Panamanian registry. The Seaboard Shipping Company's needs since the mid 1950s have been met with long-term and short-term charters of foreign flag ships.

Southern Pacific Company (SPC), 1869–1958

The story of the Southern Pacific Company is complex. Its origins lie with the Central Pacific Railroad Company, incorporated June 28, 1861, to build the western portion of the transcontinental rail line. Shortly after the completion of the rail line in 1869, the Central Pacific began a distinctive maritime career on both San Francisco Bay and the Sacramento River that would ultimately span 89 years. In 1880, the Central Pacific introduced the new river steamers *Modoc* and *Apache* on the Sacramento-San Francisco run. The Central Pacific was absorbed by the Southern Pacific Company in 1885. Up until the beginning of the 20th century, SPC made little investment in its Sacramento River steamer operation. As competition increased, it added the new steamers *Navajo* (1909) and *Seminole* (1911), and the freighter *Cherokee* (1912) to compete with the California Transportation Company and others. As had been the practice for several decades, the schedules of the SPC's river steamers and passenger trains were closely aligned. With the onset of the Great Depression, SPC reduced its losses by terminating the poorly performing Sacramento River service on January 15, 1930.

SPC also operated a fleet of ferries between Oakland (the company's major East Bay rail terminal) and San Francisco. In addition to maintaining the busy San Francisco Bay operations, SPC employed two huge train ferries on Carquinez Strait between Benicia and Port Costa. The *Salono* and *Contra Costa*, two of the world's largest vessels of their type, transported entire trains across the waterway until 1930 when the SPC opened a new railway bridge.

By the early 1920s, ferry service on San Francisco Bay was increasingly centred on transporting automobiles. By 1929 SPC and its subsidiary, Golden Gate Ferries, dominated San Francisco Bay ferry services. This monopoly changed with the opening of the San Francisco-Oakland Bay Bridge in 1936, and the Golden Gate Bridge in 1937. By 1948, only four ferry boats remained in active service. The Southern Pacific Company's veteran ferry, *San Leandro*, made her final run July 30, 1958.

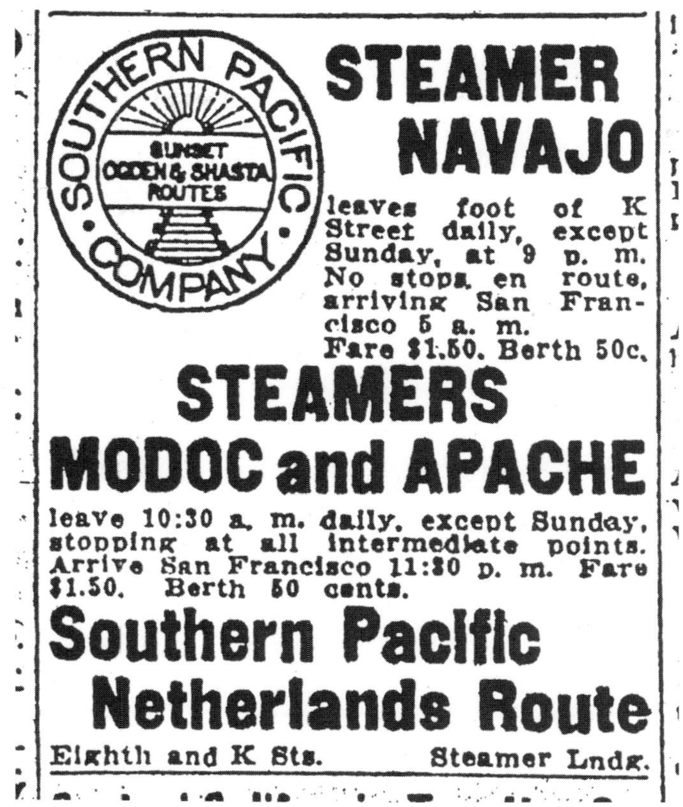
Southern Pacific Company
Sunset Ogden & Shasta Routes

STEAMER NAVAJO

leaves foot of K Street daily, except Sunday, at 9 p. m. No stops en route, arriving San Francisco 5 a. m. Fare $1.50. Berth 50c.

STEAMERS MODOC and APACHE

leave 10:30 a. m. daily, except Sunday, stopping at all intermediate points. Arrive San Francisco 11:30 p. m. Fare $1.50. Berth 50 cents.

Southern Pacific Netherlands Route

Eighth and K Sts. Steamer Lndg.

Early Southern Pacific Company newspaper advertisement. Bill Stritzel collection.

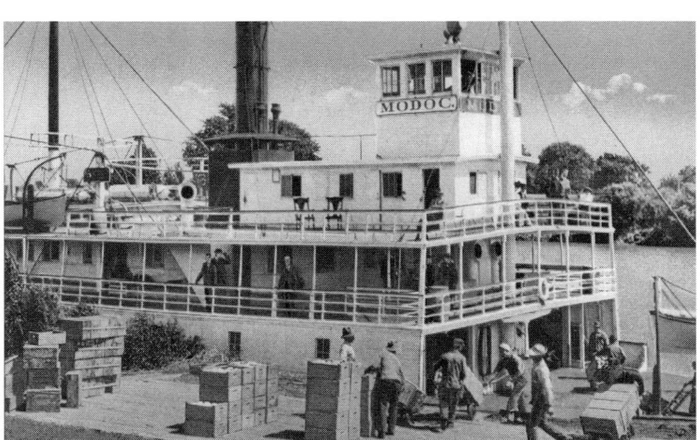

SS *Modoc* loading freight at the bank of the Sacramento River. Edward H. Mitchell photo; Bill Stritzel collection.

Standard Oil Company of California (SOC), 1906–84
Standard Oil of British Columbia, 1935–84

The Pacific Coast Oil Company was established in 1879 and purchased by the Rockefeller Standard Oil Trust in 1900. The Standard Oil Company of California was formed in 1906 to take over the marketing of Pacific Coast Oil. When the courts broke up Standard Oil in 1911, Standard Oil of California was spun off as one of the new companies. To move its oil products, SOC set up a number of shipping subsidiaries, including California Transportation Corp., the California Tanker Co. and SoCal Marine Department. During World War II, SOC ships were appropriated by the War Shipping Administration. They fuelled battleships, aircraft carriers and troop ships along the Pacific coast and towed concrete barges to the South Pacific where they acted as floating gas stations in the war zone. Over the course of its history SOC has operated 164 tankers. The first one, *Maverick*, was built for Pacific Coast Oil in 1890. The most recent is the *Chevron Mariner*, built in 1994. SOC introduced the red, white and blue chevron logo in 1931, which initially had the word "Standard" or "Calso" printed above it.

SOC expanded into Canada in 1935 as Standard Oil of British Columbia, which, in 1956, operated the MV *Standard Service* along the coast.

In 1977, SOC replaced "Standard Oil" with "Chevron", and in 1984, Standard Oil Company changed its official name to Chevron Corporation, which is still operating today.

States Steamship Company (SSS), States Line, 1928–79

The Columbia Pacific Steamship Company was incorporated in Portland in 1919.

Lumber merchant Charles E. Dant and associates bought controlling interest in Columbia Pacific in 1920; it operated to the Far East and Europe. Dant formed the States Steamship Company to handle his lumber schooners and the vessels he leased from the US Shipping Board. In 1928 he dropped the "Columbia Pacific" name and began operating everything under the name "States Steamship Co." or "States Line".

SSS entered the passenger trade in 1932 with the purchase of three United Fruit Co. steamers, *Heredia*, *Parismina* and *Cartago*, renaming them *General Pershing*, *General Sherman* and *General Lee*. The *General Pershing* sailed with passengers and cargo for Japan on December 31, 1932. On their return voyage SSS ships stopped at San Francisco. In 1936 the Marine Act was passed that required companies with passenger ships older than 20 years to make tremendous safety changes or carry no more than 12 passengers. SSS decided that it would not modernize its ships, especially when the government terminated the mail subsidy in 1937. SSS withdrew from passenger service in 1937, the *General Pershing* making its last departure from Portland on June 30. SSS stayed in business as a freighter operation, serving the Far East, until 1979. High fuel costs and a failure to move to containerization eventually caused the company to go bankrupt.

Terminal Steamship (Steam Navigation) Company (TSS), 1902–20

Using proceeds earned in the Klondike, Captain Jack A. Cates purchased the little steamer *Defiance* in 1899 and began providing passenger and freight service from Vancouver to the upper end of Howe Sound. On June 21, 1902, Captain Cates, in association with Evans-Coleman, formed the Terminal Steamship Company. The company launched its first vessel, the *Britannia*, at False Creek on July 7, 1902. TSS was reorganized and reincorporated as the Terminal Steam Navigation Company on July 13, 1908. It provided passenger and freight service to Howe Sound points, including Britannia, Woodfibre and Squamish, and operated excursions to Bowen Island.

Over its 18-year history TSS operated five passenger ships: the *Belcarra*, *Baramba*, *Britannia*, *Ballena* and *Bowena*. In 1912 Cates secured a contract with the Pacific Great Eastern Railway Company

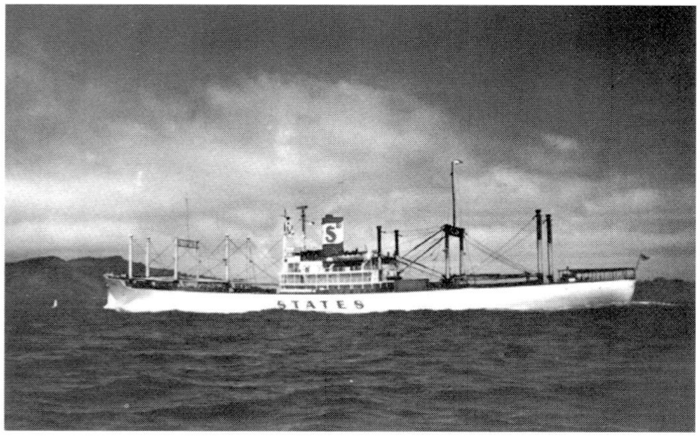

States Steamship Company white cargo liner. Mike Roberts photo.

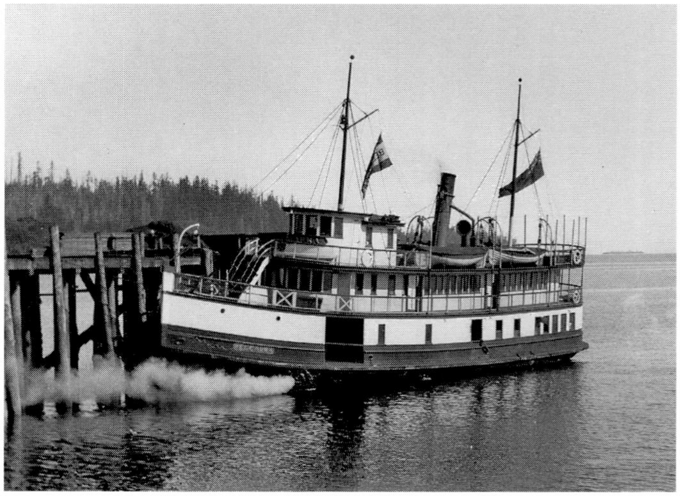

SS *Belcarra* at Sechelt, BC.

to tow a railway car barge between Vancouver and Squamish. Initially the tug *Constance* was used for this purpose, but she was replaced by a much larger tug, the *Dominion*, in 1914. In December 1920 Union Steamships of BC bought the business and Bowen Island property of the Terminal Steam Navigation Company, which included its remaining passenger steamers, *Ballena* and *Bowena*.

Union Oil Company of California (UOC), 1890–1985

The Union Oil Company of California was incorporated on October 17, 1890, by three partners: Lyman Stewart, Thomas Bard and Wallace Hardison. The headquarters for the new company was established in Santa Paula, approximately 160 kilometres southwest of Los Angeles. Recognizing oil's potential as an industrial and transportation fuel, the three partners worked to expand the uses for fuel oil in factories, steamships and railroads. UOC opened its first service station in Los Angeles in 1913 and by 1925 had more than 400 stations on the west coast. The company introduced the first 76-octane gasoline in 1930, to help motorists get better performance from their cars. In 1947, UOC placed "76" in an orange circle as a marketing symbol. To move its oil products UOC operated a tanker fleet; the fleet had eight ships in 1949.

In 1965 UOC merged with Pure Oil Company of Illinois, lifting the company from regional to national status. In 1983 Union Oil Company of California reorganized, becoming an operating subsidiary of a new holding company, Unocal Corporation. It began doing business as Unocal in 1985, and today Unocal is a leading oil and gas exploration and production company.

Union Steamship Company of British Columbia Ltd (USC), 1889–1958

The Union Steamship Company of British Columbia was incorporated in Vancouver on November 16, 1889, to provide transportation by water to outlying settlements. It began with tug and barge service, operating three tugs (*Skidegate*, *Lenora* and *Senator*) and eight scows. The company's first passenger vessel, the SS *Cutch*, began service to Nanaimo on July 3, 1890. More than 50 USC steamships followed the *Cutch*, providing scheduled passenger and freight service to

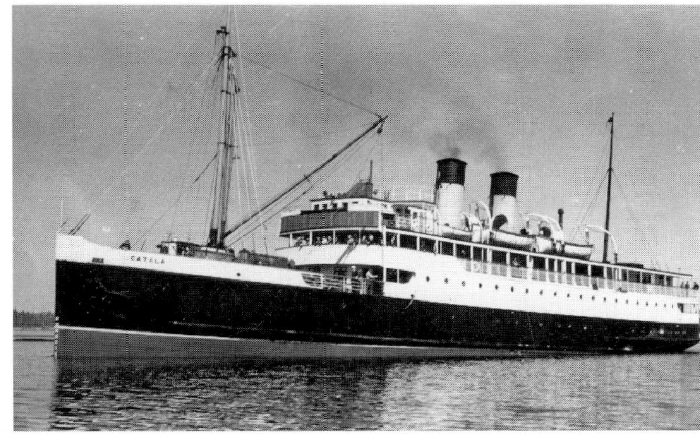

Union Steamship's flagship SS *Catala* at Bella Bella, BC. C.R. Kopas photo.

outlying settlements, sawmills, logging camps, stone quarries and mine operations along the inside waterways of BC. USC also went to Alaska and made many trips into Puget Sound. In 1925 USC purchased the Bowen Island and Selma Park estate properties to enhance its tourist-excursion business. Rising operating costs, increased competition on freight routes and the loss of a federal government subsidy for the northern passenger route forced the Union Steamship Company to close its doors. USC's fleet and floating equipment were sold in 1958 to Northland Navigation Company.

Western Canada Steamship Company (WCS), 1946–65

Western Canada Steamship Company was incorporated nationally on January 22, 1946. It was established by four Vancouver-based shipping companies and a number of prominent businessmen to purchase 21 Park-class ships from the Canadian government.

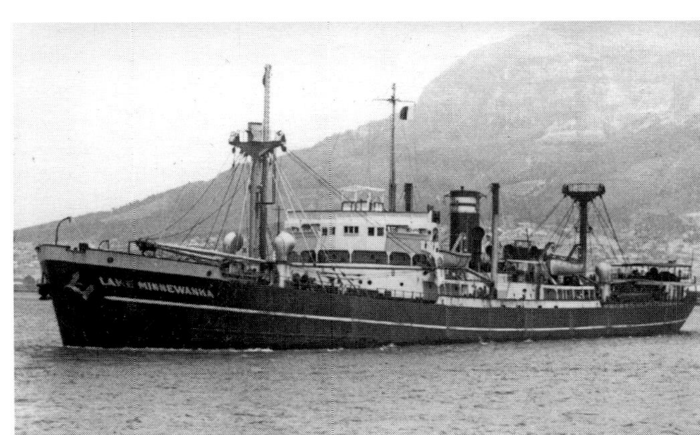

Lake Minnewanka.

WCS acted solely as a ship-owning company, chartering its ships to the shipping companies. Its ships ran on three main trade routes, between British Columbia and: the United Kingdom; the Far East; and southern and eastern Africa, returning by way of India. All ships names began with *Lake*. WCS maintained its fleet of 21 ships until 1949. It sold 6 ships that year, then 7 in 1950, 2 in 1951 and 3 in 1954. One of the company's ships was wrecked in 1957. The 2 remaining ships kept operating until 1963, when Anglo-Canadian Shipping gained control of Western Canada Steamship Company. The last ship, *Lake Atlin*, was sold in November 1965.

Western Steam Navigation Company (WSN), Vancouver Line, 1900–1907

Tacoma resident Cary W. Cook organized Cook & Company in 1897. He chartered the lower Columbia River paddle steamer *Ocean Wave,* moved it to Puget Sound and began operating it between Tacoma and Vancouver, BC, carrying passengers and freight. In 1899 Cook replaced the *Ocean Wave* with the 166-foot (51-metre) side-wheel steamer *North Pacific* purchased from the Oregon Railway and Navigation Company. He formed the Western Steam Navigation Company at Tacoma in 1900 to operate the *North Pacific* on the Tacoma-Seattle-Vancouver route. On July 31 that year WSN launched the 163-foot (50-metre) passenger steamer *Mainlander* at Tacoma. On July 18, 1903, the *North Pacific* struck Marrowstone Point and sank in deep water. A few months later the Pacific Coast Steamship Company purchased the *Mainlander* from WSN to use in its Puget Sound Service. (It was rammed and sunk in fog by the tug *Sea Lion* in 1904 while operating under the PCSS flag.)

WSN replaced the *North Pacific* and *Mainlander* with the steamers *Ramona* and *Waialeale*. Cook brought the *Waialeale* from Hawaii in 1905 for the Tacoma-Vancouver run. In 1907 the Puget Sound Navigation Company bought the *Waialeale* from WSN, bringing an end to the Western Steam Navigation Company. Cary Cook became the Pacific coast manager for the American-Hawaiian Line.

Canadian Government Services

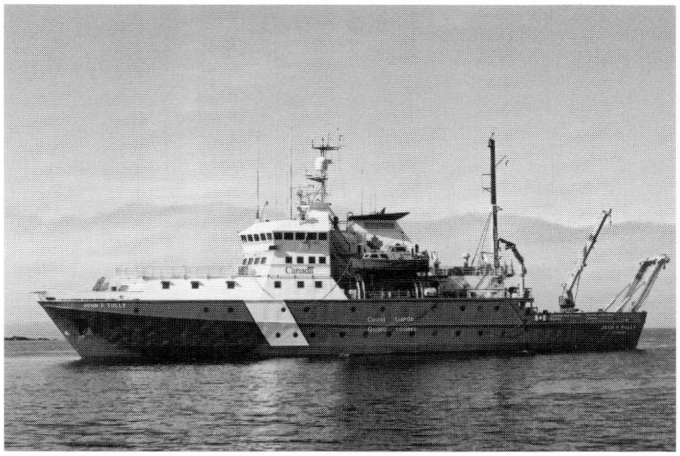

CGS *Tully*. J. Marc photo.

Canadian Coast Guard (CCG), 1962–present

The modern-day Canadian Coast Guard was created during the fifth session of the twenty-fourth parliament in January 1962, but it has been evolving since before confederation. The first lighthouse on Canadian soil was built at the fortress of Louisbourg, Nova Scotia, in 1733, and the colonial government of Nova Scotia established a lifesaving station on Sable Island in 1793. At the time of Confederation in 1867, the federal government of Canada acquired a collection of aids to navigation, lifesaving stations, canals and waterways, regulatory bodies, and associated fleets of vessels. The Department of Marine and Fisheries was established to take over all these marine responsibilities. CGS (Canadian Government Ship) prefixed the names of department ships. In 1936 the government of Canada passed the Department of Transportation Act. All modes of transportation including the marine section were consolidated under this new federal agency. The Department of Transport fleet of vessels was renamed the Canadian Coast Guard in 1962.

The job of the Coast Guard is a big one. It maintains aids to navigation, has a law enforcement role, is responsible for marine search-and-rescue operations, and looks after environmental protection. The Canadian Coast Guard left the Department of Transport and merged with the Department of Fisheries and Oceans on April 1, 1995.

Canadian Government Merchant Marine (CMM), 1918–36

The Canadian government established the Canadian Government Merchant Marine as a Crown corporation on December 30, 1918, to manage and operate a fleet of freighters that were built for the war effort. The government constructed 63 vessels between 1917 and 1921. Several were built on the west coast and put into service between British Columbia and the Orient. The name of each ship began with *Canadian*. The first west-coast ship completed, the *Canadian Raider*, began monthly service to Australia and New Zealand in 1920. The CMM briefly operated a Vancouver-India service, from 1921 to 1923. From 1920 to 1924 it operated four ships in conjunction with four ships from the Blue Funnel Line to the Orient.

The CMM operated seven ships in its west-coast operations: *Canadian Raider*, *Voyageur*, *Freighter*, *Inventor*, *Importer*, *Exporter* and *Miller*. For the first few years the CMM fleet showed an operating profit, but during 1921 freight rates dropped and foreign competition increased. This led CMM to sell off some of its ships. The *Canadian Voyageur*, last of the west coast ships, made its final voyage in 1925. Many ships on the east coast operated longer. But by 1936 the Canadian Government Merchant Marine had disposed of all of them and ceased operations.

CMM also operated the four-masted barkentine *S.F. Tolmie* on voyages to the Orient and Antipodes from 1921 to 1923. In 1924 it briefly took over operation of the former Grand Trunk Pacific ships *Prince Rupert* and *Prince George*, after GTP became part of the Canadian National Railway, and placed both on the Vancouver-Skagway run during the spring and summer.

Canadian Department of Marine and Fisheries (DMF), 1867–1927

The Department of Marine and Fisheries was created with the formation of Canada on July 1, 1867. The act authorizing its existence was passed May 22, 1868. The new department formed out of a collection of agencies acquired from the provinces joining Confederation. DMF's list of administration duties included: coast and inland fisheries, navigational aids and their maintenance, harbours and ports, steamers and vessels belonging to the government of Canada, harbour commissioners and masters, the classification of vessels,

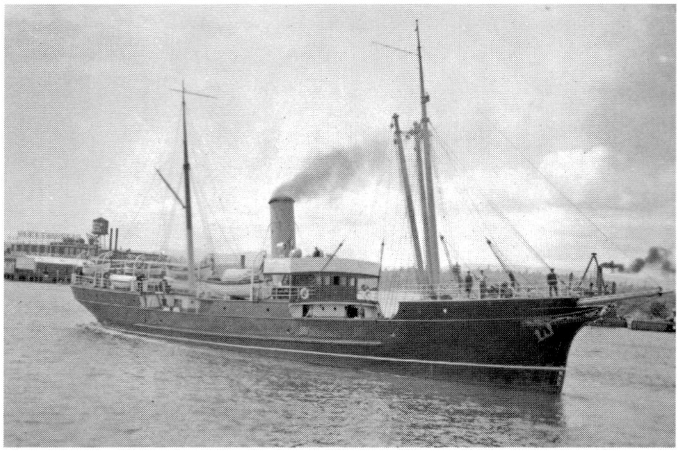

CGS *Quadra*. BC Archives A–00128.

the examination and granting of masters and mates certificates, inspection of steamboats, and shipwreck investigations. To carry out its mandate DMF operated a fleet of vessels on both coasts, principally for maintaining aids to navigation and for fisheries patrols.

The first DMF vessel on Canada's west coast was the lighthouse and buoy tender *Sir James Douglas*. It was followed by the CGS *Quadra* and others. DMF was split apart into the Department of Marine and the Department of Fisheries in 1884, but amalgamated again in 1892. The Meteorological Service was added to the DMF in 1887 and hydrographic work became another function in 1904. In 1911 the International Pelagic Sealing Treaty was signed and DMF took responsibility for the protection of the seals against poaching. It also enforced the fishery regulations in the salmon and halibut industries along BC's coast. In 1927 the Department of Marine and Fisheries once again split into two agencies: the Marine Branch and the Fisheries Branch. The Department of Transport Act of November 2, 1936, saw the Marine Branch placed under a single authority with other transportation agencies.

Royal Canadian Navy (RCN), 1910–present

The Royal Canadian Navy was created with the passing of the Naval Service Act on May 4, 1910. The first RCN vessel was the British cruiser *Rainbow,* commissioned on August 4. Its second, HMCS *Niobe,* arrived on September 6. By the outbreak of World War II, the RCN had only six destroyers and a handful of smaller ships. It expanded greatly during the war and by the end of 1945 Canada had the third-largest navy in the world, behind only the United States and the United Kingdom.

In the postwar period the RCN was reduced to a more manageable size. On February 1, 1968, the three Canadian armed services became unified as the Canadian Armed Forces. The modern Canadian navy is known as Canadian Forces Maritime Command, but still refers to itself unofficially as the Navy and maintains many RCN traditions. The Canadian Armed Forces, including the Maritime Command, performs three broad roles: protecting Canadians, defending North America in cooperation with the United States, and contributing to international peace and security. The Maritime Command is also responsible for the coordination of search-and-rescue operations in the Atlantic and Pacific regions. To carry out its role the Navy maintains a fleet of frigates, helicopter destroyers, fleet-replenishment ships and submarines. A squadron of helicopters and aircraft support the ships and carry out search-and-rescue operations.

United States Government Services

US Army Transport Service (USAT), 1898–1948

The US Army Transport Service was organized on November 27, 1898, as part of the Army Quartermaster Department to provide ocean transport of equipment and supplies in support of army operations. During the early 1900s the US Army operated a large trans-Pacific operation in support of US troops in the Philippines. In World War I the USAT sent 50 ships to support the American Expeditionary Force in France. At the peak of its operations during the war, the USAT operated 168 vessels larger than 1000 tons. The USAT was absorbed into the US Army's Transportation Corps, becoming part of the Water Division in 1942.

During the war, four government agencies competed to utilize the commercial merchant marine. This proved cumbersome, so after studying the situation, the Secretary of Defence announced on December 15, 1948, that all military sea transport, including army transports, would be placed under navy command in a new organization known as the Military Sea Transportation Service (MSTS). On March 1, 1950, the 72 ships of the ATS based in the continental United States were redesignated from US Army Transports to US Naval Ships. Four months later, on July 1, the ATS offices and assets based at Bremerhaven (Germany), San Juan (Puerto Rico), Balboa (Panama), Honolulu and Tokyo were tuned over to the navy. The last portion of the army's fleet, 10 ships stationed in Alaska, joined MSTS on November 1, 1950. The Army Transportation Corps still operated a substantial fleet of small craft.

US Bureau of Fisheries (USBF), 1903–39

The United States Commission of Fish and Fisheries was established in 1871 to protect, study, manage and restore fish populations. In 1903 the Commission's activities were transferred to the newly established Bureau of Fisheries, under the Department of Commerce and Labor. The USBF retained the scientific responsibilities of the Commission of Fish and Fisheries, and took on more functions, such as the jurisdiction, supervision and control of Alaskan fur seals. In its 36-year existence the USBF built and operated fish hatcheries all over the continental US. The first hatchery on the west coast was established in 1905 on a lake at Yes Bay in southeastern Alaska. To move fish, the USBF operated a number of fish distribution railroad cars. Biological research laboratories were built all over the country to study fish. The first permanent station was built at Woods Hole, Connecticut, in 1885. Since 1871, Fisheries Commission/Bureau has had many research vessels sail out of Woods Hole.

From 1915 to 1921 the RV *Albatross* conducted considerable research off the coasts of Oregon, Washington and California, including tuna studies of southern California and Baja California. The USBF has always carried out law enforcement, but an official Division of Law Enforcement wasn't set up until 1930. To enforce Pacific salmon regulations in the 1930s, the USBF employed 228 agents using 24 vessels and its first airplane. On June 30, 1940, the Bureau of Fisheries was transferred to the US Department of the Interior, Fish and Wildlife Service. In 1956 the US Fish and Wildlife Service established the Bureau of Commercial Fisheries. In 1970, the Bureau of Commercial Fisheries amalgamated with other agencies to form the National Marine Fisheries Service, part of the National Oceanic and Atmospheric Administration.

US Coast and Geodetic Survey (USCS), 1878–1970

The Survey of the Coast was established through legislation on February 10, 1807, to undertake a survey of the coasts of the United States. The survey conducted its early activities under the Treasury Department, where it shared vessels with the Revenue Cutter Service. In 1878, the Survey of the Coast's name was changed to Coast and Geodetic Survey to reflect the role of land surveying in the agency's work. Over its 92-year history, the USCS established 13,000 geodetic benchmarks across the continent, developed and produced tide-prediction tables, carried out hydrographic surveys, charted US airways, produced nautical and aeronautical charts, and published books for coast pilots containing navigation information not portrayed on charts. During World War II it operated and maintained 9 major ships to carry out its work; by the 1960s its fleet had increased to 14. With the formation of the National Oceanic and Atmospheric Administration (NOAA) in 1970, the Coast and Geodetic Survey was renamed the National Ocean Survey.

The USCS operated survey ships on the west coast as early as 1914. The survey ship *Patterson* assisted in the search for the revenue cutter *Tahoma* lost on September 26, 1914. On May 30, 1929, the USCS steamer *Surveyor* attended the scene of the *Aleutian* sinking.

US Coast Guard (USCG), 1915–present

The history of the United States Coast Guard is complicated because it comprises five separate federal agencies. On January 28, 1915, the US government combined the Life Saving Service and Revenue Cutter Service to form the modern-day Coast Guard. It added the Lighthouse Service on July 1, 1939, and the Bureau of Marine Inspection (formerly Steamboat Inspection Service and Bureau of Navigation) on July 16, 1946.

The USCG reported to the Treasury Department (except during the two world wars, when it reported to the US Navy) until 1967, when it moved to Department of Transportation. The USCG has many responsibilities, including maintaining aids to navigation, enforcing the law in US waters, conducting marine search-and-rescue operations, and looking after preventative safety and environmental protection. To carry out these duties the US Coast Guard maintains a fleet of ships and aircraft.

US Lighthouse Service (USLS), 1852–1939

The first lighthouse in the United States was established in 1716 by the British on Little Brewster Island in Boston Harbor. The costs and responsibility for the construction and maintenance of early lighthouses were borne by local governments and states. The first federal authority to manage lighthouses was established in 1789 and placed in the Treasury Department. The responsibility for the lighthouse system was transferred between the secretary of the treasury and commissioner of revenue several times between 1802 and 1820. The United States Lighthouse Service was established on October 9, 1852, to take over responsibility of 331 lighthouses and 42 lightships in the lighthouse system.

The USLS carried out light inspections, managed the distribution of materials and supplies to stations and vessels, built new facilities and maintained existing ones, and established light-keeper duties and guidelines. By 1910 there were 11,713 types of navigational aids and the USLS could no longer manage these effectively. It was replaced by the Bureau of Lighthouses on June 17, 1910, which was staffed primarily by civilians. The Bureau of Lighthouses became part of the US Coast Guard on July 7, 1939.

US Marine Hospital Service (USMH), 1870–1902

The Marine Hospital Service was created in 1870 to provide care to sick and injured merchant seamen and was the precursor to today's US Public Health Service. The USMH was created by converting a loose network of locally controlled marine hospitals into a centralized organization, with its headquarters in Washington, DC. John Maynard Woodworth was the first Supervising Surgeon (later called the Surgeon General) in 1871. Woodworth created a cadre of mobile service physicians who could be assigned as needed to the various marine hospitals. The National Quarantine Act of 1878 conferred authority for quarantine activities to the USMH; in 1891, it became responsible for conducting medical inspections on arriving immigrants to prevent disease from entering the country. The USMH also conducted the first physicals at US Coast Guard Academy. As a result of its broadening responsibilities the name of the Marine Hospital Service was changed, in 1902, to the Public Health and Marine Hospital Service. This was shortened to Public Health Service in 1912.

US Maritime Commission (USMC), 1936–50

The United States Maritime Commission was established by the Merchant Marine Act of June 29, 1936. It succeeded the US Shipping Board and the US Shipping Board Bureau in the Department of Commerce. Its purpose was to develop and maintain a merchant marine for the promotion of US commerce and defence. The USMC was authorized to regulate US ocean commerce, supervise freight and terminal facilities, and administer government funds to construct and operate commercial ships.

When it entered World War II in 1942, the United States established the War Shipping Administration as part of the USMC. Its task was to purchase and operate civilian ships needed to fight the war. Numerous west coast shipping companies, including American Presidents Line, acted as agents for the War Shipping Administration, overseeing the handling, manning, equipping, overhauling and repair of cargo and passenger vessels. The USMC was abolished on May 24, 1950, and its functions were transferred to the Department of Commerce, where they were assigned to the Federal Maritime Board and the Maritime Administration.

US National Oceanic and Atmospheric Administration (NOAA), 1970–present

The National Oceanic and Atmospheric Administration was established on October 3, 1970, under the US Department of Commerce. It combined the Environmental Science Services Administration, Bureau of Commercial Fisheries, Bureau of Sport Fisheries and Wildlife, National Sea Grant Program (from the National Science Foundation), elements of the US Lake Survey (from the Department of the Army) and the National Oceanographic Data Centre (from the Department of the Navy). The NOAA has evolved into a science agency with conservation management and regulatory responsibilities. Today, through its various departments, the NOAA forecasts weather and sea conditions, charts the ocean, monitors the environment, manages the fisheries resource and protects National Marine Sanctuaries. In 1970 the NOAA acquired the 14 ships from the Coast and Geodetic Survey.

US Navy (USN), 1775–present

The United States Navy is a branch of the US Armed Forces responsible for conducting naval operations. Its stated mission is to maintain, train and equip combat-ready naval forces capable of winning wars, deterring aggression and maintaining freedom of the seas. The US Navy traces its origins to the Continental Navy, which was established October 13, 1775, during the American Revolutionary War, but was disbanded in 1790. Attacks against American shipping by Barbary-coast pirates spurred the US Congress to pass the Naval Act of 1794, which initiated the construction and manning of six frigates and the creation of the first federal navy. These six ships were the *United States*, *Constellation*, *Constitution*, *Chesapeake*, *Congress* and *President*.

When the US declared war on Great Britain in 1812, it dispatched its navy to do battle against the British warships. Naval actions during the US Civil War mostly consisted of blockades by the North against southern ports. After the civil war, the US Navy fell into a short period of decline; it shrank as its ships were broken up or sold. A modernization program beginning in the 1880s brought the US Navy into the first rank of the world's navies by the end of the century. When the Spanish-American War was declared in April 1898, the Asiatic fleet attacked and defeated the Spanish fleet in the Philippines. A few weeks later, the Atlantic fleet destroyed Spanish ships in the Caribbean.

In the early 1900s, under Theodore Roosevelt's administration, the US Navy added many more ships to its fleet. Seeking to demonstrate growing US military power, Roosevelt dispatched a fleet consisting of four battleship squadrons and their escorts on a circumnavigation of the world from December 16, 1907, to February 22, 1909. The US stayed out of World War I until 1917, so the US Navy's role was mostly limited to escorting convoys and transporting troops. Following the American entry into the World War II, after Pearl Harbor, the US Navy grew tremendously. By war's end in 1945, the US Navy had added hundreds of new ships, including 18 aircraft carriers and 8 battleships. The US Navy was a major participant in the Vietnam War and blockaded Cuba during the Cuban Missile Crisis. Today it maintains a sizeable presence (278 ships and 4000 aircraft) in the world, deploying in such areas East Asia, southern Europe and the Middle East.

US Revenue Cutter Service (USRC), 1790–1915

On August 4, 1790, the US Congress authorized the Secretary of the Treasury, Alexander Hamilton, to create a maritime service to enforce laws governing the collection of customs and tonnage duties. In later years the Revenue Cutter Service also worked to suppress smuggling, piracy and the slave trade, it assisted ships and removed navigational hazards, and it enforced quarantine regulations, neutrality laws and laws prohibiting the importation of Chinese coolies; after 1867 it enforced regulations in Alaska concerning the hunting of fur-bearing animals, the protection of fisheries, and the traffic of firearms, ammunition and liquor.

The USRC was supervised by collectors of customs from 1791 to 1871. A new Revenue Marine Division of the Treasury Department was established in 1871. It became officially known as the Revenue Cutter Service on July 31, 1894.

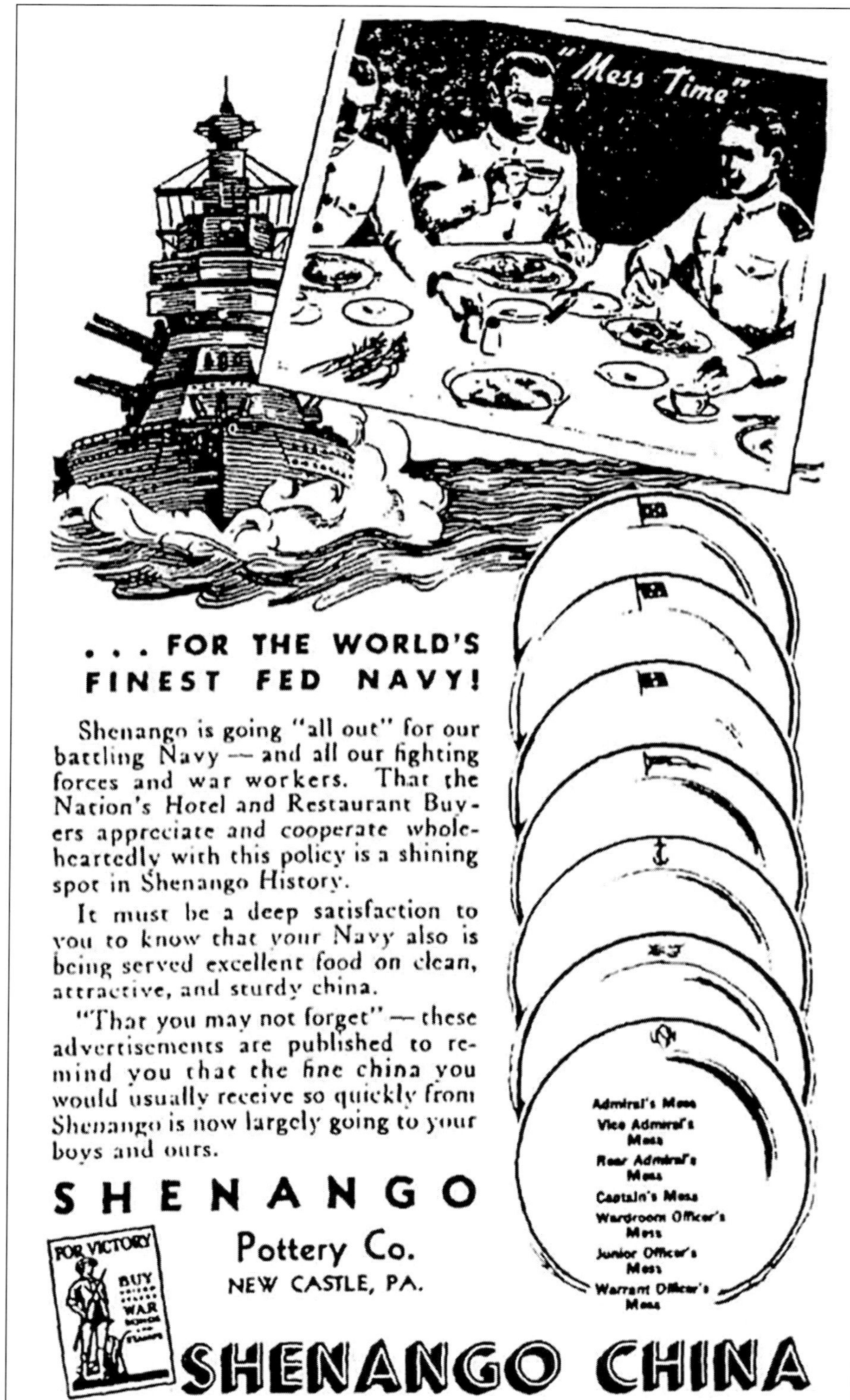

Shenango China advertisement.

No cutters were assigned to the northwest coast until September 1854, when the topsail schooner *Jefferson Davis* arrived in Puget Sound. In 1861 she was replaced with the *Joe Lane*, another topsail schooner; sometimes the *Shubrick*, a steam-powered lighthouse tender, supplemented the *Joe Lane*. The cutter *Lincoln* took over *Joe Lane*'s duties in 1866. Perhaps the most famous revenue cutter to serve in territorial waters was the *Oliver Wolcott*. She patrolled the coast from 1875 to 1897, often dealing with vessels attempting to smuggle opium and Chinese labourers from Canada into Washington. In 1908 the life-saving and salvage tug *Snohomish* arrived on the coast. It operated out of Neah Bay and Port Angeles. In 1915, the Revenue Cutter Service merged with the Lifesaving Service to form the modern-day Coast Guard.

US Shipping Board (USSB), 1917–33

The United States Shipping Board was organized on January 30, 1917, to develop and maintain a strong merchant marine and to regulate ships involved in foreign and interstate commerce. With the entry of the United States into World War I in April 1917, it became necessary for the USSB to assemble a large fleet of vessels to transport men and materials to Europe. The US government authorized the board to seize enemy-owned ships and to requisition ships from within the jurisdiction of the United States. The USSB equipped, manned and operated these ships. Under the Merchant Marine Act of June 5, 1920, the USSB was given broad authority to study the economics of shipbuilding and operation, to lend money for private construction of vessels, to determine what steamship lines should be opened in foreign and domestic commerce, and to recommend to Congress the shipbuilding programs necessary for the maintenance of an adequate merchant marine.

USSB vessels were operated by various US lines, including Pacific Steamship, Pacific Mail, American Mail and Dollar Steamship on the west coast. A class of USSB passenger vessels known as 535s (in reference to their length) was named after American presidents. In its latter years the USSB was responsible for the liquidation of the government-owned merchant fleet that it had amassed. An executive order on June 30, 1933, directed that the United States Shipping Board be abolished and its functions transferred to the newly formed US Shipping Board Bureau.

5. Identifying Ship China

How do you identify which steamship company owned a particular piece of china and where that piece was used? In some circumstances shipping companies top marked their china with their house flag. These flags, unique to each shipping company, were flown at the company offices and aboard ship when entering or leaving a harbour. Learning shipping company house flags is a good way to identify what line a piece of china is from. Drawings of house flags for shipping companies discussed in this book can be found in Appendix 4. To identify other flags, consult *Flags, Funnels and Hull Colours* by Colin Stewart or *Brown's Flags and Funnels, Shipping Companies of the World*, compiled by J.L. Loughran.

It is important to note that most shipping company house flags are rectangular, with or without swallowtails. There are a few exceptions: pennants are normally from yacht clubs, but the Canadian National Steamships and Alaska Commercial Company both used pennants. The US Navy commissioning pennant is another exception.

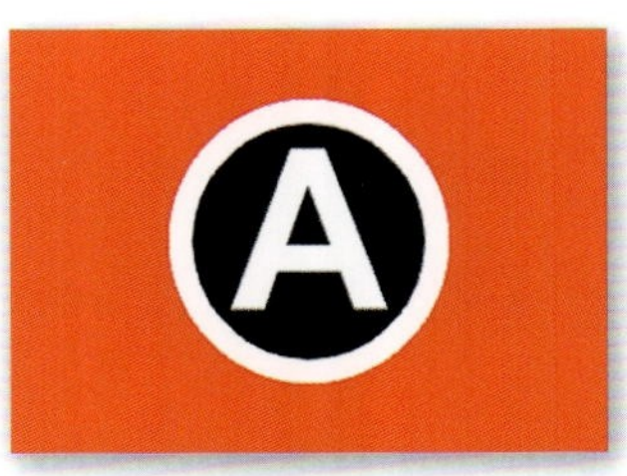

Alaska Steamship flag.

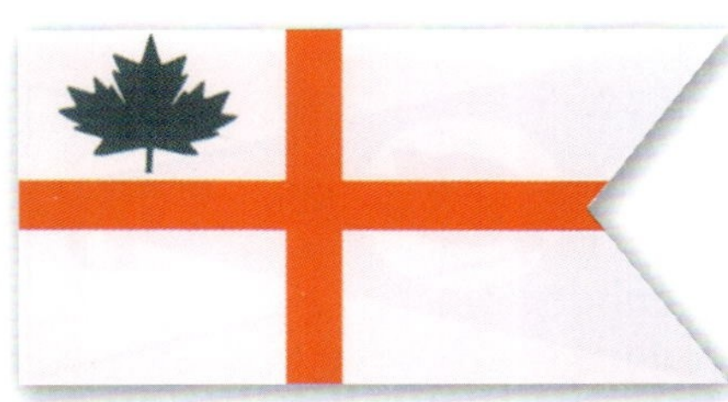

Western Canada Steamship swallowtail.

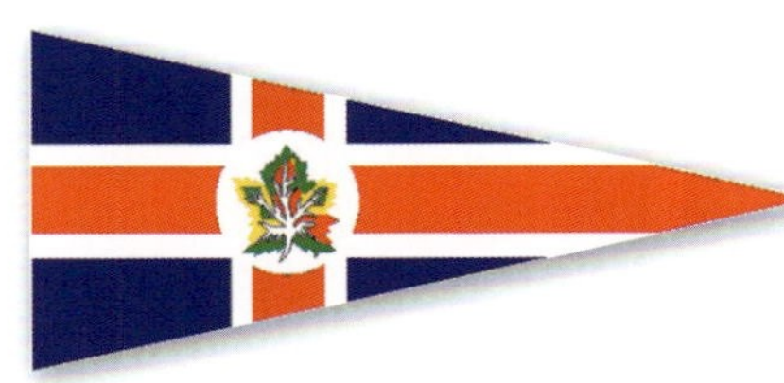

Canadian National Steamships pennant.

In some circumstances (especially in first class), shipping lines used high-quality, beautifully decorated china that was not top marked. In this case the line name can often be found on the bottom.

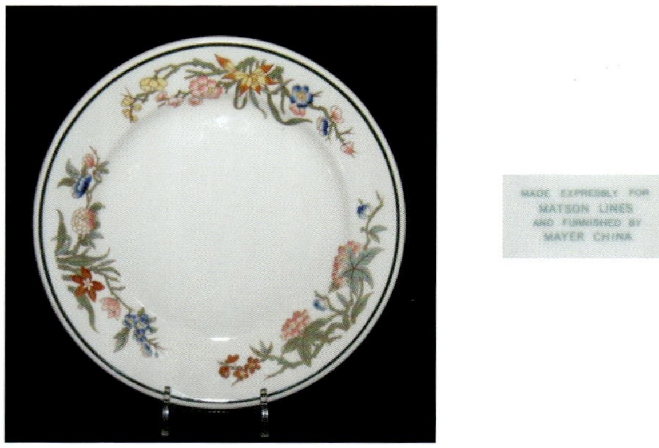

A Matson Bombay pattern (MNC-3, see page 112), back stamped with the line's name.

In some cases companies used symbols, intertwined lettering or a corporate logo in place of their house flag. These methods of labelling pose problems, because no comprehensive source exists for their identification. But the presence of the letters "SS" often indicates steamship use.

When investigating more modern patterns it is possible to write to the manufacturer to learn who ordered the china. I recently found a new pattern of china made by Homer Laughlin. The badge reminded me of one used by the Matson Navigation Company. The date code read 1992, yet Matson got rid of its last passenger vessel in 1972. I contacted Homer Laughlin, who said that Matson Navigation Company had placed an order for the china in 1992. (See pattern MNC-4 on page 113.)

One of the best sources of identification for company logos can be a company's ephemera (schedules, passes, brochures and menu covers). Because these items are often dated, they can indicate when a company started using a corporate logo. For example, the earliest brochure that uses the Alaska Steamship Company slogan "The Alaska Line" is dated 1927, suggesting that any china top marked with this slogan post-dates 1927. China back-stamp information confirms this hypothesis, as the earliest date code on "The Alaska Line" china is early 1930s. Another confirmation comes from china recovered from the SS *Aleutian*, which sank in 1929, bearing the old diamond logo pattern without the slogan.

It is also possible to find china stamped with a ship's name. The Matson Matsonia pattern has been observed with the names of the SS *Maui* and SS *Matsonia* imprinted on the back. The SS *Vadso* of the Boscowitz Steamship Company had china top marked with the ship's name.

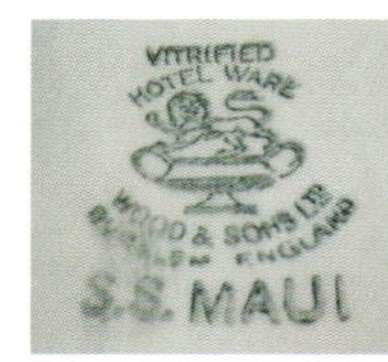

SS *Maui* back stamp.

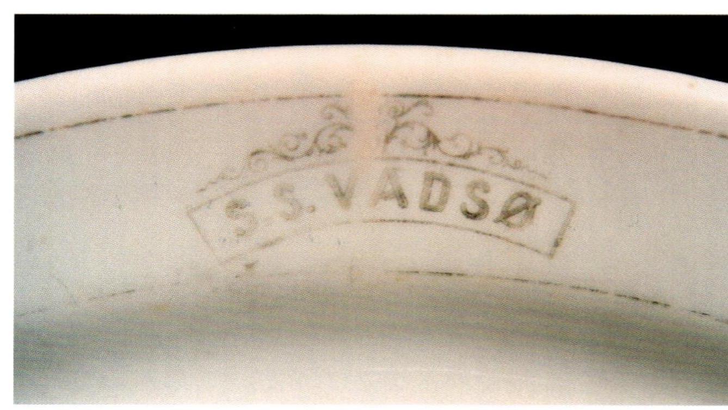

SS *Vadso* top mark (see pattern BSC-1, page 80).

Most modern bone-china pieces bearing the picture of a ship and its name were not used aboard the ship. They are either souvenir items sold aboard the ship or given to VIPs by line executives to mark the launching of a ship, or promotional items sent to shippers and travel agents.

6. Using Decorations and Design Elements to Establish Origin

Pacific coast nautical china patterns use many of the same decorative elements. These elements often have significance and can assist in identifying the origin of the china.

Shipping company house flags occur on 71 patterns of china by themselves or with other elements, making this the most common type of decoration. The benefit of this is that most house flags can be found in published flag and funnel books, assisting with the identification of unknown patterns.

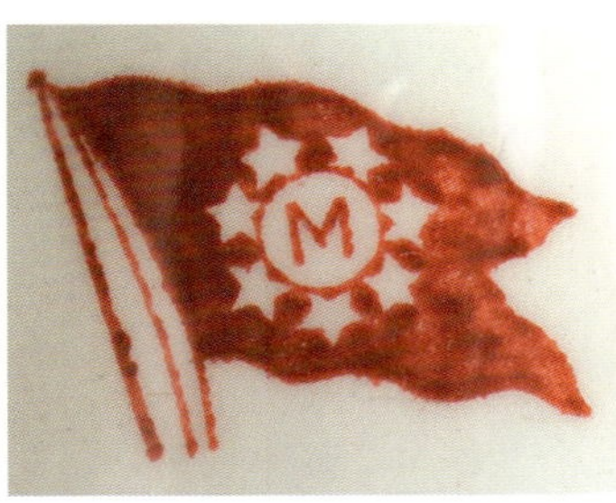

Matson Company house flag.

The second most frequently seen design element is the garter or garter belt. The garter occurs on 25 patterns of china. The garter is most commonly associated with military insignia, royal coats of arms and crests. Its origins go back to 1348 when King Edward III established the Most Noble Order of the Garter, and issued garters embroidered with the motto "*Honi soit qui mal y pense*" (evil to him who thinks it evil) to his knights. Over time the use of the garter broadened and became incorporated in the trademark and logos of private companies. Canadian Pacific Steamships made extensive use of the garter. From a diagnostic standpoint, garters generally appear on early (pre-1930) patterns of china.

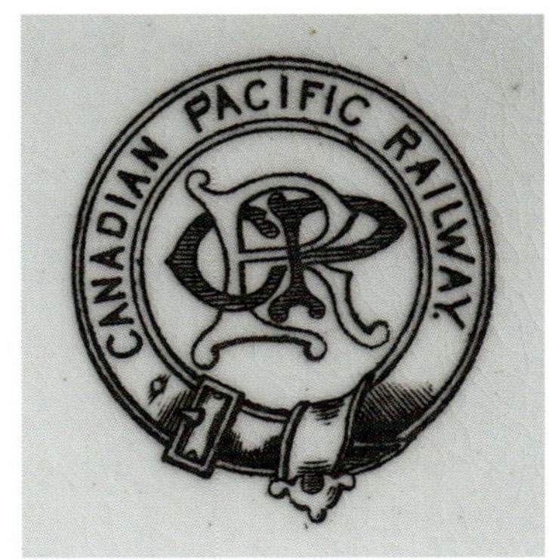

Canadian Pacific garter.

Ring logos appear almost as frequently as the garter. On the Pacific coast, garters were used primarily by Canadian companies, while ring logos were used more by American companies. In most cases, the name of a shipping line arches within the ring logo. Pacific Coast Steamship Company's Coos Bay and American Mail's Island Mail patterns are good examples of a ring logo. A variant on the ring logo is the life-ring logo used by the Pacific Steamship Company and California Transportation Company. See patterns PSS-1, -2 and -3 and CTC-1 to CTC-5.

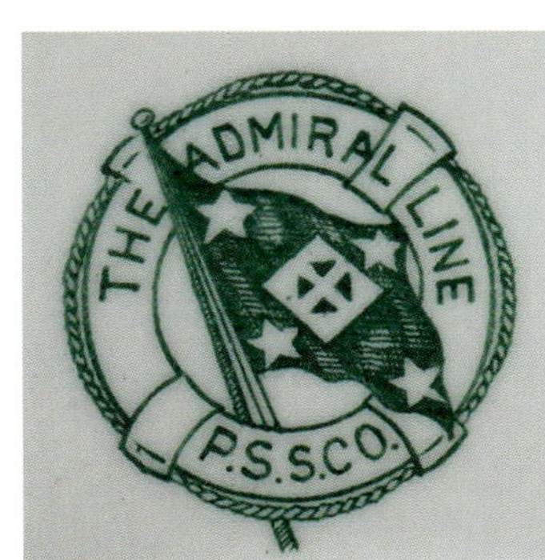

Pacific Steamship life-ring logo.

Superimposed lettering or script initials occur on 22 patterns. In almost all cases the lettering forms the initials of the steamship company. Many also include the letters "SS", meaning "steamship". The San Francisco and Portland Steamship Company's Bear pattern and Pacific Coast Steamship's Santa Rosa patterns are good examples.

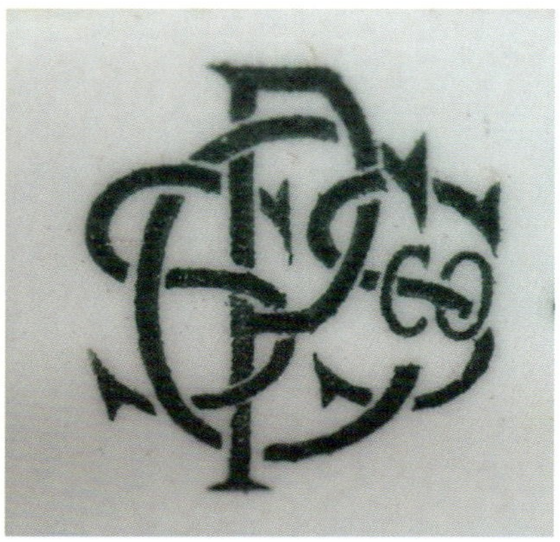

Pacific Coast Steamship Company's superimposed lettering.

Anchors appear frequently (26 patterns), almost always linked to Canadian and American government services. The anchor has been used by the US and Canadian navies, the Canadian Department of Marine and Fisheries and the US Coast Guard.

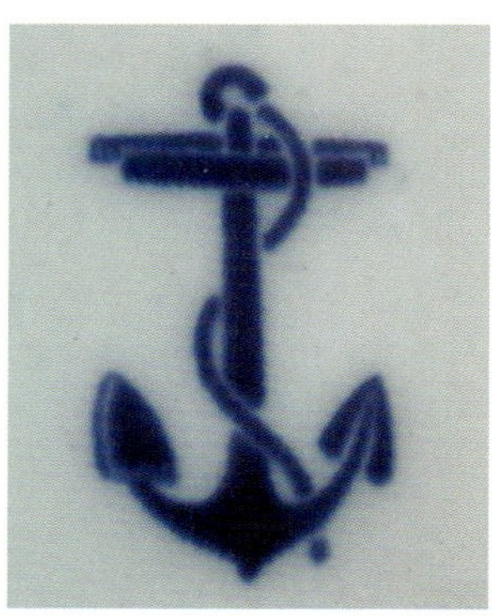

US Navy anchor.

Shields occur on 15 patterns, most used by the US government. Examples include early Revenue Cutter Service and the US Shipping Board. The CPR also used a shield to top mark and back stamp some patterns of china between 1929 and 1958.

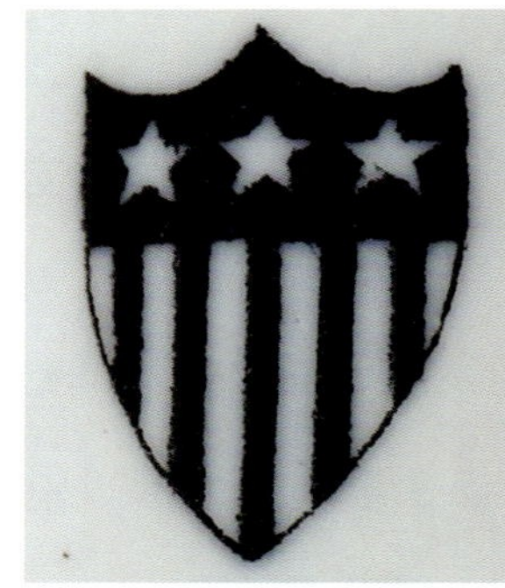

US Shipping Board Shield.

Crowns occur on 14 patterns, all but one associated with the Canadian government. The exception is the CPR crown pattern, the promotional symbol of the Empress Hotel. The style of crown found on military china can be used to help date china. The Imperial (Tudor) crown was used on RCN china before 1955, and from 1956 onward the St Edward's crown.

St Edward's crown.

Maple leaves occur on 25 patterns, all associated with Canadian companies or the Canadian government. Canadian National used the maple leaf on the Toronto and Bonaventure patterns. CPR used the maple leaf on its Blue Maple Leaf, Brown Maple Leaf and Foliage patterns. The maple leaf is also found on the Canadian Navy and Coast Guard badges.

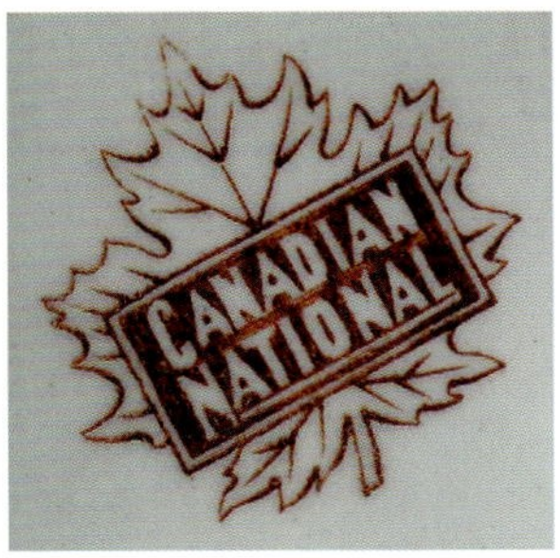

CN maple leaf.

Floral transfer prints were also popular on shipping china. Several shipping companies used handsomely decorated china for their first-class dining and for afternoon tea service. Matson is known for its popular tourist class Bombay pattern. The beautiful Copeland-Spode Heron pattern was used in first class aboard the Canadian Pacific Empress ships.

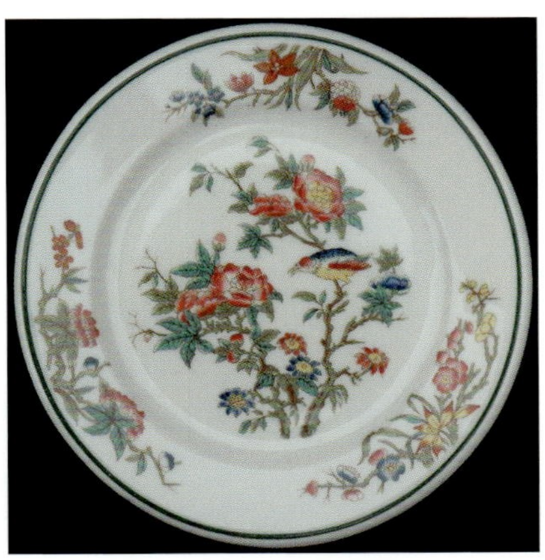

Matson Bombay pattern china.

Distinctive dolphin patterns can be found on the china used by the US Navy submarine service and the Canadian Coast Guard.

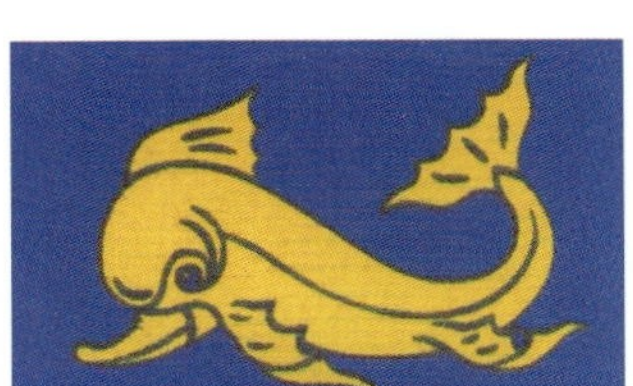

Canadian Coast Guard dolphin.

Eagles are featured on several US Navy patterns, one Coast & Geodetic Survey pattern and on the American President Lines' Eagle pattern.

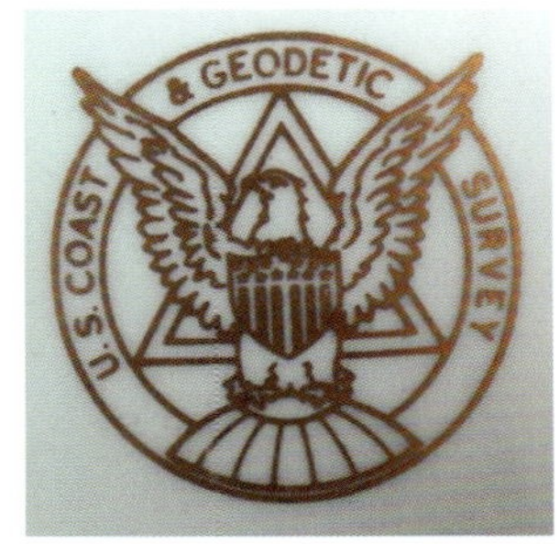

Coast & Geodetic eagle.

The Panama Mail Line Columbia pattern features a paddle steamer and "1849", the year that Pacific Mail Line started service. Panama Mail Line grew out of the ashes of Pacific Mail Line in 1925.

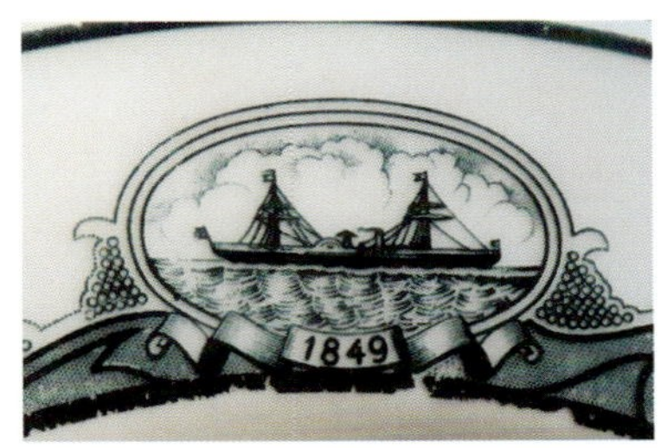

Panama Mail paddle steamer.

The earliest US Lighthouse Service pattern features a lighthouse on a rock, symbolic of the service it represents.

US Lighthouse Service badge.

7. Dating Ship China

Dating steamship china can be a challenge, but the markings stamped or incised on the bottom of the china should provide the necessary clues. These markings provide manufacturing, date-code and distributor information. Other information that will help date china is shipping company histories, ephemera and the quality and style of china. China recovered from shipwrecks also provides accurate dating.

Manufacturer's Marks

The first step in dating china starts with learning the manufacturer marks and date codes stamped on the bottom of the china. Manufacturer back stamps have evolved over time and most changes have been recorded in published works. A number of good books have been written on china back stamps and provide the dates that the logos were in use. For American manufacturers, see *Lehner's Encyclopedia of US Marks on Pottery, Porcelain and Clay* by Lois Lehner, or *Restaurant China*, Volume 2: *Identification and Value Guide for Restaurant, Airline, Ship and Railroad Dinnerware* by Barbara Conroy. For British marks,

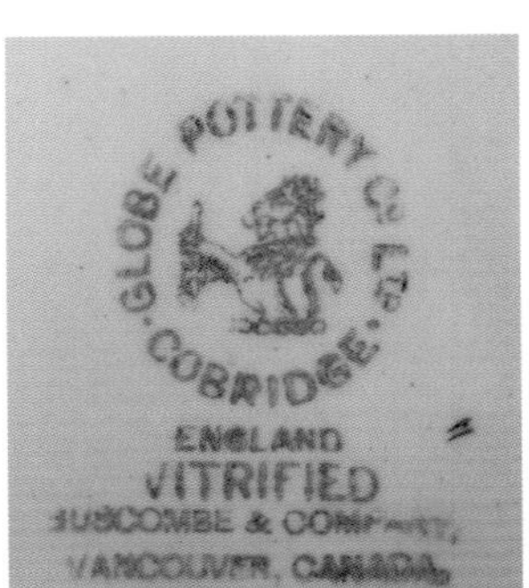

Globe Pottery back stamped "Cobridge".

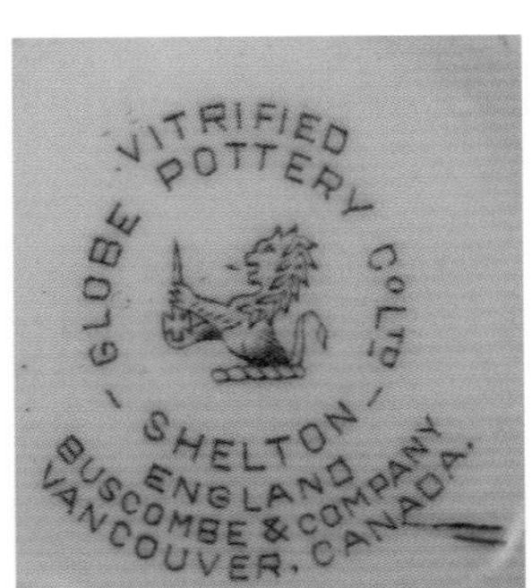

Globe Pottery back stamped "Shelton".

consult *Encyclopedia of British Pottery and Porcelain Marks* by Geoffrey Godden, or *Dictionary of Pottery and Porcelain Marks* by Terry and Ralph Kovel.

A company could change its back stamp, replacing it with a completely new one, or simply adding, dropping or replacing a word. For example, when dating Union Steamship china made by Globe Pottery, the word "Cobridge" dates the china to between 1930 and 1934: Globe Pottery started operations in 1930 in Cobridge, then moved to Shelton in 1934; all Union Steamship china stamped "Shelton" was manufactured after 1934.

In some instances a specific back stamp may have been used for decades, as was the case with English manufacturer John Maddock & Sons. The prevalent green lion back stamp was used on marine china from 1896 to 1946, during which many shipping companies came and went.

Other features in the manufacturer's back stamp that can help with dating. British manufacturers added "Limited" or "Ltd" in the 1880s, and then added "England" after 1891 to comply with the American McKinley Tariff Act. "Made in England" applies to wares made in the 20th century.

Appendix 2 contains a list of manufacturers, their dates of operation and descriptions of selected manufacturing companies.

Date Coding

Many china manufacturers stamped products with date codes in addition to their corporate back stamp. These codes consist of letters, numbers or symbols indicating the year, and sometimes the month, in which the china was produced. The books named at the beginning of this chapter contain information on manufacturer date codes. Unfortunately not every manufacturer employed date codes.

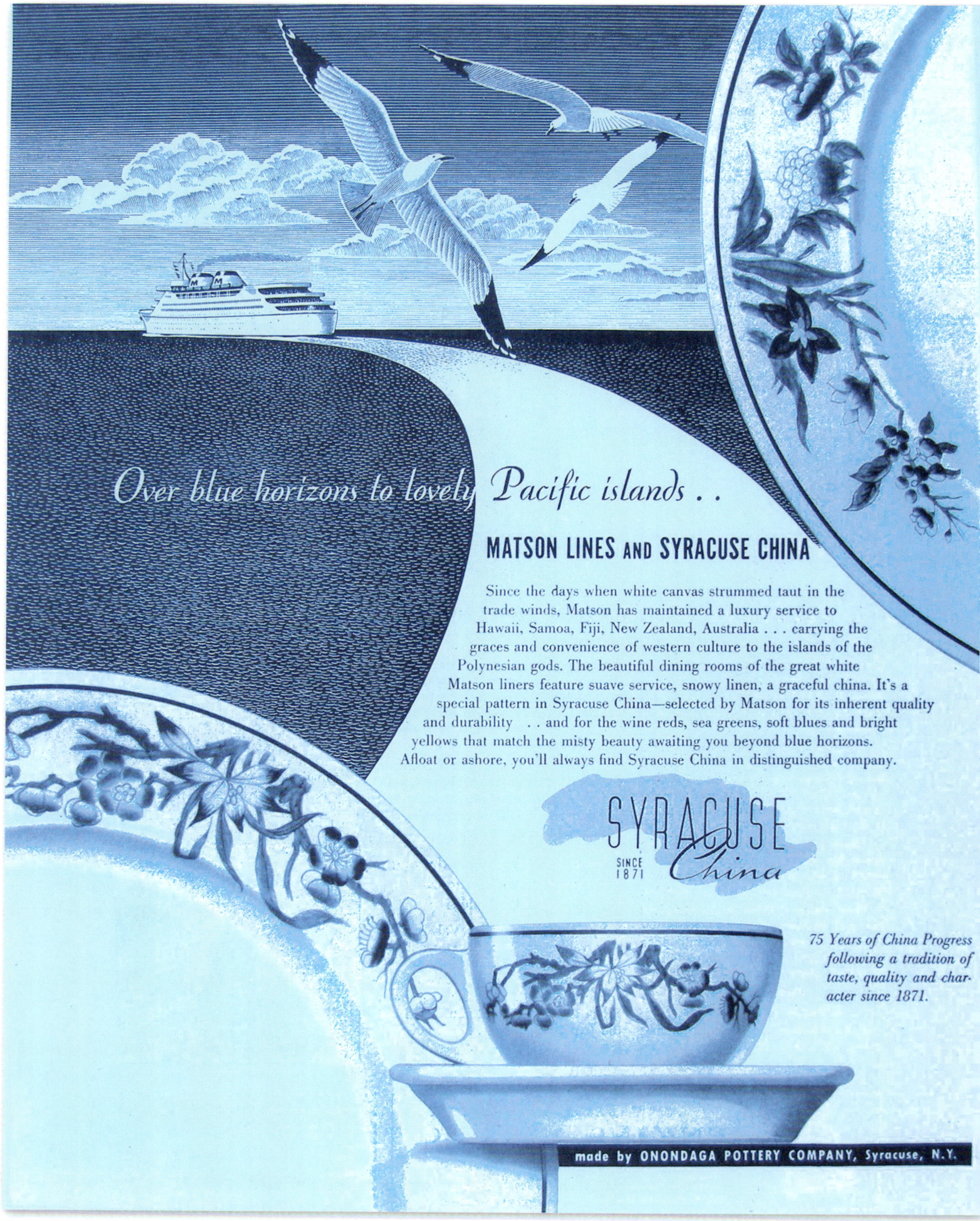

Onondaga Pottery Company advertisement for Matson Navigation Company china. The pattern shown is Bombay (MNC-3; see page 112).

Date codes can be incised in the china body or stamped in ink on the bottom. An incised code often indicates when the body of the china was made, not when it was decorated. An ink stamp was normally applied at the time the piece was decorated and fired, so provides a more accurate date. Some china pieces are bottom marked with the year of manufacture. This is especially true for military and shipping board china. Date-code information for the most prolific manufacturers can be found in Appendix 2.

British Registry Marks

Some china patterns manufactured in England also bear a registration number from the British design-registry office. By referring to the registration number in the appropriate chart, you can determine the date that a pattern was copyrighted. English patent law established that articles manufactured must be marked with the identification number and distributed for use within six months of the registration date. This provides a relatively accurate date for the start of service for a design with a registry mark. See Appendix 1 to identify the dates associated with registry marks. The only problem with this method of identification is that it sometimes took a year or two before a pattern became registered, as was the case with the Canadian Pacific Empress pattern.

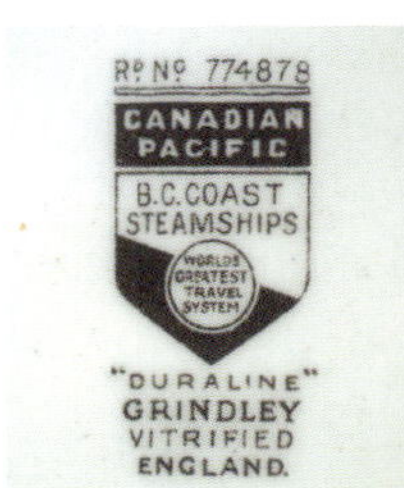

A Canadian Pacific Empress Pattern back stamp showing the registration number 774878 (1932).

Distributors and Importers

Many shipping companies purchased their china through distributors or importers, and these companies often had their name added to the china back stamp at the time of production. Learning when a particular distributor or importer operated can often help narrow down the time period of manufacture. Distributor's marks are difficult to date. The best method for dating them is to research their incorporation date and subsequent operating period. You can check with the registrar of companies to identify the years a company was in business. City directories can also be used to identify the years a company was in business. This will cover a period of years during which the china would have been manufactured. Barbara Conroy, in her book about restaurant china, provides some operating periods for distributors. I have substantially expanded her list with my own list of distributors who supplied nautical china and their known dates of operation – see Appendix 3.

Shipping Line History

Another avenue of dating china lies in the name changes of shipping companies. Since most ships' china was top marked with the line name, a name change often meant new china was ordered. Tracing corporate name changes can be a challenge. The easiest way is to go to published works that chronicle a company's history, such George Musk's *History of Canadian Pacific.*

In Musk's book, we learn that the name of the Canadian Pacific shipping division began in 1891 as Canadian Pacific Railway Co. In 1915 it changed the name to Canadian Pacific Ocean Services Ltd, then in 1921 to Canadian Pacific Steamships Lines Ltd and in 1968 to CP Ships. These name changes appear on the company's dinnerware.

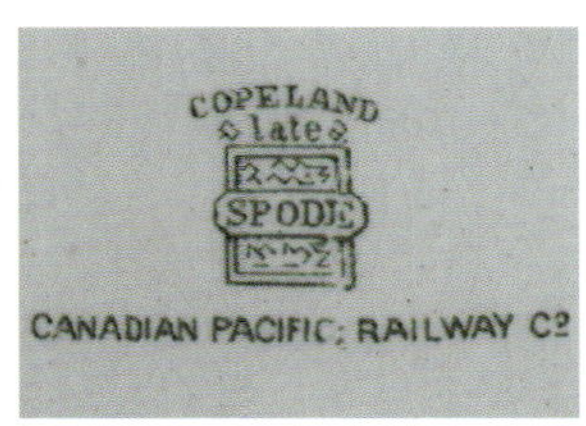

Canadian Pacific Railway back stamp, 1891–1915.

Canadian Pacific Ocean Services back stamp, 1915–21.

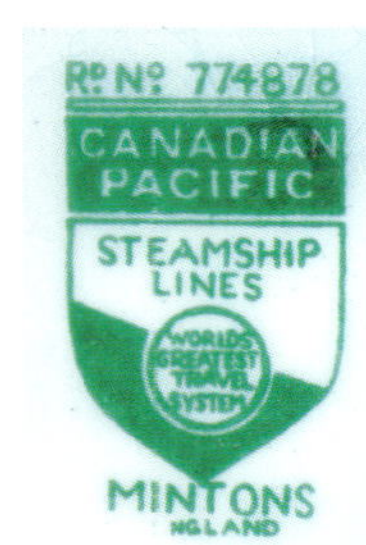

Canadian Pacific Steamship Lines back stamp, 1921–68.

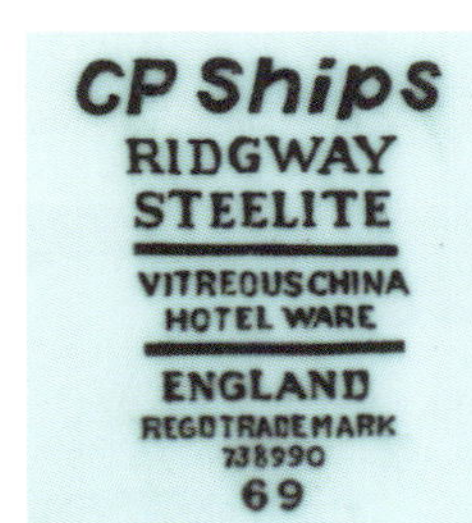

CP Ships back stamp, 1968–present.

One of the earliest CPR patterns, Black Crest, is back stamped Canadian Pacific Railway Co., suggesting that it predates 1915. A pattern called Heron, which was registered in 1913, is back stamped both Canadian Pacific Railway Co. and Canadian Pacific Ocean Services. The Empress pattern that was registered in 1932 is back stamped Canadian Pacific Steamship Lines. The Halifax pattern is back stamped CP Ships.

If the history of a corporate line has not been published, then it may be necessary to gather information from a museum or archives in the company's home port. There you might find useful booklets, brochures or news clippings. The registrar of companies is another method for tracking individual companies.

Corporate Logos

House flags, funnel markings and logos often changed over the corporate life of a shipping company. Learning when these changes occurred can help date china. Shipping schedules, brochures, advertisements and stationary are good sources for determining when logos changed. Once a new logo appeared on the company's printed material it often appeared on the line's china. For example, States Steamship Co. began operating out of Portland in 1920. Initially the company's house flag was red with a black swastika. Early silver-plated items from the States Line exhibit this logo. In 1940 the company changed its logo to a blue wheel with cross in it because Germany had adopted the swastika. In 1954 the States Line changed its house flag and corporate logo to a red sea horse. From this, we can surmise that all States Line china with the sea-horse logo postdates 1954.

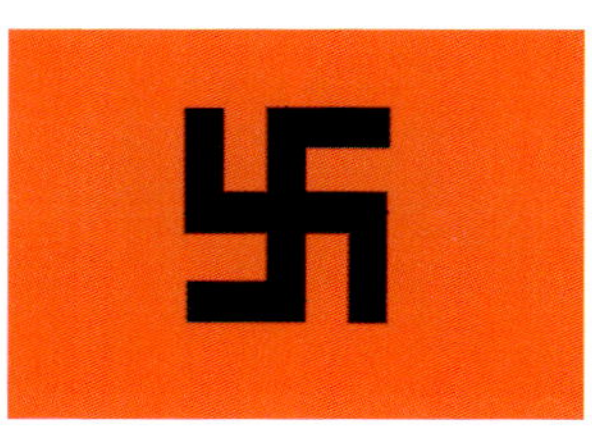

States Line flag with swastika logo, 1928–40.

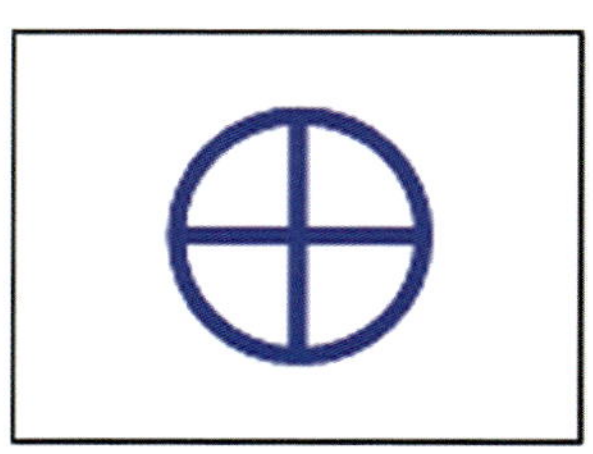

States Line flag with blue circle-and-cross logo, 1941–53.

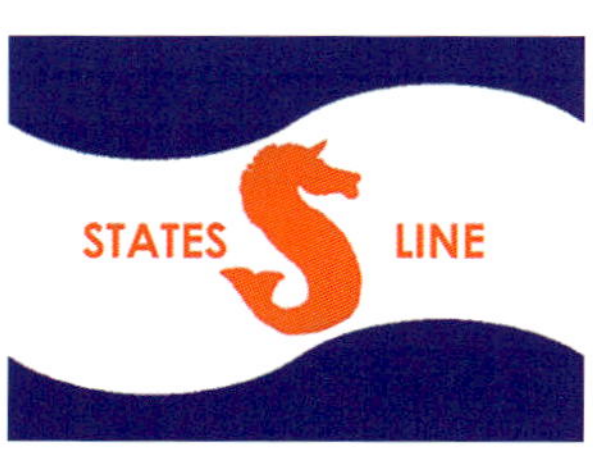

States Line flag with sea horse, 1954–73.

Ephemera

Shipboard photos, company brochures, advertising material, postcards and other ephemera that show china in service can also help identify when a pattern may have been used. Paper items are easy to date, because almost all have printing dates or post marks on them. If you can make out a china pattern in a picture, it stands to reason you will also have a firm date. However, use ephemera with caution as a dating method. A 1959 Canadian National Steamships brochure that clearly shows the Prince George china pattern in use. But this is highly unlikely, as this china pattern was last produced in 1948. Just because the brochure was produced in 1959, it doesn't mean that the photos were taken in 1959. To make matters worse the same brochure shows the Canadian National Bonaventure pattern also being used. This is likely the actual pattern used in 1959.

Similar problems have been encountered with ephemera such as menus when identifying shipping logos. Shipping companies often had menu covers mass-printed to save money, and then used them for five or six years, so the logos and dates on menus don't always correspond. A classic case is the Macouillard watercolour menu covers produced for Matson. When Matson sold its ships to the Pacific Far East Line it must have included all shore stock. PFE was still using the Macouillard Matson menu covers in the mid 1970s.

Quality, Style, Shape and Decorative Elements

The quality and style of china has changed with time. Early ships were equipped with white or grey heavy ironstone china with few decorations. At the beginning of the 20th century this gave way to vitrified restaurant ware with hand-applied transfer prints and hand-painted pinstripes. By the 1920s and '30s, higher quality china, including some bone china with colourful transfer prints, was being used on ships. Examples of these include the Matson Bombay pattern, the Canadian National Jasper pattern and the Canadian Pacific Coloured Heron pattern. By the 1970s, in the waning years of ship-china production, techniques had progressed to a point where most china was highly refined and most decorations were applied decals (e.g., Seaboard Shipping). While it is not possible to accurately date china using quality and style, it is possible to make an educated guess at the era a piece of china is from, based on these characteristics.

Decorative trends have also influenced china shapes. For example, art deco cube tea service sets were produced for Los Angeles Steamships and Canadian National Steamships in the late '20s and early '30s.

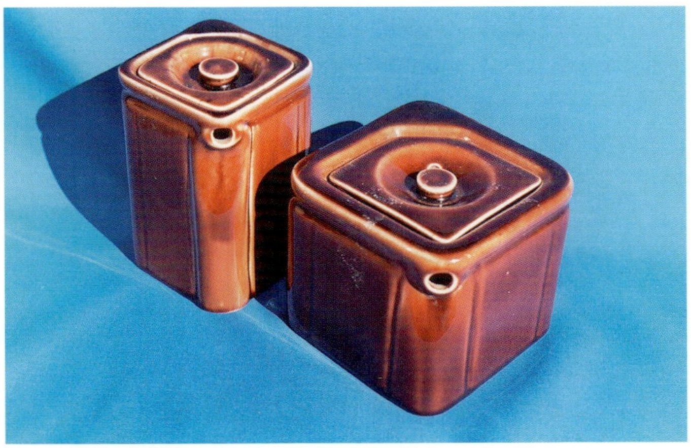

Los Angeles Steamship art-deco-era teapot and hot-water pots. Daniel Krummes photo.

As table fashions changed, some specialized pieces of china fell out of use. Crescent-shaped salad dishes and individual butter pats seemed to drop out of favour by the 1930s.

Shipwrecks

Ship china goes down with a sinking ship, so shipwrecks act as time capsules. China recovered from a wreck can tell us what patterns were being used at the time of its sinking.

There can be some surprises. Many collectors presumed that the earliest Pacific Coast Steamship pattern was the green interscroll pattern. Divers recovered this pattern from the SS *State of California*, which sank in Gambier Bay, Alaska, in 1913. Date coding substantiated that it was actually the most recent pattern of the four known top-marked Pacific Coast Steamship patterns.

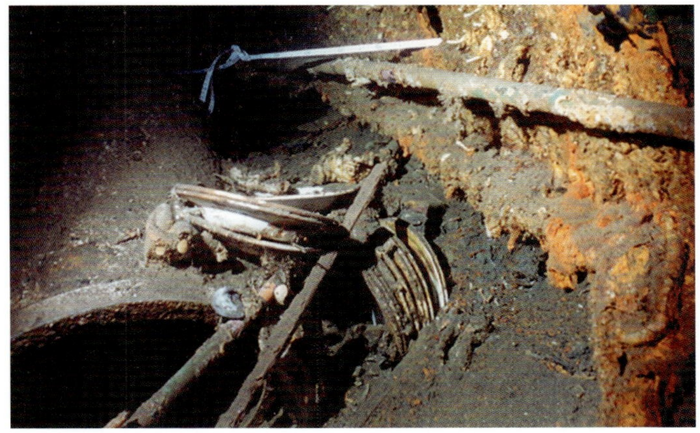

China buried in sediment on the wreck of the *Princess Sophia*. Bruce Lanham photo.

Period of Manufacture

In order for the previous sections to be useful, we have to be able to apply our new-found knowledge to a piece of china. Dating china is easy if the manufacturer has stamped the china with a date code that can be referenced. But if a date code is absent or unintelligible, then you can use the following approach to help establish the period of manufacture:

1. Make a list of attributes or characteristics associated with the china that can be dated (see previous section). Each characteristic must have known starting and ending points (years) to be of use.
2. Make a table showing the time line associated with each attribute.
3. When you have completed all possible time lines, add an extra row to the bottom of the table to record the likely period of manufacture based on the evidence.

Seldom, if ever, will you be able to establish the exact date of manufacture.

Example 1. CPR-302 Tremblant BCCSS

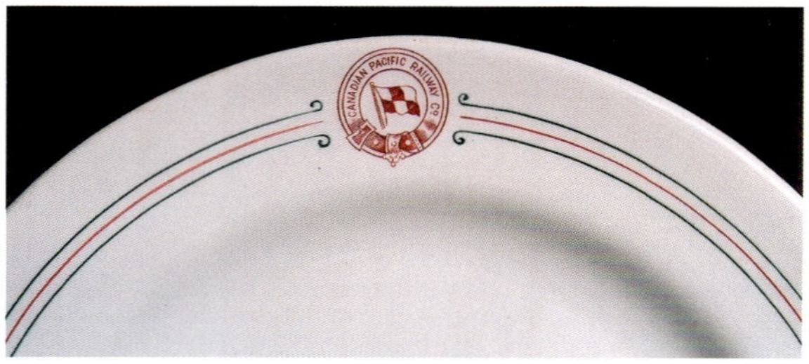

Tremblant pattern top mark.

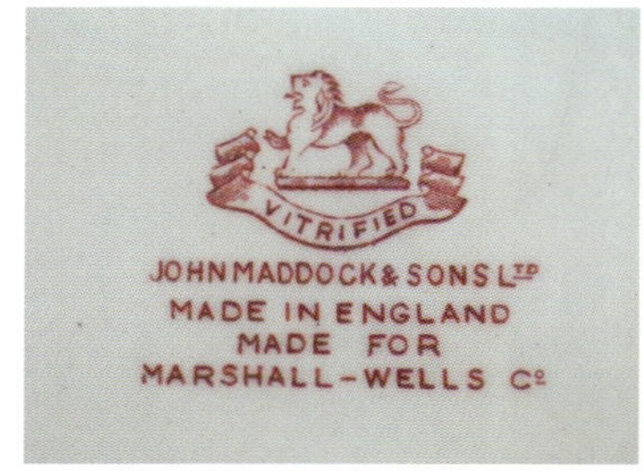

John Maddock & Sons back stamp.

The attributes with potential for establishing a time line include the house flag, company name, manufacturer's mark and distributor's mark. Research reveals that Canadian Pacific introduced the checkered flag in 1891 and retired it in 1968. BC Coast Steamships was formed in 1903 and ceased operations in 1974. John Maddock and Sons introduced the red lion back stamp in about 1945, and Marshall Wells was formed in 1899 and operated until 1957.

Based on the time lines below, it appears that this piece of china was manufactured between 1945 and 1957. A record of impressed date codes for this pattern, with this back stamp and distributor, indicate that the china was actually produced in 1947 and 1948, well within the range established by the time-line analysis.

Pattern: CPR-302 Tremblant BCCSS	1890	1910	1930	1950
CPR house flag 1891–1968				
BC Coast Service 1903–74				
Manufacturer's mark: John Maddock & Sons, after about 1945				
Distributor: Marshall Wells, 1899–1957				
Period of manufacture, about 1945–57				

Example 2. PCS-3 Santa Rosa

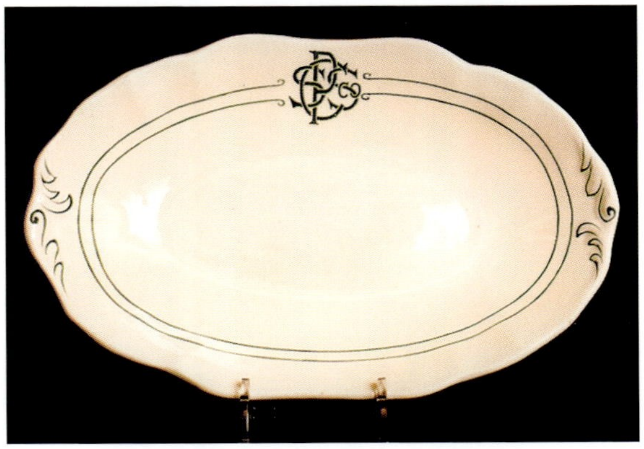

Santa Rosa pattern top mark.

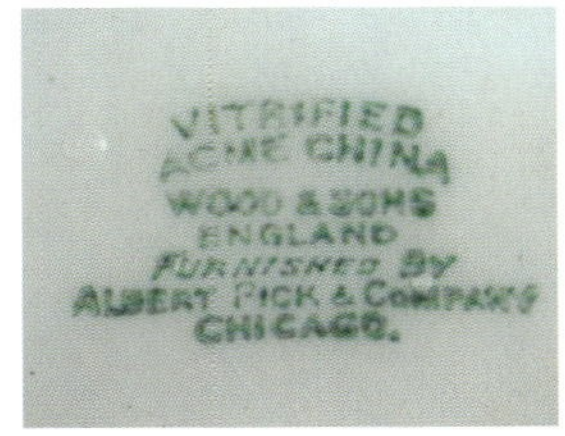

Wood & Sons back stamp.

The attributes with potential for establishing a time line include the company name, manufacturer's mark and distributor's mark. An added bonus for dating this piece is that it was recovered from the wreck of the steamer, *State of California*. Research reveals that Pacific Coast Steamships was formed in 1877 and ceased operations in 1916. Wood and Sons used this back stamp from around 1907 to 1910, and Albert Pick & Co. of San Francisco operated from 1858 until 1926. The *State of California* was launched in 1879 and sank in 1913.

Based on the time lines below, this piece of china appears to have been manufactured between 1907 and 1910. A record of impressed date codes for this pattern, with this back stamp and distributor indicate that the china was actually manufactured in 1910 and 1911. Its use on board the *State of California* confirms that it was still in service in 1913.

Pattern: Santa Rosa PCS-3	1890	1910	1930	1950
Shipping line history: 1877–1916				
Manufacturer's mark: Wood & Sons, 1907–10				
Distributor: Albert Pick & Co., San Francisco, 1858–1926				
Shipwreck: on board *State of California*, wrecked 1913				
Period of manufacture, about 1907–10				

Example 3. CPR-502 White Ware

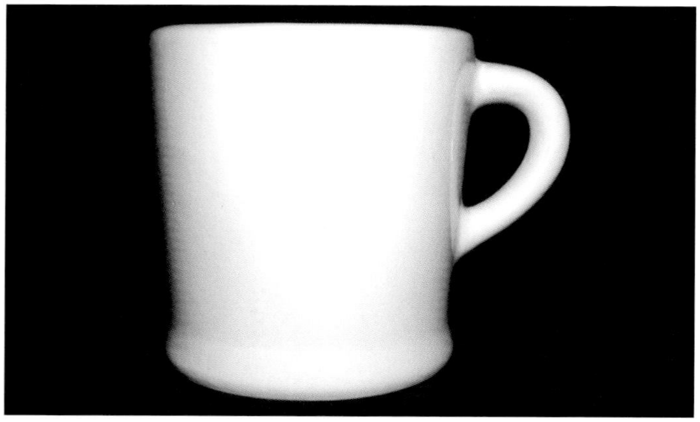

White Ware.

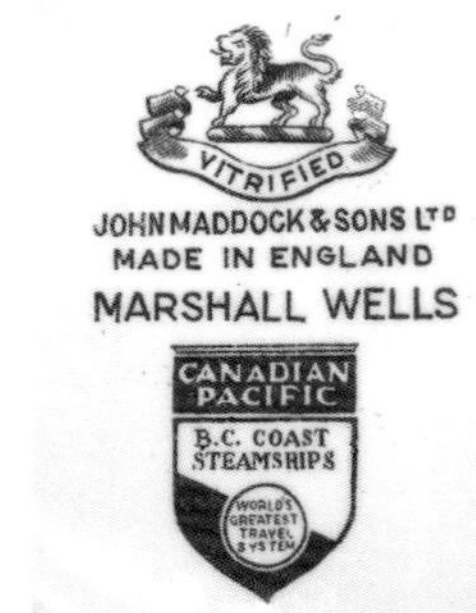

John Maddock & Sons back stamp.

The attributes with potential for establishing a time line include the company name, manufacturer's mark, distributor's mark and advertising shield. BC Coast Steamships was formed in 1903 and ceased operations in 1974. John Maddock & Sons introduced the refined green lion back stamp in about 1945, and Marshall Wells operated from 1899 until 1957. CPR used the shield logo and advertising slogan, "The World's Greatest Travel System", between 1929 and 1946.

Based on the time lines below, this piece of china was manufactured in 1945 or 1946. This coincides with date codes collected off other white-ware pieces made by other manufacturers.

Pattern: CPR-502	1890		1910		1930		1950	
BC coast service 1903–74								
Manufacturer's mark: John Maddock & Sons, after about 1945								
Distributor: Marshall Wells 1899–1957								
Advertising shield 1929–46								
Period of manufacture, about 1945–46								

8. Value and Rarity

This book is not a value guide for Pacific coast ship china. Rather, it is meant to help collectors and institutions identify and date china. I have not assigned value to the pieces in this guide, because:

1. value is dependent on factors such as rarity, condition, quantity and type of piece;
2. the asking price for ship china varies greatly from source to source and region to region; and
3. the value of china fluctuates with the number of pieces available in the marketplace and as the economy improves or declines.

Still, most collectors want to know how rare or common their find is. Consequently, I have developed a rarity guide. As most collectors know, there can be a correlation between rarity and value. Generally the rarer a piece, the more valuable it is. As with value, the rarity of particular patterns is influenced by many factors.

The number of years a company operated is a key factor in a pattern's rarity: a company that only operated for a short time generally has much less china available than one that operated for a long time. Age also plays a role: if the company operated at the turn of the 20th century, far fewer pieces will have survived, making them less common. Small companies would have produced less china than larger companies with more ships. The availability of information on the Internet has shown that some patterns once considered rare are actually more prevalent.

I have assigned a rarity factor to the patterns in this book based on the number of pieces observed on the open market and recorded in personal and museum collections. The scale corresponds with that used by other writers.

Very common	★	frequently available at shows
Common	★★	
Uncommon	★★★	available but seldom seen
Rare	★★★★	
Very rare	★★★★★	five or fewer pieces known

This rarity factor applies to patterns only and not to individual pieces within a pattern. Certain shapes will be scarcer as fewer pieces would have been made; for example, egg cups and creamers are rarer than cups, saucers and plates.

The Matson Bombay Pattern is an example of a two-star pattern. It is readily available at most transportation shows and on-line auction sites. The American Yukon Navigation Company's Yukon pattern is an example of a five-star rating. Most stock patterns have been assigned a one-star rating as they often were produced for many transportation customers including hotels and restaurants.

9. Identifying and Dating Ship China

I have organized china patterns alphabetically by shipping company name, then in date-of-manufacture order. It builds on the work of Spence, Luckin, Conroy, Krummes and others.

Numbering China Patterns

Pattern numbers provide a means of cataloguing china patterns and give collectors and scholars a reference for identification and discussion purposes. Pattern numbers have been assigned by a number of authors of different publications over the years. As no one had ever done an in-depth study of Pacific-coast marine patterns, the cataloguing is inconsistent. The same pattern has different numbers depending on which publication you open, and many previously undescribed patterns have no numbers at all.

For this reason, I have a new numbering system for this book. Consistent with other authors, I use a three-four letter code to identify the shipping company. The letters are the initials of the shipping company (e.g., "PCS" for Pacific Coast Steamship). For clarity, I use the initials of the registered company, not the line name. A number follows the lettering, beginning at one. This practice is consistent with Luckin, Conroy, Spence and Krummes. Because time lines have been established for most of the Pacific-coast patterns, I have numbered patterns in chronological order, with one designating the earliest recorded pattern for a particular steamship line. Pattern variants are identified by a lowercase letter after the number; for example, the number for the Canadian Pacific Tremblant pattern is CPR-202 and one variant with a gold border stripe is CPR-202a.

Canadian Pacific operated three separate services on the west coast: Canadian Pacific Steamship Lines ran the Pacific Empress ships to the Orient, British Columbia Coastal Steamship Service operated the Princess ships in coastwise service, and BC Lakes and River Service operated paddlewheelers and tugs on lakes and rivers. CPR also operated ships on the Atlantic and the Great Lakes. Some china patterns were exclusive to one service or another, while others were used simultaneously on all services. To cope with this problem I have used a 100-series numbering system to reflect the decade in which each pattern of china was manufactured:

001+	patterns manufactured before 1900.
100+	patterns manufactured in 1900–1909
200+	patterns manufactured in 1910–19
300+	patterns manufactured in 1920–29
400+	patterns manufactured in 1930–39
500+	patterns manufactured in 1940–49
600+	patterns manufactured in 1950–59
700+	patterns manufactured in 1960–69

So that readers can relate the patterns in this book to those described in other published works, I have cross referenced the numbers assigned by other authors where they exist. To save space only the author's last name is used.

Naming China Patterns

The pattern names in this guide have come from a variety of sources:

1. if known, the shipping company's name for the pattern;
2. if the shipping company's name is unknown, then the manufacturer's name;
3. if neither shipping company's nor manufacturer's names can be found, then the name previously assigned to the pattern by another author; or
4. if previous authors have disagreed on a pattern name or where a new pattern has been identified, I have assigned a pattern name.

When assigning a name to a new pattern, the first task was to determine the earliest date code recorded for the pattern. On the assumption that new china patterns were produced for special occasions, such as the launch or acquisition of a new vessel, the china was named after the vessel with the closest launch or purchase date.

In some instances, collectors regularly refer to specific patterns by name. Where this was the case, I continue using this name (e.g., CPR Tremblant pattern). In a few cases, names have been attributed to china based on the transfer print or decorations.

Some pattern names used by others have been changed to reflect new information. A good example of this is Canadian Pacific Heron pattern. Previous authors have called this pattern Blue Floral or Yule. The Copeland-Spode pattern name is, in fact, Heron.

Shipping companies ordered some patterns by name. Where this is the case that is the pattern name assigned. For example, the British Columbia Ferry Corporation purchases its china from the Dudson Company in England. The order name for the recent blue-and-red pattern is Spirit. The pattern was introduced aboard ship with the launching of the new Spirit class ferries in 1993.

In some cases, the china manufacturer labelled the china with their name for the pattern (e.g., the Canadian Pacific Newport and Colonial stock patterns).

Description Guide

The top line shows the pattern number, pattern name and rarity (five stars indicates very rare and one star very common – see page 71). The china description follows, with the provenance, if known. The last lines, in order, list the manufacturers ("Mfr") that produced the china and the documented dates of manufacture, the cross references ("Ref.") to pattern numbers used by other authors, and a credit for the photograph if it was taken by someone other than me.

If I was unable to collect enough date-code information to provide an accurate date for china, then I give the approximate time period using manufacturer's back stamp (e.g., ca 1925). See Chapter 4 for the list of abbreviations for shipping company names and Appendix 2 for a list of manufacturer's abbreviations.

Shipping Company Patterns

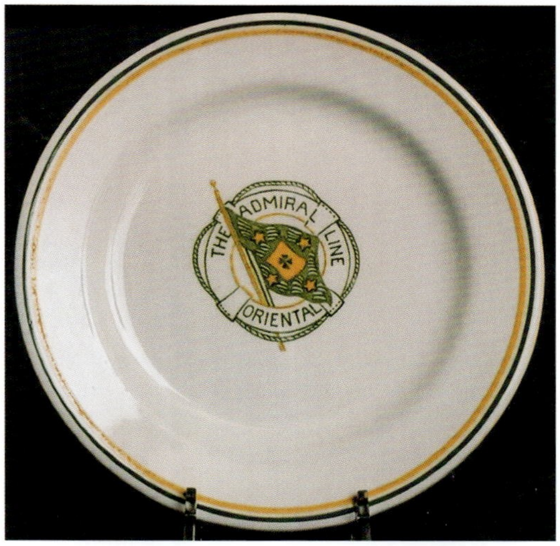

Admiral Oriental Line

AOL-1 Oriental ★★★

White china with the AOL logo and trimmed with two medium border pinstripes, the outer green and the inner yellow. The logo features a green life ring with "The Admiral Line" arched left to right within it. The word "Oriental" arches downward beneath the house flag.
Mfr: NPT 1922–26.

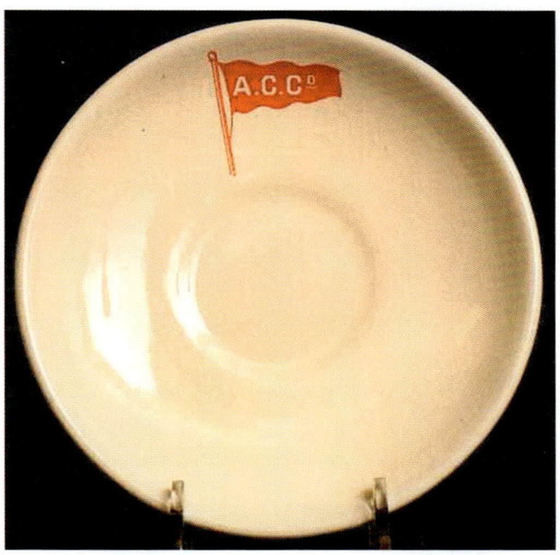

Alaska Commercial Company

ACC-1 Roanoke ★★★★

White china top marked with red/orange pennant containing the letters "A.C. Co." This was the Klondike Gold Rush pattern.
Mfr: WLK unknown dates, MDK 1896–1901.
Ref.: Conroy ACC-1, Spence ACC-1.

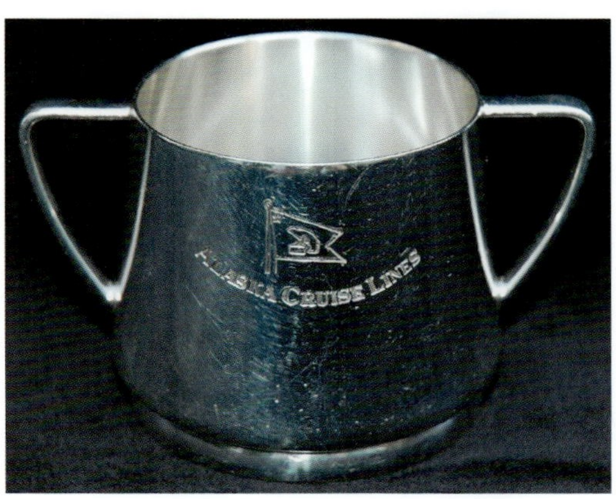

Alaska Cruise Lines

ACL-1 Yukon Star ★★★

Silver plate hollowware top marked with the ACL house flag. The flag features a polar bear on a rock looking at the north star. This logo found on silver plate items only.
Mfr: Morton-Parker 1958–70.

ACL-2 **West Star** ★★★

White china top marked with a red Naviera Cruceros Company house flag. The house flag features a white star in the top left corner and line initials NC in lower right corner. ACL acquired the *Cabo Izarra* from the Spanish company, Naviera Cruceros, in 1970 and renamed it *West Star.* The Naviera Cruceros china aboard the *West Star* remained in service after purchase.
Mfr: GAM 1967–70.

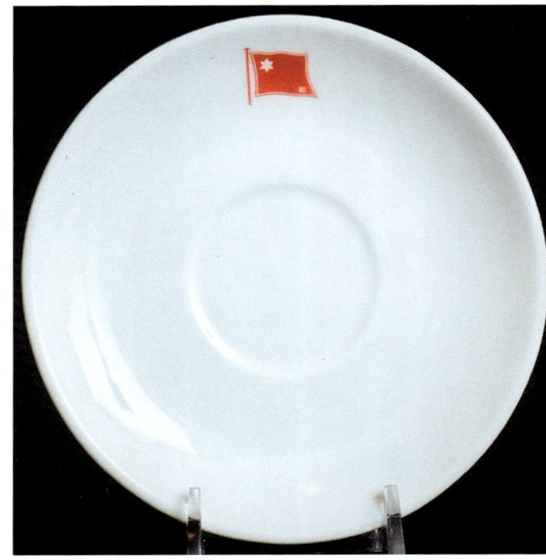

Alaska Steamship Company

ASC-1 **Alaska** ★★★

White china top marked with the ASC black-diamond logo. Inside the diamond, "Alaska Steamship Company" encircles the house flag, which is red with a black ball and the letter "A" on it. A pale blue band flanked on either side with thin teal pinstripes trims the outer edge of china. The black-diamond logo first appears on advertising in 1916. This pattern was still in service on the SS *Aleutian* when it sank in 1929.
Mfr: MDK 1923, OPC 1923–24, BWK 1925–27, GLY 1927.
Ref.: Spence ASL-2.

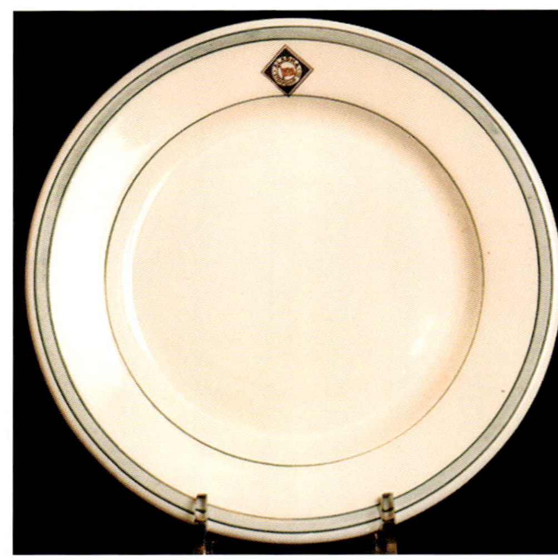

ASC-2 **Yukon** ★★

White china top marked with a black ASC diamond logo and trimmed with a pale blue border band flanked on each side by teal pinstripes. "The Alaska Line" appears under the logo in black type; these words were added to advertising material in 1927. Exact dating inconclusive for all manufacturers.
Mfr: SGO 1930s, MYR 1940s, BUF 1941–42 and 1955,
Ref.: Conroy ALS-1, Spence ASL-1.

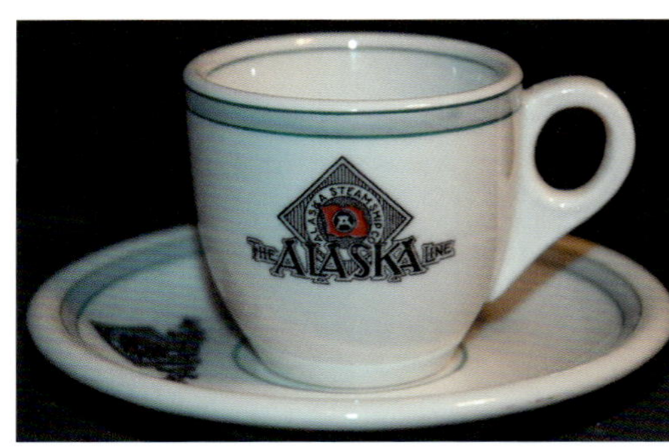

ASC-3 **Glendale** ★

White china with stylized orange-and-grey tulip border. This pattern was copyrighted by OPC in 1923. It was recovered from the SS *Aleutian*, which sank in 1929.
Mfr: OPC 1925–26.
Ref.: Conroy (vol. 2, p. 479).

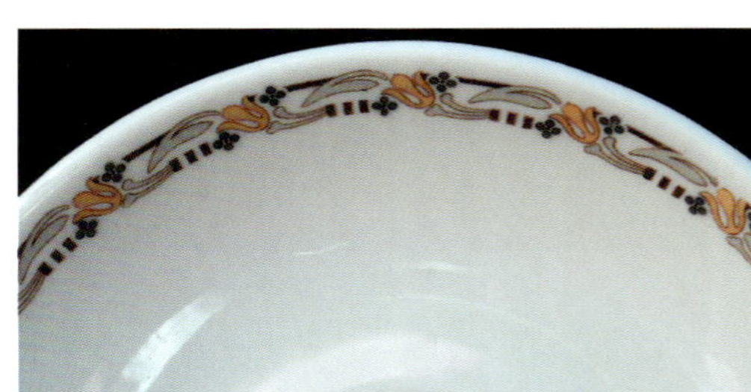

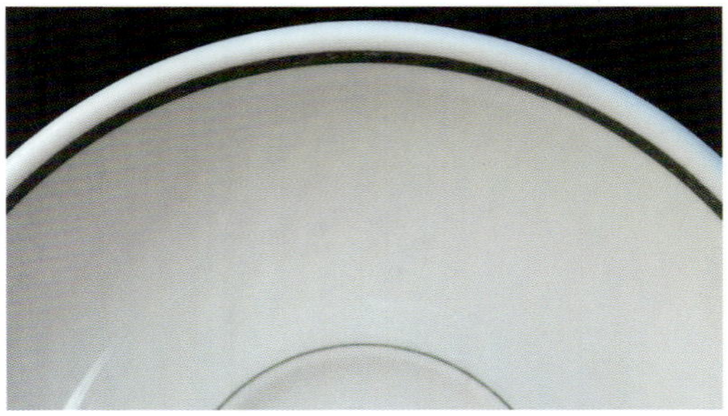

ASC-4 Calvert ★

White china with a forest-green stripe around the perimeter and a thin green pinstripe around the cup well. This pattern recovered from the SS *Aleutian*, which sank in Uyak Bay, Alaska, in 1929. The Mayer China name for this green-stripe pattern is Calvert.
Mfr: OPC 1926–27.

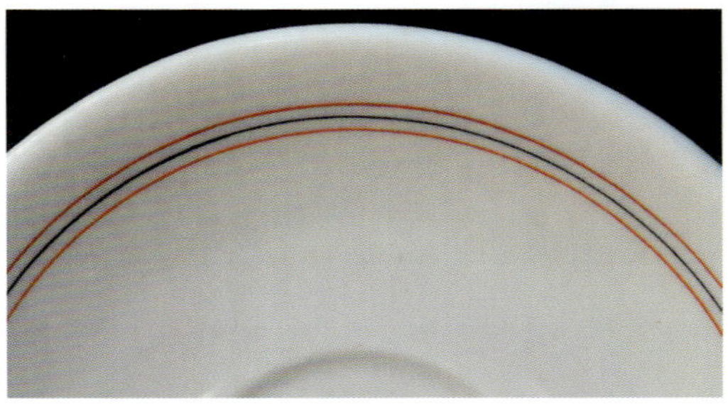

ASC-5 Aleutian ★

White china with alternating brown, black and brown pinstripes around the perimeter. This stock pattern was recovered from the SS *Aleutian*, which sank in Uyak Bay, Alaska, in 1929.
Mfr: SGO 1927–29.

American Mail Line

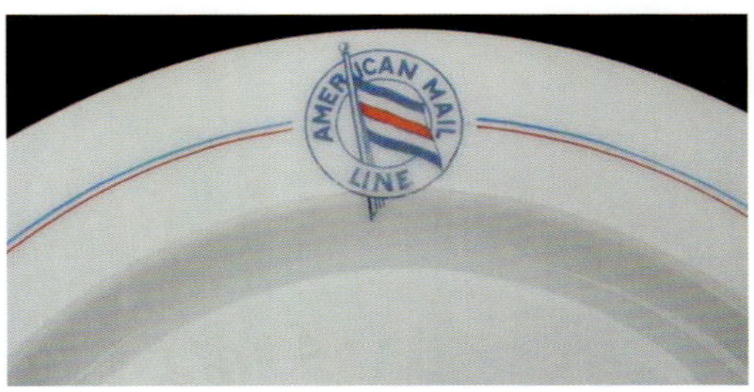

AML-1 Island Mail ★★

White china top marked with the AML logo and trimmed with two side-by-side border pinstripes, the outer blue and the inner red. The logo consists of the AML house flag encircled with "American Mail Line" within two blue rings. The *Island Mail* was launched in 1941.
Mfr: Schwarzburg (Germany) 1926–32, BUF 1941–50.
Ref.: Luckin AML-1, Spence AM-1.

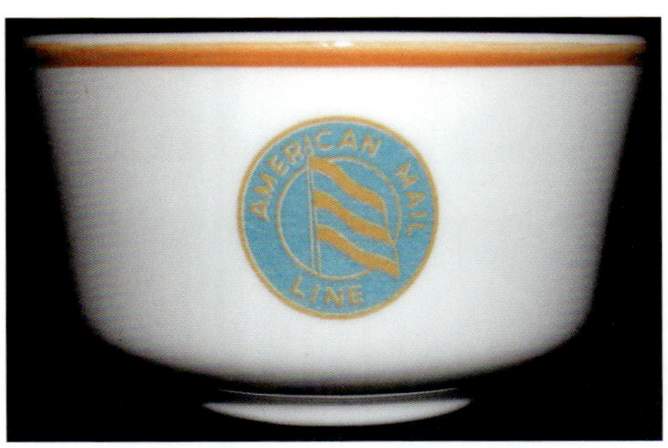

AML-2 Washington Mail ★★★

White china top marked with a green AML logo and trimmed with a beige border stripe. The logo consists of beige AML house flag encircled with "American Mail Line" within two beige rings. This pattern visible in Mariner class ship dining room postcard from the 1960s. The earliest date code coincides with the launching of the C-4 class ship *Washington Mail* in May 1961.
Mfr: MYR 1961.

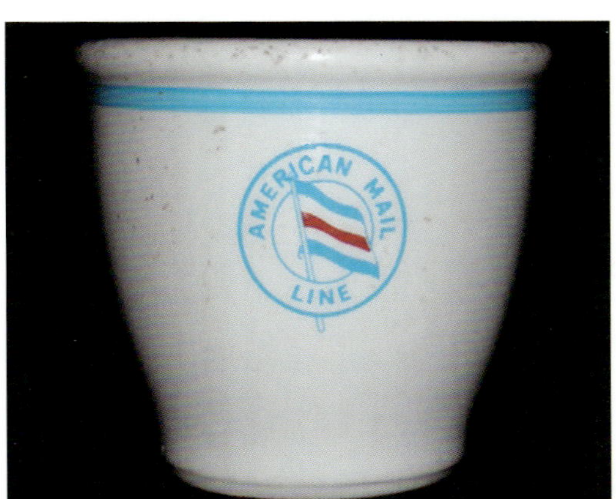

AML-3 Alaska Mail ★★★

White china top marked with the AML logo and trimmed with a pale blue border stripe. The logo consists of the AML house flag encircled with "American Mail Line" within two blue rings. The earliest manufacturing date coincides with the launching of the C-5 class ship *Alaska Mail* on April 1, 1968.
Mfr: MYR 1968–69.

American President Lines

APL-1 Brown Floral ★★

Beige china decorated with brown floral wreath transfer print and trimmed with a brown border stripe. A postcard showing SS *President Wilson* dining room shows this pattern in use about 1948. Pieces back stamped "US Maritime Commission". APL managed ships on behalf of the Maritime Commission during WW II.
Mfr: WAL 1947.
Ref.: Conroy MCOM-1.

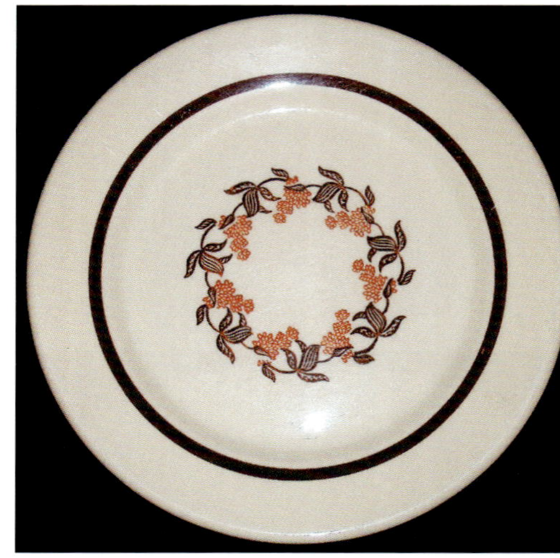

APL-2 President Wilson ★★★

Cream-coloured china top marked with a red APL logo and trimmed with two stripes (a thick inner and a thin outer). Badge consists of a red shield with script letters "APL" inside. The shield is topped with a ship and nine stars form an arch over the ship. This is a post-WW II pattern.
Mfr: SGO pre-1957.
Ref.: Conroy APL-1, Luckin APL-1.

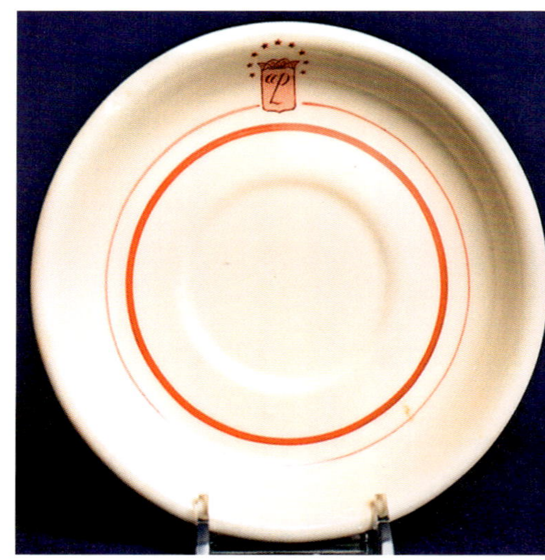

APL-3 President Cleveland ★★★★★

Cream-coloured china with an oriental theme in the well of the plate. The plate is trimmed with a grey border band flanked with thin maroon pinstripes on either side. The band is broken at 12 o'clock to allow for the placement of the APL eagle. Depicted in the well of the plate are the Taj Mahal (India), a pagoda (China), a bridge and bonsai trees (Japan), and a hut on stilts (Indonesia). Former APL steward Ted Marcollo confirms that this pattern was produced for use on board the *President Cleveland* (1947) and *President Wilson* (1948).
Mfr: WKR 1942–54.
Daniel Krummes photo.

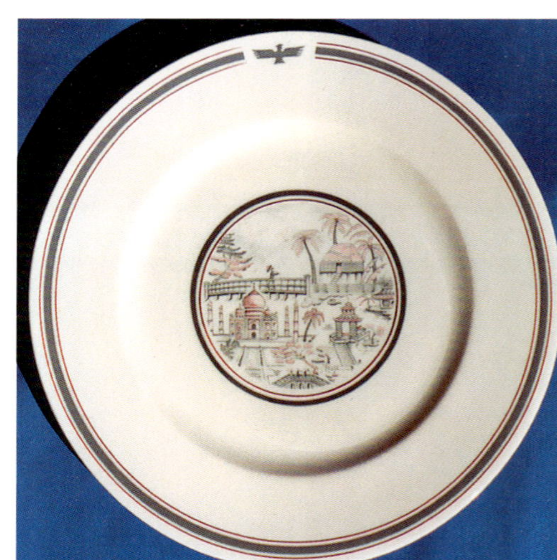

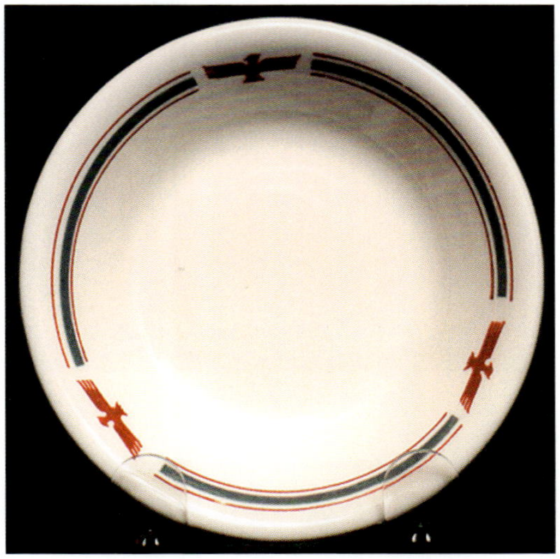

APL-4 **Eagle** ★

White china with a grey border band flanked with thin maroon pinstripes on either side. The band is broken at 4, 8 and 12 o'clock to allow for the placement of a red eagle.
Mfr: SGO 1956, WKR 1957–62, STR 1968–69, WEL 1959–69.
Ref.: Conroy APL-2, Luckin APL-2, Spence AP-1.
APL-4a features a more stylized eagle and is made by Corning.

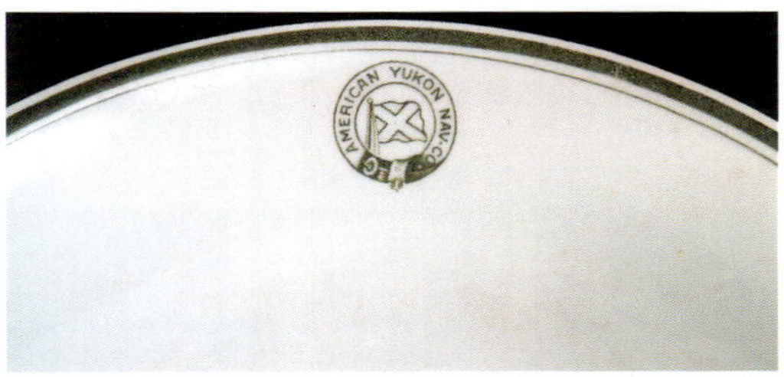

American Yukon Navigation Company

AYN-1 **Yukon** ★★★★★

White china decorated with a dark green garter containing "American Yukon Navigation Company" and the house flag, and trimmed with a dark green border stripe, flanked on the inside with a green pinstripe. The SS *Yukon* was launched in 1914 to compete with Northern Navigation Company. Only one piece known is known to exist.
Mfr: COK 1914–15.

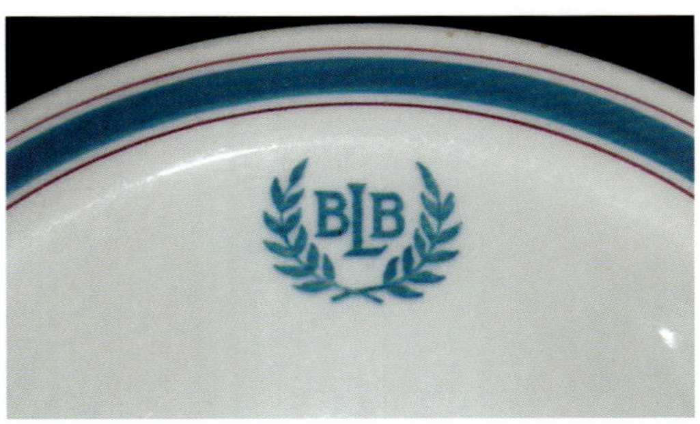

Black Ball Lines – Canada

BBL-1 **Chinook** ★★★

White china top marked with the BBL logo and trimmed with a five-millimetre medium blue band flanked on each side with thin burgundy pinstripes. Logo consists of the blue letters "BLB" between two blue fern branches.
Mfr: SGO 1953–58.

Bonners Ferry and Kaslo Transport Company

BFT-1 Bonners Ferry ★★★★★

White china trimmed with red lettering that reads "Bonners Ferry and Kaslo Nav. Co." Only two pieces are known to exist.
Mfr: KTK 1893.
Nelson Museum of Art and History photo.

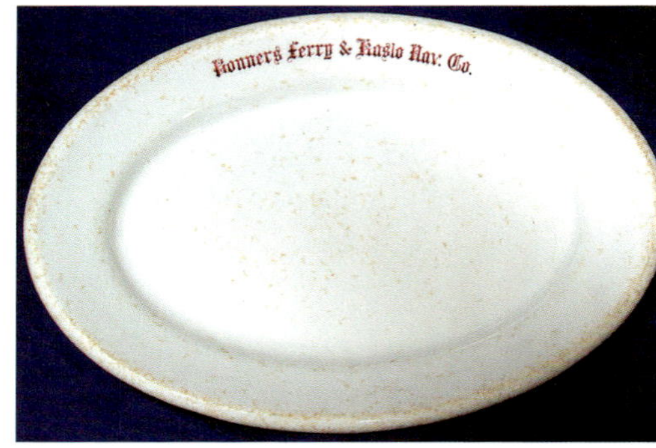

Boscowitz Steamship Company

BSC-1 Vadso ★★★★★

White china with a black banner containing the words "SS *Vadso*" and trimmed with a black border pinstripe. The *Vadso* served on the Christiana-Vadso route in Sweden before being purchased by BSC. This china arrived on the west coast aboard the SS *Vadso* in 1907.
Mfr: Stettin (England) 1906.

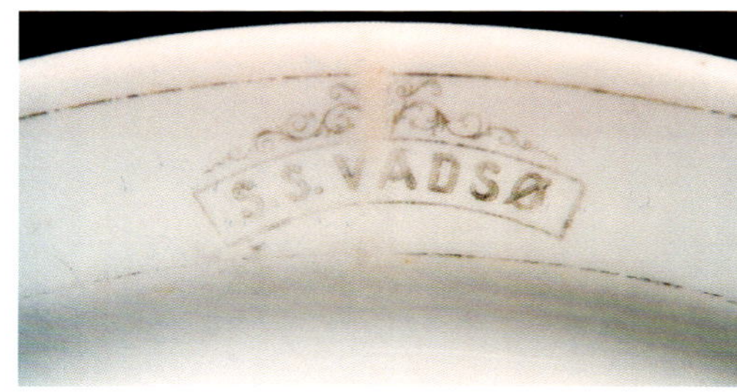

BSC-2 Boscowitz ★★★★★

White china decorated with a blue garter containing the BSC name and house flag. The flag is a red flag with a black "V" superimposed on a white horizontal diamond.
Mfr: FUR ca 1908.

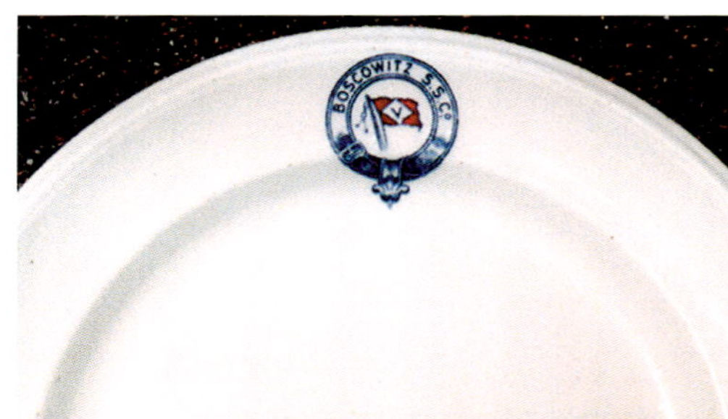

BSC-3 Venture ★★★★★

White china top marked with the BSC house flag in green. Beneath the house flag is a banner containing "B.S.S.C." The china it trimmed with two green pinstripes centred on plate rim. This pattern named for the SS *Venture*, launched on February 16, 1910.
Mfr: Sneddon & Sons ca 1910.

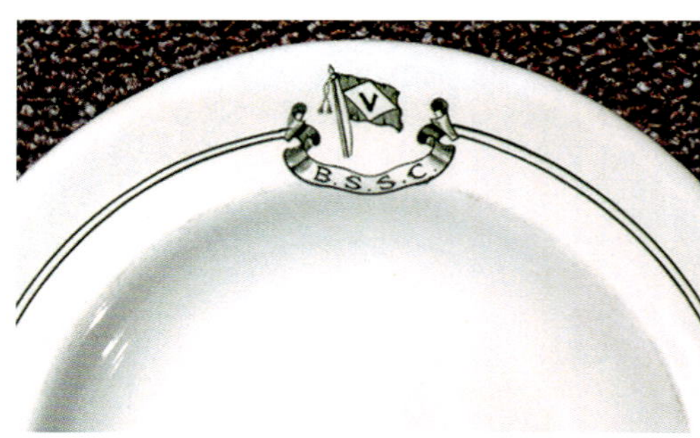

British Columbia Ferries

BCF-1 Malaspina ★

White china trimmed with a five-millimetre yellow band flanked on either side with thin orange pinstripes. A stock pattern not top marked or bottom stamped. The name is as it appeared on BCF requisition forms.
Mfr: HTH 1971–75, RID 1977–85.

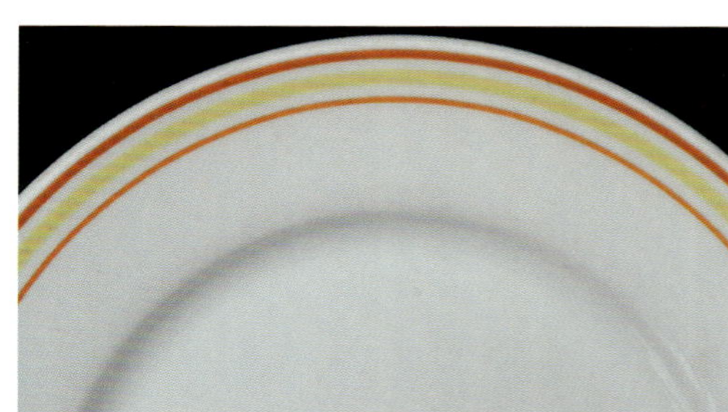

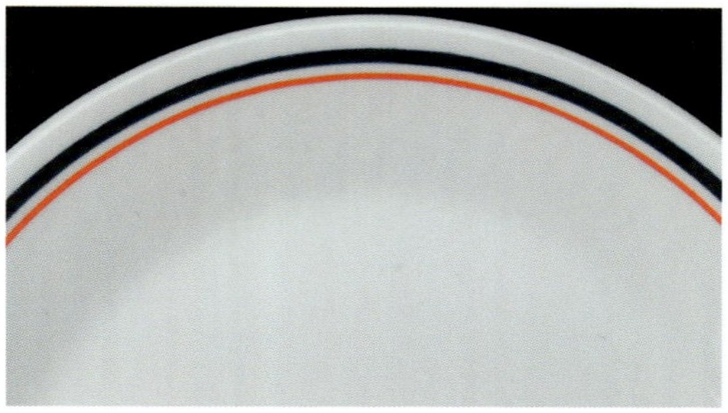

BCF-2 **Spirit** ★

White china trimmed with a blue outer stripe flanked by an inner red pinstripe. A stock pattern not top marked or bottom stamped. It was commissioned with the launch of the Spirit class ferries in 1993. The name is as appears on BCF requisition forms.
Mfr: DUD 1993–2005.

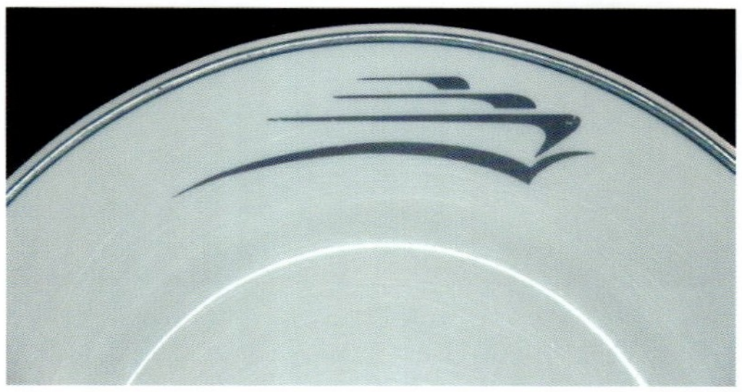

BCF-3 **Fast Cat** ★★★

White milk glass dinnerware top marked with a blue stylized vessel motif. This pattern was produced for the launch of the BCF high-speed catamarans in 1998. It is not top marked or bottom stamped with line name.
Mfr: ACP 1998.

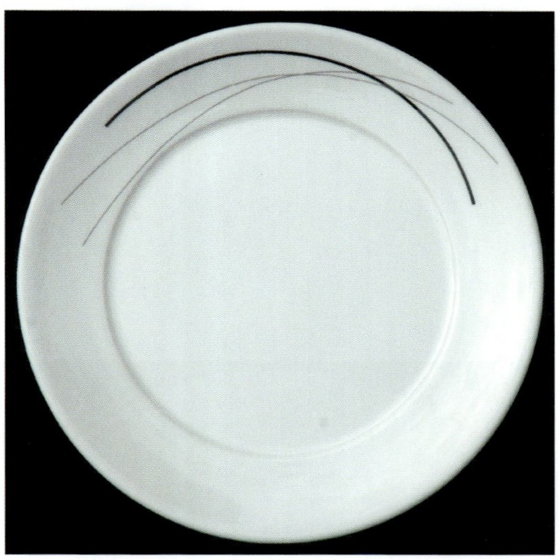

BCF-4 **Renaissance** ★

White china decorated with a black stripe that arches around the top third of the plate rim and is underlined with two criss-crossed pinstripes. Produced for the *Coastal Renaissance* ferry and her sister ships built in 2007. Pieces of this china are already showing up in second-hand shops.
Mfr: DUD 2007+.

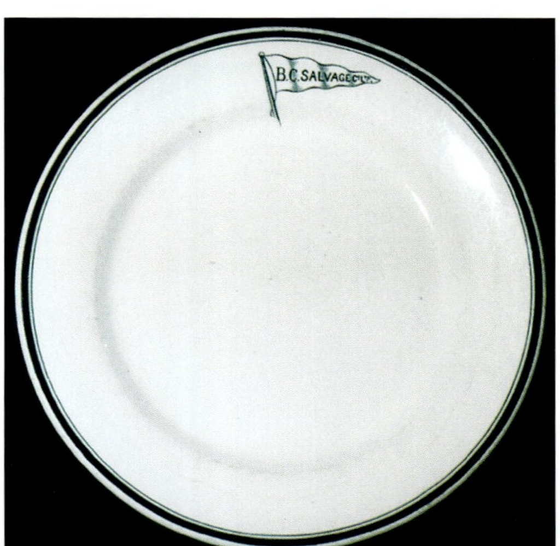

British Columbia Salvage Company

BCS-1 **Salvor** ★★★★

White china top marked with the BCS house flag and trimmed with a green border stripe and inner pinstripe. The house flag is a white pennant with "BC Salvage Co." in green type. The SS *Salvor* was commissioned in 1905.
Mfr: MDK 1907.

British Columbia Steamship Company (1975)

BSS-1 **Regent** ★

White china with a red rectangular border design. This stock pattern was used on the *Princess Marguerite* and *Vancouver Island Princess* when they were operated by BSS. The manufacturer's name for this pattern is Regent.
Mfr: RID 1976–86.

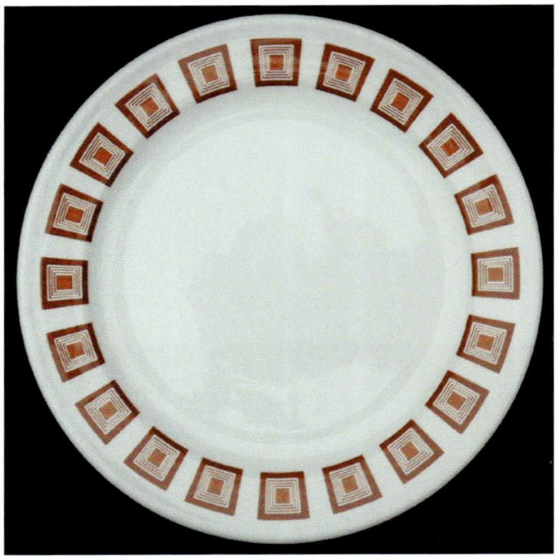

California Navigation & Improvement Company

CNI-1 **J.D. Peters** ★★★★

White china with CNI house flag and three red border pinstripes. The house flag has a thick blue border and the flag staff has two tassels hanging from it. Most likely introduced to J.D. Peters when the passenger trade increased in the early 1900s.
Mfr: MDK ca 1909.
Daniel Krummes photo.

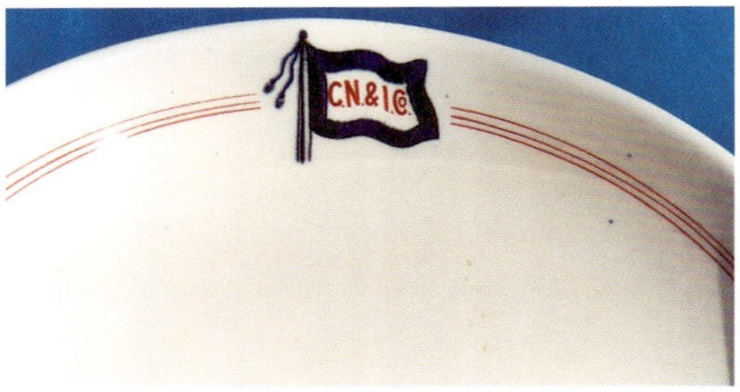

CNI-2 **Stockton** ★★★

White china with CNI house flag and three red border pinstripes. The house flag is white with a thin blue border and contains "C.N. & I. Co" in red.
Mfr: MDK ca 1910–22.
Ref.: Luckin & Conroy CN&I-1, Spence CNI-1.

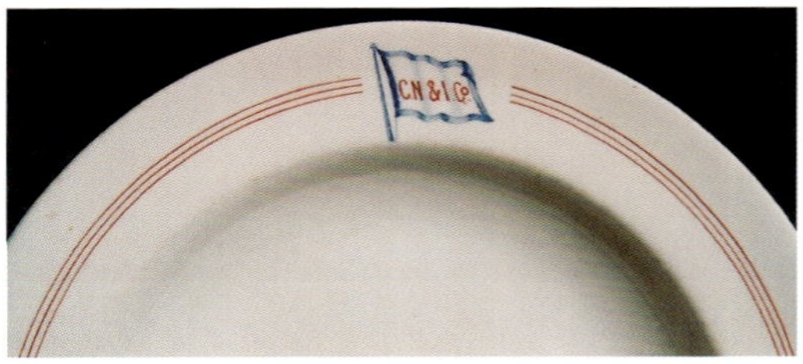

California Transportation Company

CTC-1 **Iselton** ★★★★★

White china top marked with a blue silhouette life ring containing the red CTC house flag and trimmed with two dark green pinstripes. The ring is simple with no shading and no rope around the perimeter. The flag is wholly within the ring. The passenger steamer *Isleton* entered service in 1902.
Mfr: MYR unknown dates.
Lester Silva photo.

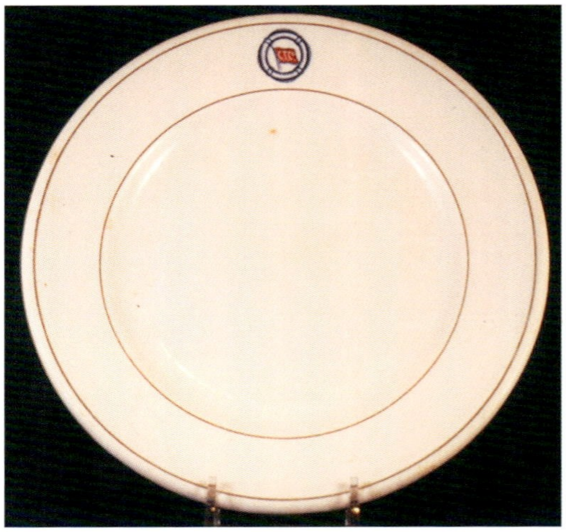

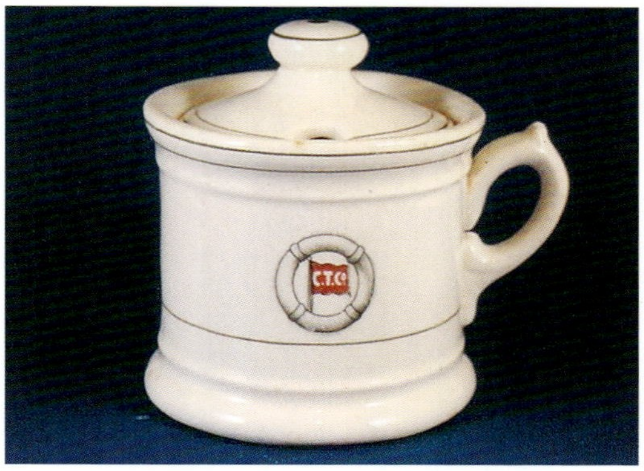

CTC-2 Reform ★★★★★

White china top marked with a green silhouette life ring containing the red CTC house flag and trimmed with a single dark green pinstripe. The ring is simple with no shading and no rope around the perimeter. The flag is wholly within the ring. The steamer *Reform* was launched in 1906.
Mfr: RID 1906–10.
Lester Silva photo.

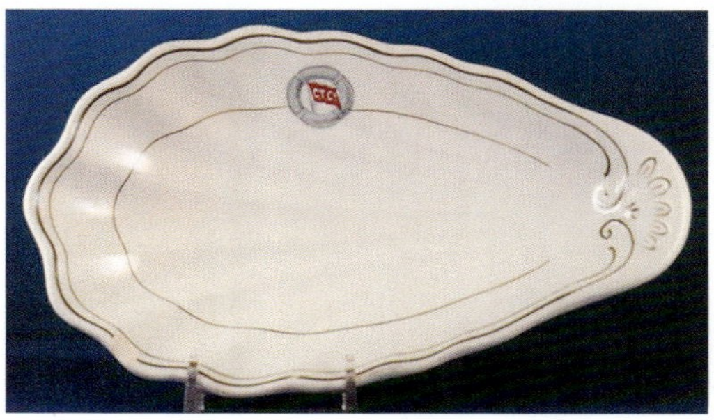

CTC-3 Gray Life Ring ★★★★★

White china top marked with a grey-shaded life ring containing the red CTC house flag and trimmed with dark green pinstripes. The ring is shaded and has a rope perimeter. The house flag cuts across the life ring.
Mfr: WOD 1907–10.
Lester Silva photo.

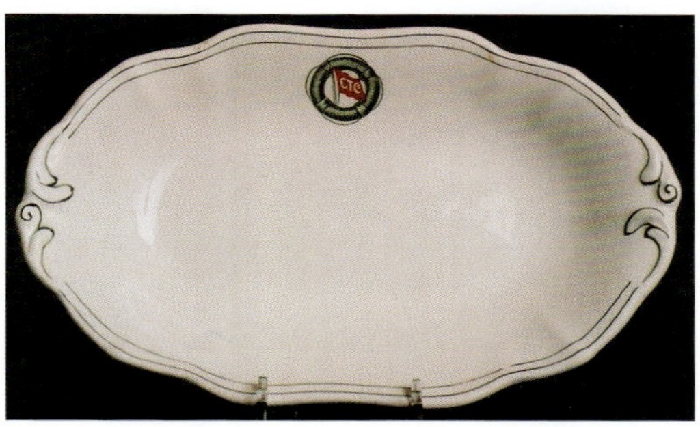

CTC-4 Capital City ★★★★

White china top marked with a green life ring containing the red CTC house flag and trimmed with two green border pinstripes. The steamer *Capital City* was placed into service in 1910.
Mfr: MDK 1909– .

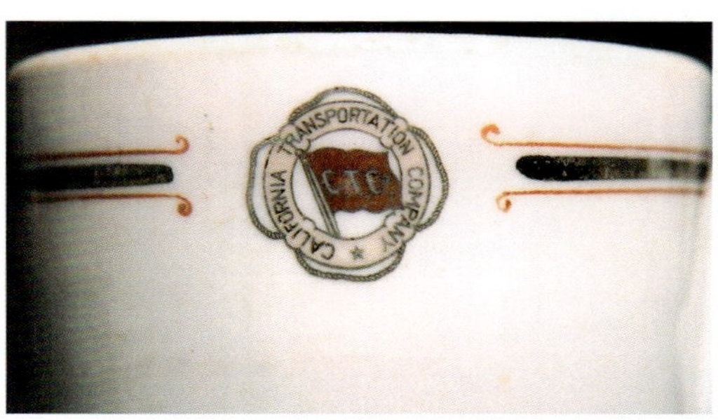

CTC-5 Delta King ★★★

White china top marked with a beige life ring containing the red CTC house flag and trimmed with a black stripe flanked on each side by a brown pinstripe. The text within the life ring reads "California Transportation Company". The steamers *Delta King* and *Delta Queen* entered service in 1927.
Mfr: IRQ 1929–30s.
Ref.: Spence CAL-1.

Canadian National Steamship Company

CNS-1 Flag & Leaf ★★★★

White china top marked with the early CNS house flag. The house flag is reddish orange with a green maple leaf on a white circle and "CNR" on the maple leaf. This pattern is similar to GTP-1.
Mfr: GLY 1923.
Ref.: Luckin CN-4.1.

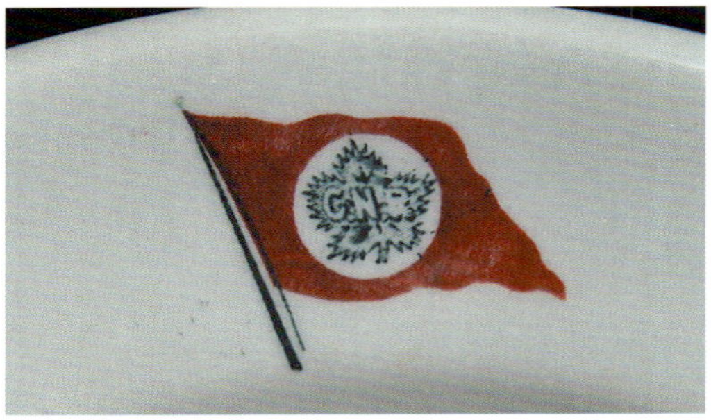

CNS-1a Flag & Leaf ★★★★

This pattern is the same as CNS-1 except that the "CNR" has been replaced by "CNS" on the maple leaf. This logo occurs on CNS advertising starting in 1926.
Mfr: GLY 1926–29.

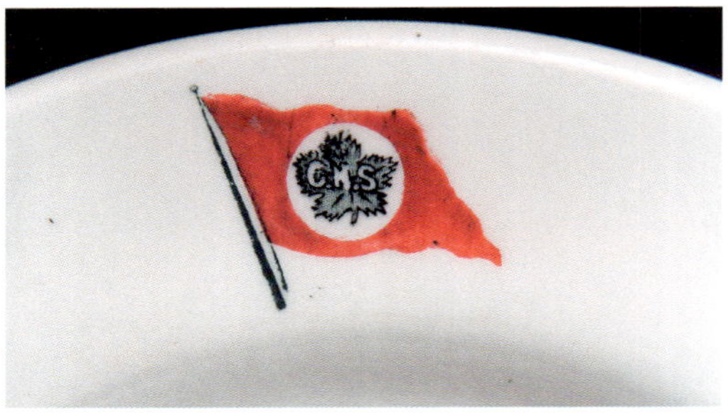

CNS-2 Prince George ★★★

White china with the CNS house flag and navy-blue-and-gold border design. Back stamped "Canadian National Steamships" or "Canadian National System".
Mfr: RLD 1930–38, GLB ca 1934–36.
Ref.: Luckin & Spence CN-16, Laister p. 32.

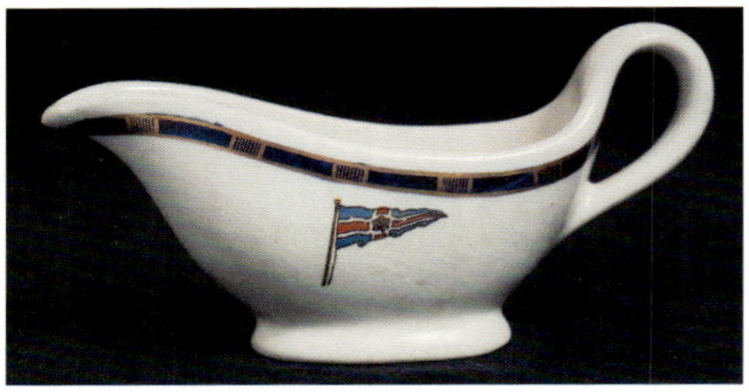

CNS-2a Prince George ★★★

Same as CNS-2, except that the gold in border design has been replaced with mustard yellow enamel.
Mfr: GLY 1938–42.
Ref.: Conroy & Luckin CN-16.

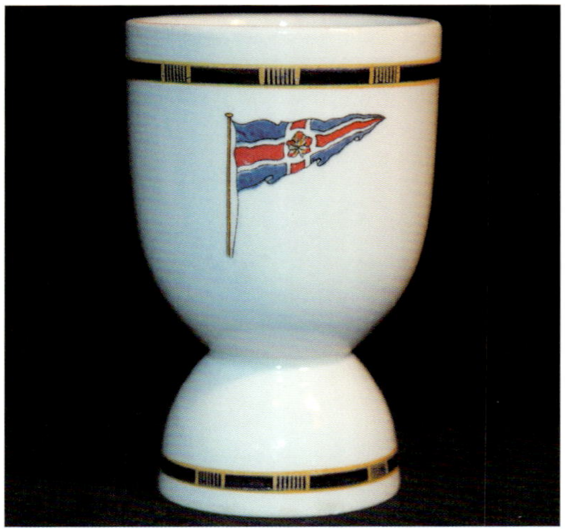

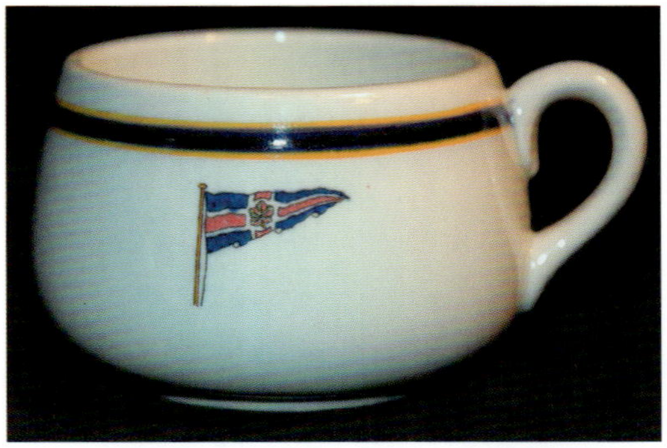

CNS-3 **Transition** ★★★

White china top marked with the CNS house flag and trimmed with a cobalt blue border stripe flanked by yellow pinstripes. The back stamp reads "Canadian National System". This pattern differs from CNS-2 in that it has border stripes not a border design.
Mfr: GLY 1942–47.

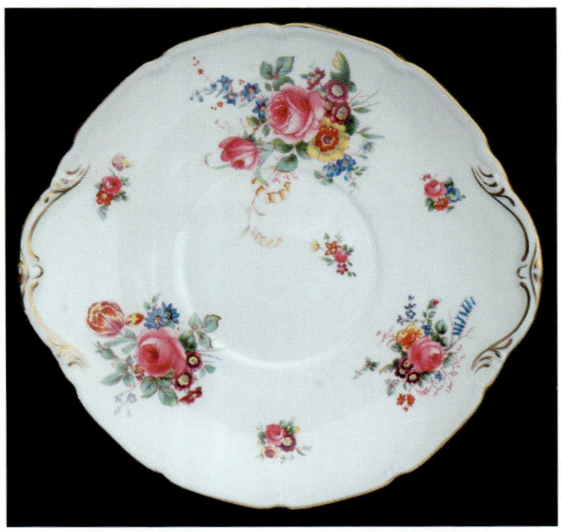

CNS-4 **Jasper** ★★

White china decorated with a multicoloured floral design and trimmed with gold. Back stamped "Canadian National Steamships", "Canadian National System" or "CNR". This is a Royal Doulton stock pattern, CN 3079, introduced in 1930. The Royal Doulton name of this transfer print is the "Bristol".
Mfr: RLD 1930–41 ("Steamship" back stamp 1930–33, "System" back stamp 1940–41).
Ref: Luckin CN-8, Laister p. 33.

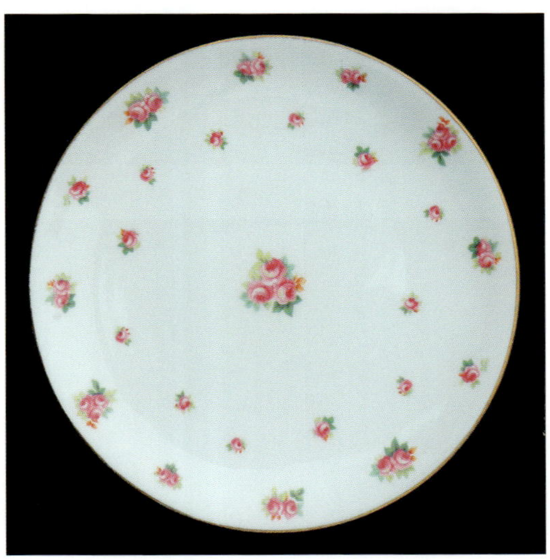

CNS-5 **Rosebud** ★★★

White china decorated with a pink rosebud transfer print and trimmed with gold on the rim. Back stamped "Canadian National Steamships" or "Canadian National System". Royal Doulton has no record of this pattern. This design, trimmed with a green stripe, was used in the 1930s by the Royal Mail Line, Union Castle Line and Bank Line.
Mfr: RLD 1930–39 ("Steamship" back stamp 1930–38, "System" back stamp 1936).
Ref.: Laister p. 33.

CNS-6 **Green Wreath** ★★★★★

White china top marked with a green CNS logo and trimmed with a green border stripe flanked on each side by a green pinstripe. The badge consists of "Canadian National Steamship" surrounded by a wreath of green leaves. The badge on this china is typically seen only on silver-plated items. Only two pieces are known to exist.
Mfr: RLD ca 1930–38.

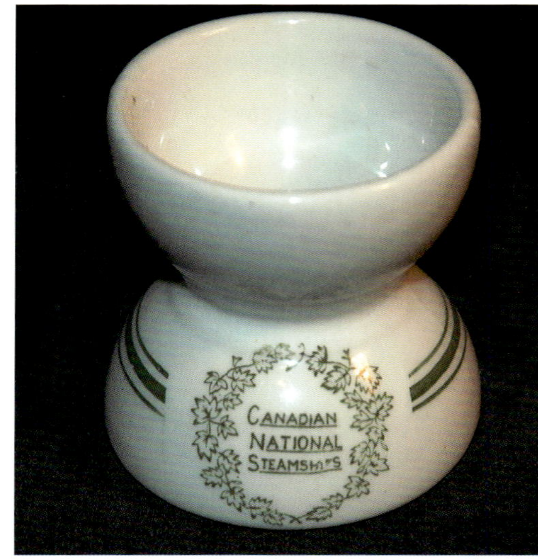

CNS-7 **Truro** ★★

White china with a brown CNS canted box logo, trimmed with an aquamarine border stripe, flanked on each side by pinstripes of the same colour. The text in the canted box reads "Canadian National System". This pattern was first introduced by the Canadian National Railway in about 1923, when the canted box contained "Canadian National Railways". It was reintroduced in the late 1940s with "Canadian National System" in the box. All dive-recovered pieces read "Canadian National System".
Mfr: SOV 1947–51, SGO 1951–60, WHG 1950s.
Ref.: Luckin CN-22 and 22a.

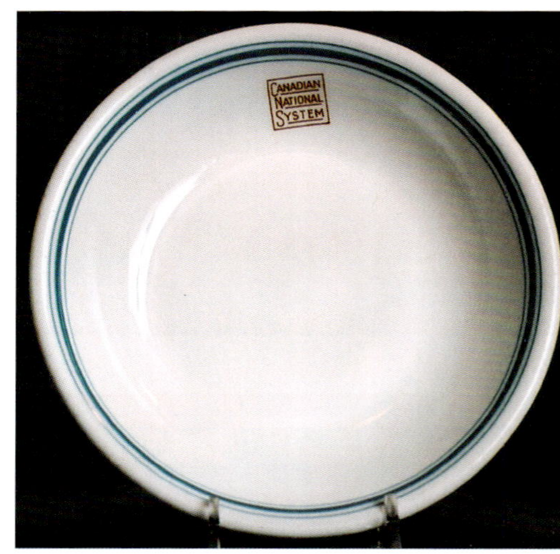

CNS-8 **Rupert** ★★

Cream-coloured china top marked with a brown CNS canted box logo. Text in box reads "Canadian National System".
All dive-recovered pieces made by Sovereign Potters.
Mfr: MED ca 1940s, SOV 1947–54.
Ref.: Luckin CN-20.

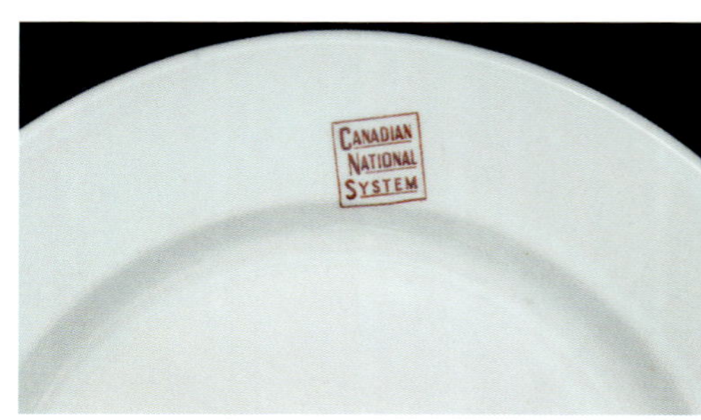

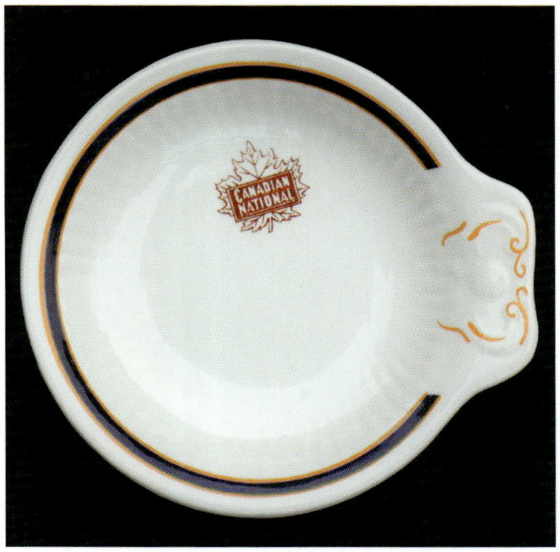

CNS-9 Bonaventure ★★

White china top marked with the CNS maple-leaf logo and trimmed with a royal blue border stripe flanked on each side by a single mustard yellow pinstripe. Text in the logo reads "Canadian National". This pattern is often found underwater at dock sites.
Mfr: GLY 1946–54.
Ref.: Luckin CN-3.

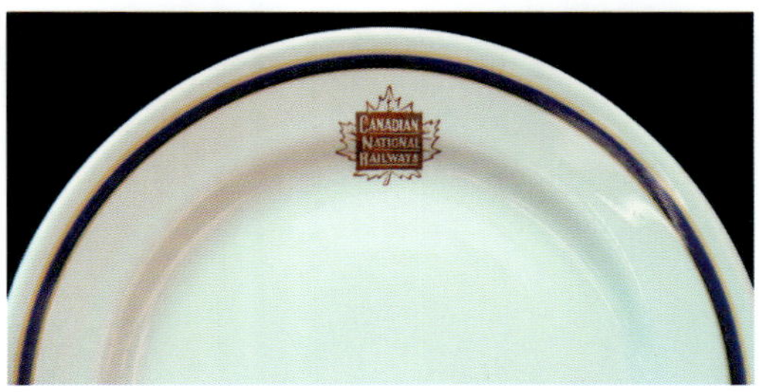

CNS-9a Bonaventure ★★

Same as CN-9, except for the text in the maple-leaf logo, which reads "Canadian National Railways". A few pieces of this variant have been found underwater, but it is generally considered a railway pattern.
Mfr: GLY 1954–75.
Ref.: Luckin CN-3a.

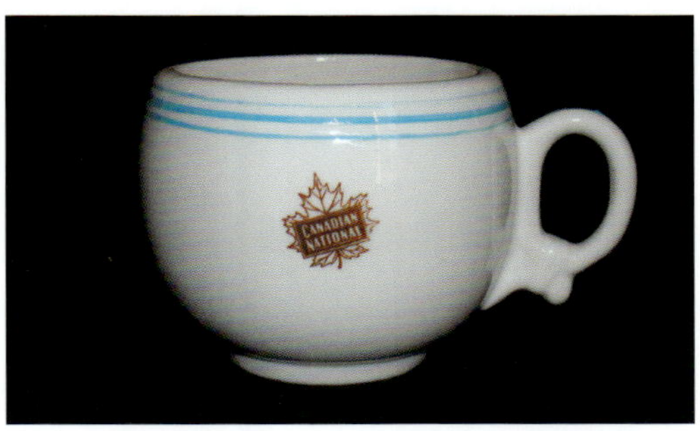

CNS-10 Toronto ★★

White china with a brown CNS maple-leaf logo, trimmed with a baby blue border stripe flanked on each side by a pinstripe of the same colour. The text within the maple leaf reads Canadian National. Pieces of this pattern have been retrieved from CNS dock sites. Steven Sandknop *(Nothing Could be Finer)* called this pattern Maritime.
Mfr: SOV 1947-54.
Ref.: Luckin CN-21.

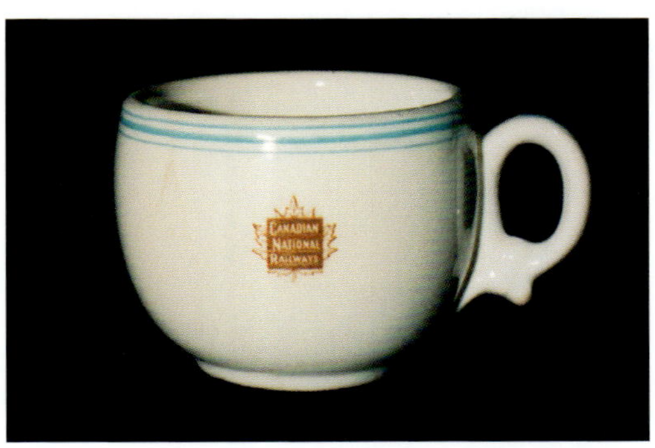

CNS-10a Toronto ★★

Same as CNS-10, except that the text in the maple-leaf logo reads "Canadian National Railways". Pieces of this pattern were retrieved from the galley of the west-coast railway ship SS *Canora* when it was retired from service in 1967.
Mfr: SOV 1954–62.
Ref.: Luckin CN-21a.

Canadian Pacific Navigation Company

CPN-1 Charmer ★★★★★
White china top marked with the CPN house flag (designed by Captain Irving in 1883) and beneath it "C.P.N. Co." Only one piece of this pattern is known to exist.
Mfr: unknown ca 1890.

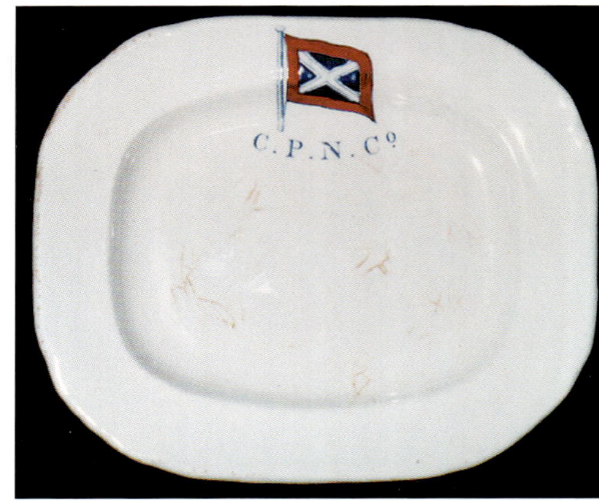

CPN-2 Islander ★★★★
White or grey china with superimposed black letters "CPNCo." Pieces of this pattern were recovered from the wreck of the SS *Islander* when it was salvaged in 1934.
Mfr: FUR 1890–95, MKN 1895–1901.
Ref.: Spence CPN-1.
Brian Fardoe photo.

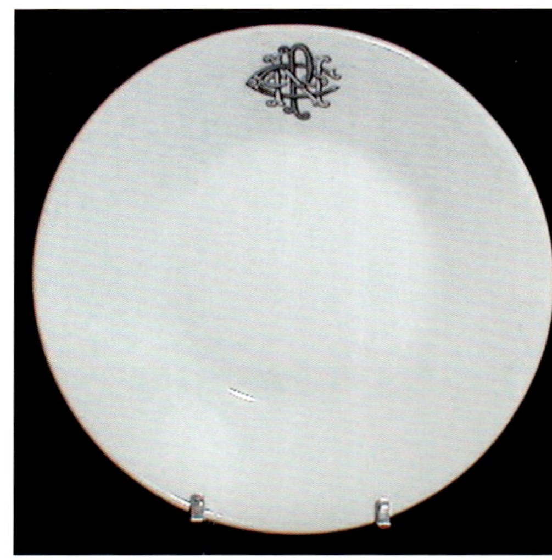

Canadian Pacific Railway Company

CPR-001 Black Crest ★★★★
White china top marked with a black garter containing the superimposed letters "CPR". "Canadian Pacific Railway" forms an arch inside the garter and two small birds are perched on a roost under the garter. This pattern dates to about 1900. Pieces have been found under water at Prince Rupert, in Swanson Bay and in Kootenay Lake.
Mfr: GDA unknown dates, CAF 1894–unknown date, HCO unknown dates.
Ref.: Luckin CP-14.

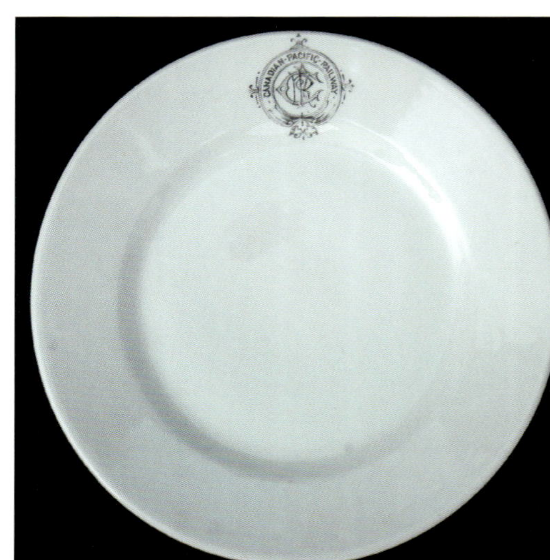

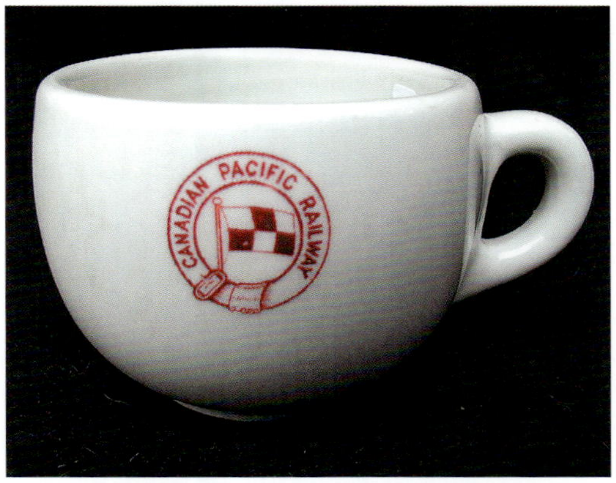

CPR-002 Moyie ★★★★

White china top marked with a red garter containing the CPR red-and-white checkered house flag. The garter text reads "Canadian Pacific Railway". This pattern has been dive-recovered in Kootenay Lake. It was most likely on board the SS *Moyie* and SS *Minto*, launched in the fall of 1898. This pattern also saw extensive service aboard the Great Lakes steamer *Keewatin* in the 1950s. The *Keewatin* china rates two stars, because is readily available at shows and on the Internet.
Mfr: CAF 1898–1906, WGC 1928–32, SOV 1954–62, GLY 1946, 1962–69.
Ref.: Luckin CP-21a, Krummes CP-1, Laister p. 38.
David Smith photo.

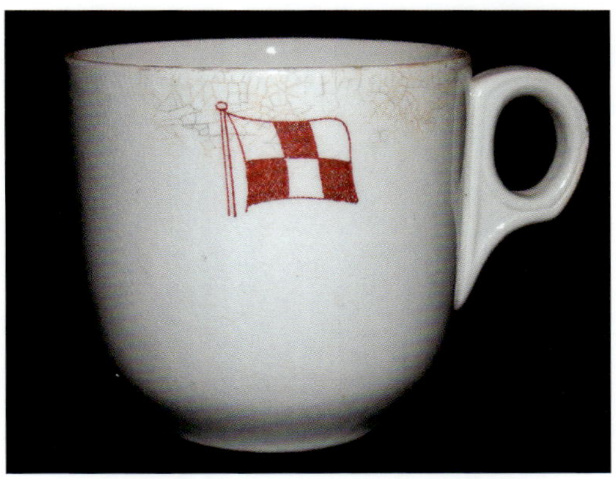

CPR-101 Checkered Flag ★★★★★

White china top marked with the CPR red-and-white checkered house flag. This badge was used on early silverware around 1906. The piece shown was dive-recovered from Victoria's Inner Harbour.
Mfr: unknown ca 1904–06.

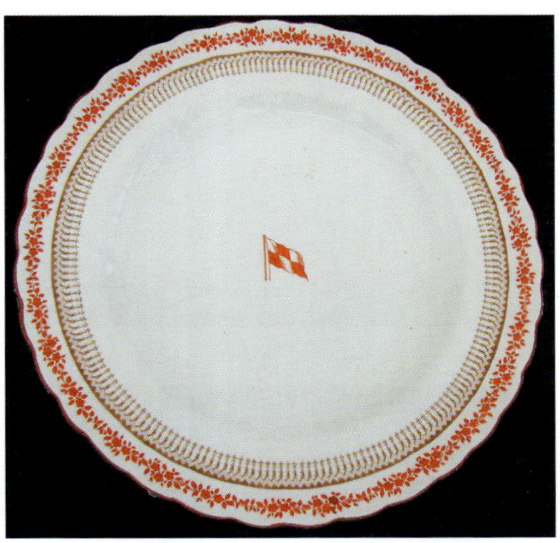

CPR-102 Empress of India ★★★★★

White china top marked with the CPR red-and-white checkered house flag and trimmed with a red-and-brown border design. Registration no. 90067 dates the pattern design to 1896. The flag was introduced in 1891 with the launch of the *Empress of India.*
Mfr: COP (impressed back stamp) 1904.
Ref.: Laister p.35.
Peter Laister photo.

CPR-103 Orange Garter ★★★★

This pattern is the same as CPR-002, except that the garter text reads "Canadian Pacific Railway Co." and the badge is slightly more orange.
Mfr: MDK ca 1908–10.

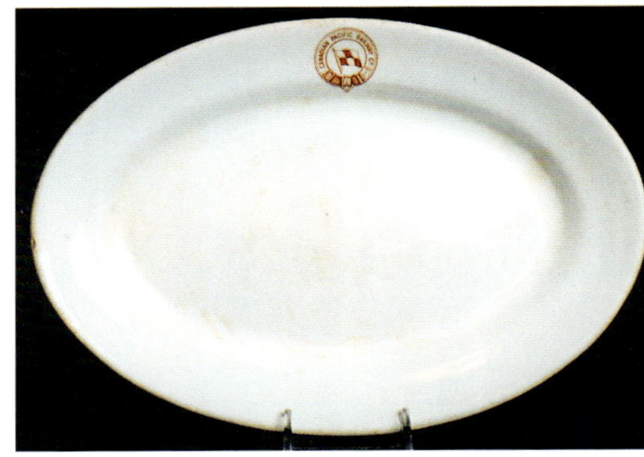

CPR-104 Green Interscroll ★★★

White china top marked with a simple green CPR interscroll. This pattern has been dive-recovered from New Castle Island Resort and other CPR dock sites.
Mfr: RID 1906–10, COP 1913, GLY ca 1922.

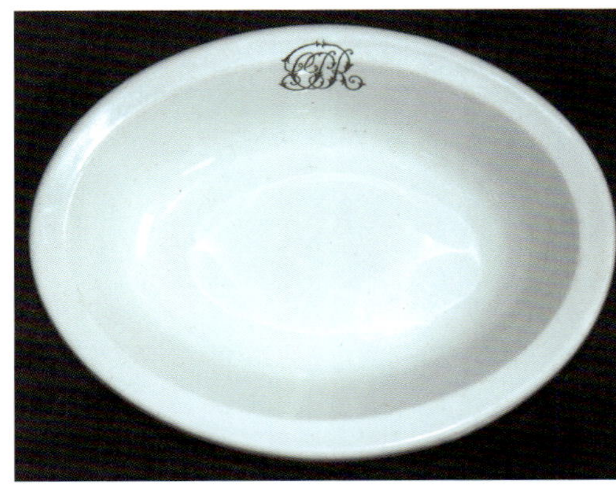

CPR-105 Red Stripe ★★★★

White china top marked with the CPR garter logo containing the house flag and trimmed with a red border stripe flanked inside by a red pinstripe. The garter text reads "Canadian Pacific Railway Co." This pattern was introduced to BC coastal service in 1906 and was still in service on the *Princess Sophia* in 1918.
Mfr: RID 1906–07, COP 1906, MKN unknown dates, MIN 1908.
Ref.: Luckin CP-21c.

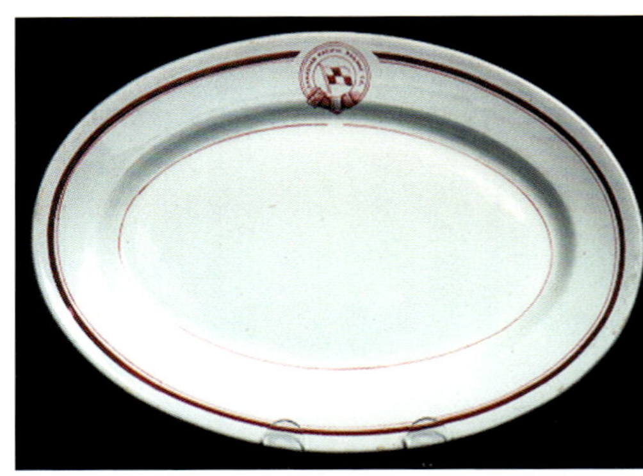

CPR-105a Red Stripe VIP ★★★★

This pattern is the same as CPR-105, but includes the addition of a gold highlight on the rim. This pattern was most likely used for the captain's table or VIP service.
Mfr: RID 1906–07.

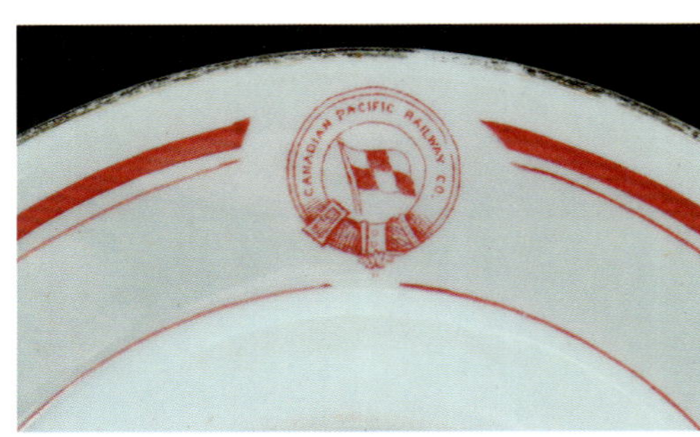

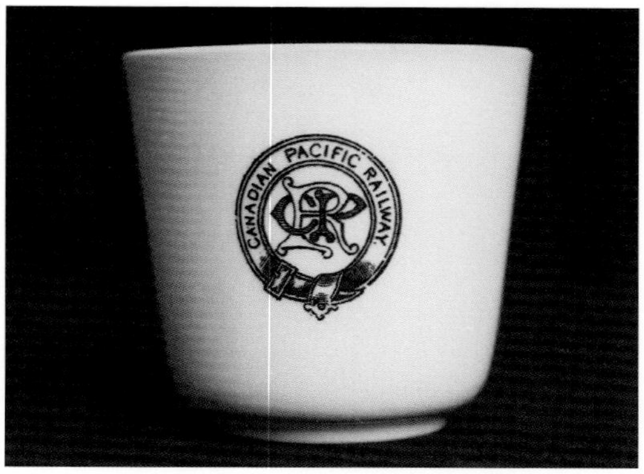

CPR-106 Black Garter ★★★

White china top marked with a black garter containing the CPR name and superimposed black initials. The garter text reads "Canadian Pacific Railway". This pattern has been recovered from Barkley Sound, but it is most commonly associated with the *Empress of Ireland*, third class.
Mfr: MIN 1906–14.
Ref.: Luckin CP-3.1, Krummes CPS-5.
Mike Paris photo.

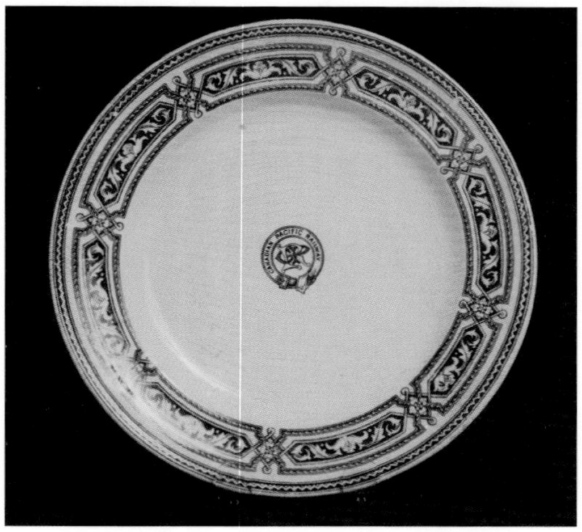

CPR-107 Alton ★★★

White china top marked with a black garter containing the superimposed black initials "CPR" and trimmed with an ornate black border design. The garter text reads "Canadian Pacific Railway". This pattern was used on *Empress* ships and has been found in Victoria's Inner Harbour and at west-coast dock sites.
Mfr: MIN 1906–14
Ref.: Luckin CP-22, Krummes CPS-4, Laister p. 37.
Mike Paris photo.

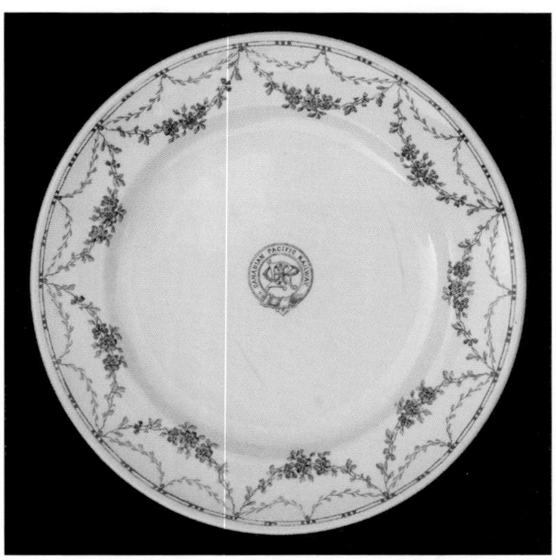

CPR-108 Fontenay ★★★

White china top marked with a green garter containing the superimposed green initials "CPR" and trimmed with an ornate green-and-pink border design. The garter text reads "Canadian Pacific Railway". West-coast use has not been verified. This pattern was used on the *Empress of Ireland* in first class.
Mfr: MIN 1906–14.
Ref.: Luckin & Spence CP-11.11, Krummes CPS-3, Laister p. 36.
Mike Paris photo.

CPR-109 Denmark ★★★★

White china top marked with an orange garter logo containing superimposed "CPR" and trimmed with ornate oriental border design. The garter text reads "Canadian Pacific Railway". The rim of china is trimmed with gold. The provenance of this china is not known. It is thought to have been used aboard the *Pacific Empress* ships, given that the oriental motif and date coding is consistent with CP-106 and CP-107.
Mfr: MIN 1906–14.
Ref.: Luckin CP-6.
David Smith photo.

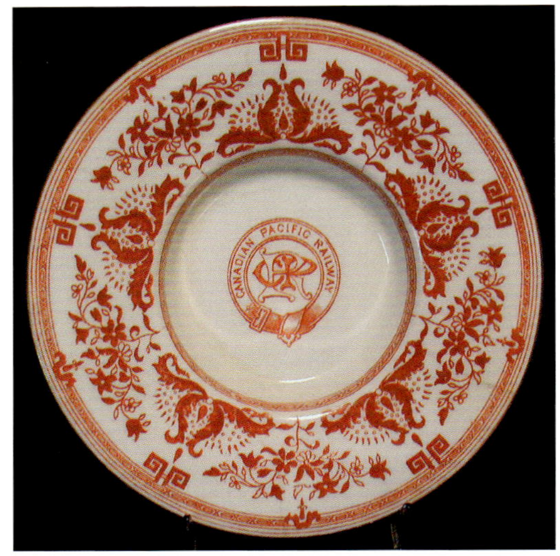

CPR-110 Faucille ★★★★★

White china top marked with a green garter containing "CPR" superimposed in green, and trimmed with an ornate green border design. The garter text reads "Canadian Pacific Railway". The edge is trimmed with gold. West coast use has not been verified. This pattern has been dive-recovered from *Empress of Ireland*'s first-class saloon.
Mfr: MIN 1906–14.
Steve Gatto photo.

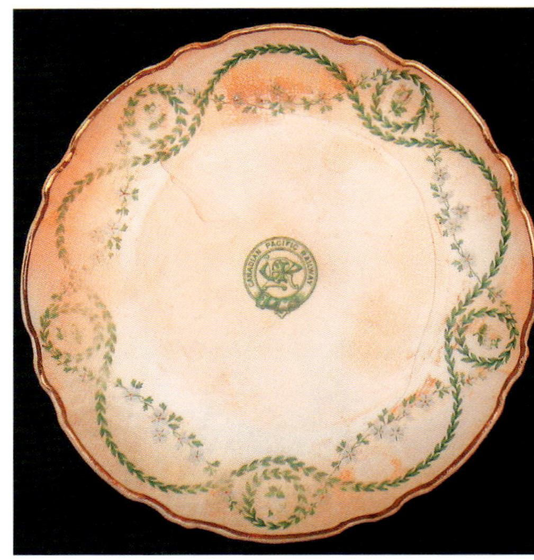

CPR-111 Epping Fruit ★★★★★

White china top marked with a black garter containing "CPR" superimposed in black, and trimmed with an ornate black border design. The edge is trimmed with gold. The garter text reads "Canadian Pacific Railway". West coast use has not been verified. This pattern has been dive-recovered from *Empress of Ireland*'s first-class saloon.
Mfr: MIN 1906–14.
Phil Beaudry photo.

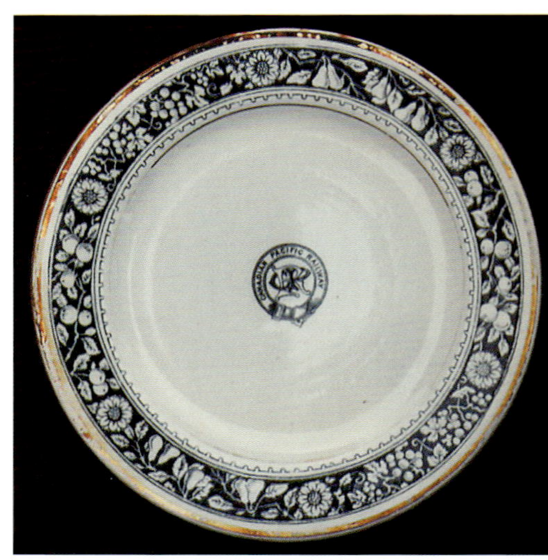

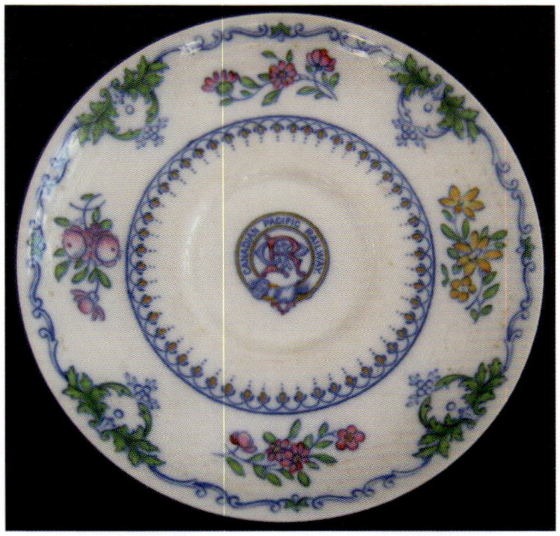

CPR-112 Tri-colour Floral ★★★★★
White china top marked with a blue garter containing "CPR" superimposed in red and blue, and decorated with a pink-and-yellow flower motif. The garter text reads "Canadian Pacific Railway". Thought to be a marine pattern although exact provenance is unknown.
Mfr: COP ca 1913–15.
Ref.: Laister p. 35.
Peter Laister photo.

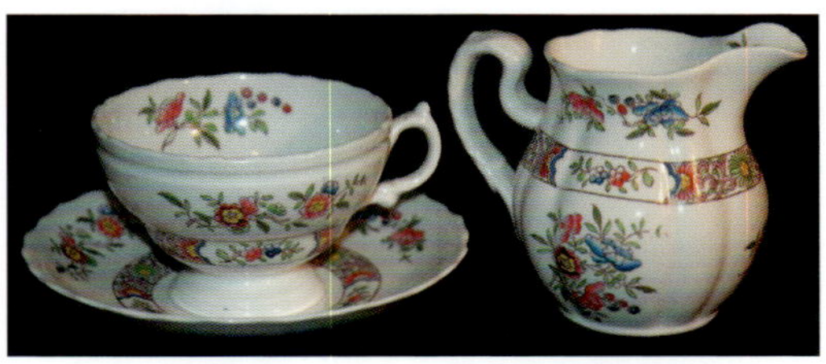

CPR-113 Nigel ★★★★★
White china decorated with red and blue flower transfer print. Back stamped "Canadian Pacific Railway Co." Some pieces marked with British registration number 512488, which dates to 1907. Pattern also known as Naran or New Nigel. Provenance unknown.
Mfr: COP 1907–15.

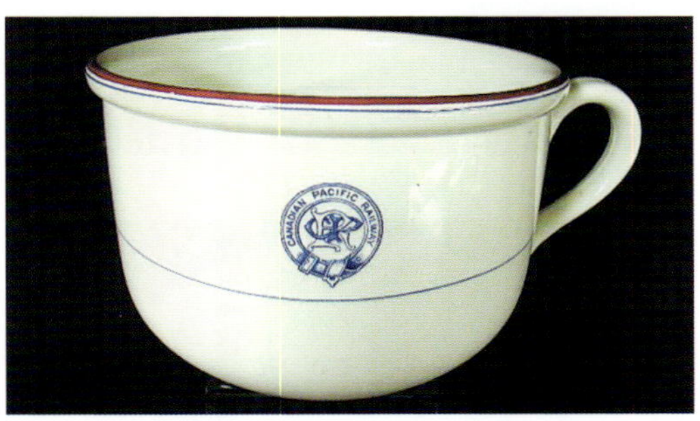

CPR-114 Blue Garter ★★★★★
White china top marked with a blue garter containing "CPR" superimposed in blue. The rim is trimmed with red and flanked with an outer blue pinstripe. A second blue pinstripe rings the body about halfway down. The garter text reads "Canadian Pacific Railway". Back stamped "Canadian Pacific Railway Co." Chamber pots of this style and colour scheme were used by the Allan Line up until the formation of CP Ocean Services in 1916.
Mfr: COP ca 1913–15
Ref.: Luckin CP-3.3.
David Smith photo.

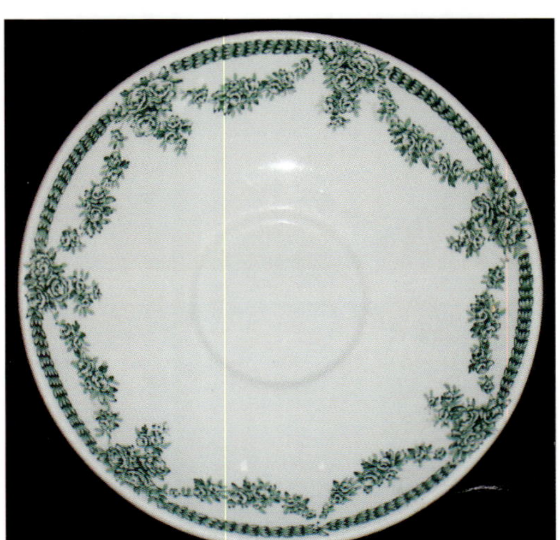

CPR-115 Newport ★★
White china decorated with a green ornate border design. This is a stock pattern found at numerous CPR dock sites and recovered from the wreck of the *Princess Sophia*. The name "Newport" is often included in the manufacturer's back stamp.
Mfr: MDK 1907–18.

CPR-116 Empress Crown ★★★★★

White china top marked with a large green crown. This pattern is attributed to the Empress Hotel in Victoria, BC, but has been found at dock sites along the BC coast. Three pieces are known, all chamber pots.
Mfr: MDK 1908–10.

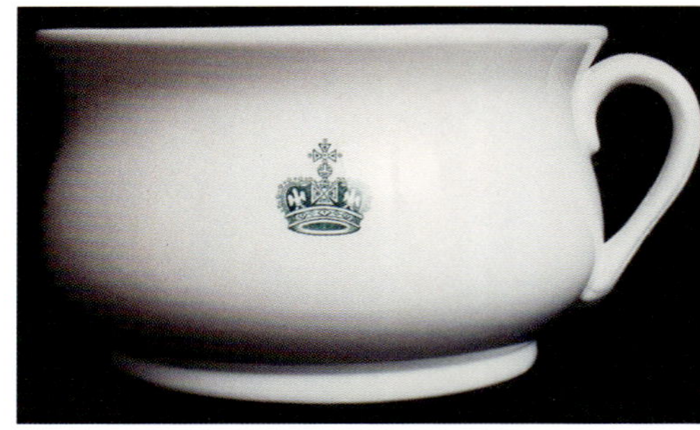

CPR-116a Crown & Stripes ★★★

White china top marked with a large green crown and trimmed with three border pinstripes, red with a green on each side. This pattern is attributed to the Empress Hotel in Victoria, BC, but has been found at several dock sites along the BC coast. It was most likely a filler pattern on ships during World War I.
Mfr: MDK 1908–10.

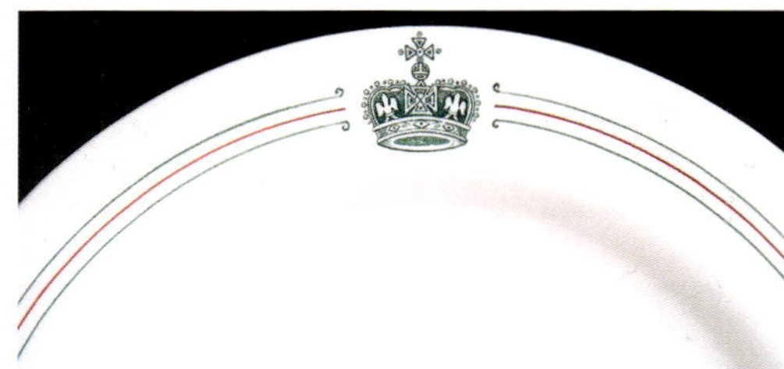

CPR-117 C.P.R. ★★★★★

Dark green china with a white circle badge containing "C.P.R." in white. One piece known, dive-recovered from Victoria's Inner Harbour.
Mfr: unknown (most likely British) ca 1900–1910.

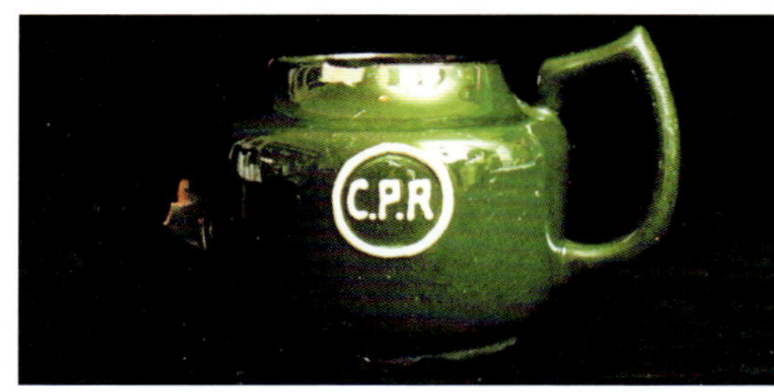

CPR-201 Colonial ★★

White china decorated with a border design of green bows and beads. This is a stock pattern found at numerous BC dock sites and recovered from the wreck of the *Princess Sophia*. The name "Colonial" is often included in the manufacturer's back stamp.
Mfr: MDK 1910–22.

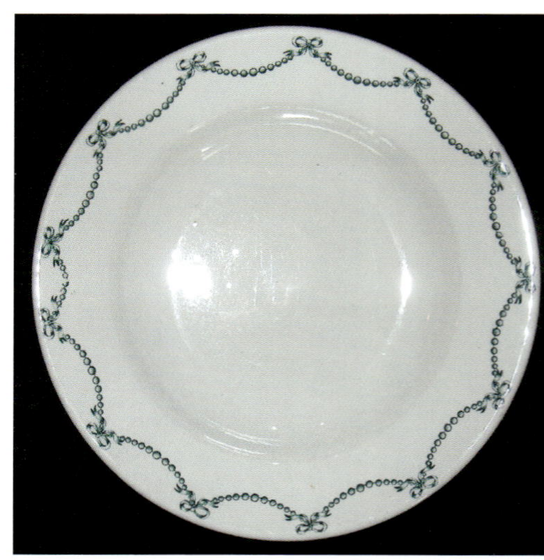

CPR-202 Tremblant ★★

White china top marked with a red garter containing the CPR chequered house flag and trimmed with green, red and green border pinstripes. The garter text reads "Canadian Pacific Railway Co."
Mfr: MDK 1910–18, GLY 1922–28, MDK 1928–35, GLY 1939–45.
Ref.: Luckin CP-21.

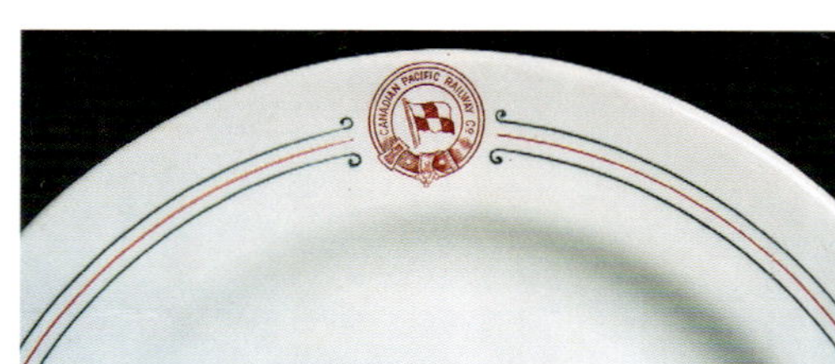

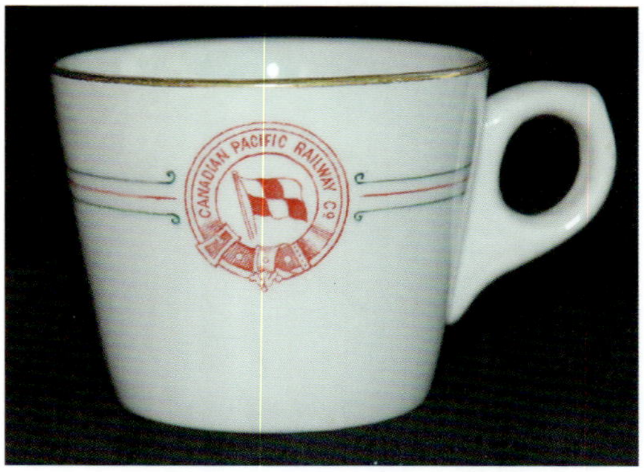

CPR-202a Tremblant VIP ★★★★
Same as 202 but with gold trim on the rim. This pattern was most likely used at the captain's table or for VIP service, as it occurs in limited quantity.
Mfr: MDK 1910–12, GLY 1922–23.

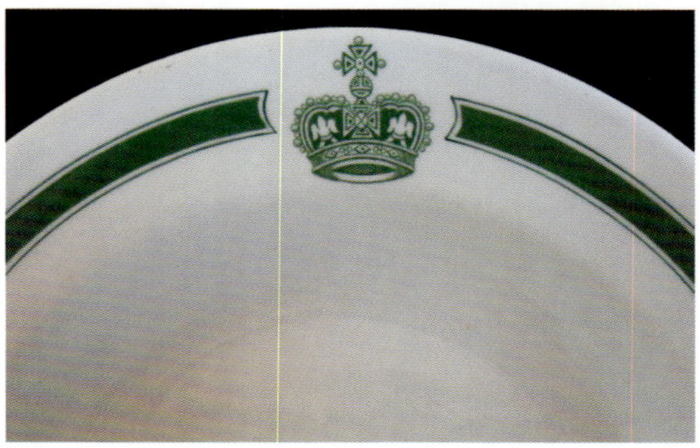

CPR-203 Green Band Crown ★★★
White china top marked with a large green crown and trimmed with a green band flanked on each side by green pinstripes. This pattern is attributed to the Empress Hotel in Victoria, BC, but has also been found at dock sites along the BC coast, and many soup plates were found on the wreck of the *Princess Sophia*.
Mfr: MDK 1910–12.

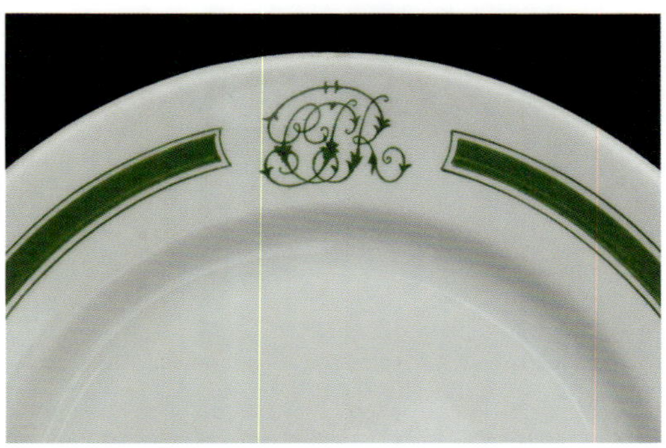

CPR-204 Green Band Interscroll ★★
White china top marked with a simple green CPR interscroll and trimmed with a green band, flanked on either side by green pinstripes. This appears to have been a multipurpose pattern, seeing service on ships and trains and in hotels. It is CPR's longest running pattern, spanning 43 years.
Mfr: RID 1910; MDK 1910–15; GLY 1924–31, 1946-50, 1955; MED 1951–53.
Ref.: Luckin CP-15b, Spence CP-2, Krummes CP-3, Laister p. 37.

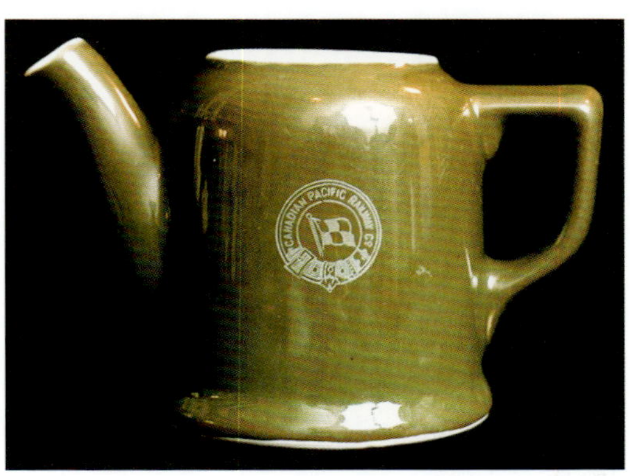

CPR-205 Jade Teapot ★★★★★
Green china etched with a white CPR garter logo containing the house flag. The garter text reads "Canadian Pacific Railway Co." This pattern is known to exist in Chicago- and Boston-style teapots only.
Mfr: HAL ca 1910–30.

CPR-206 Blue Heron ★★★
White china decorated with a cobalt blue floral transfer print. Back stamped "Canadian Pacific Railway Co." or "Canadian Pacific Ocean Services". The British registration number 616247 dates to 1913. Numerous photographs exist showing this pattern in use on *Pacific Empress* ships.
Mfr: COP 1913–21 (CPR back stamp 1913–15, CPOS back stamp 1916–21).
Ref.: Luckin CP-7.1, Krummes CPS-1, Laister p. 36.

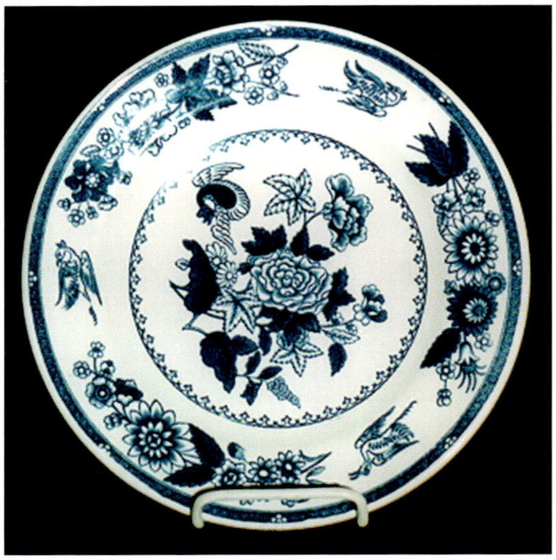

CPR-207 Coloured Heron ★★★
White china decorated with colourful floral transfer prints. Back stamped "Canadian Pacific Railway Co." or "Canadian Pacific Ocean Services". The British registration number 616247 dates to 1913. Photographs show this pattern in use on *Pacific Empress* ships.
Mfr: COP 1913–21 (CPR back stamp 1913–15, CPOS back stamp 1916–21).
Ref.: Luckin CP-25, Spence CP 1, Conroy CP-12.1, Laister p. 36.

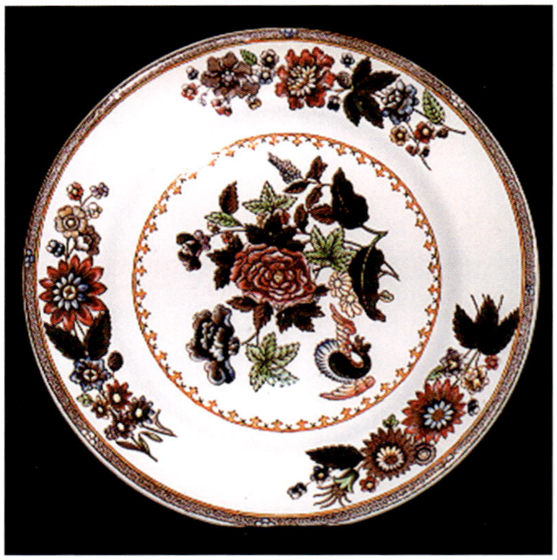

CPR-208 Butterfly Floral ★★★★
White china with a raised flower relief and decorated with a colourful transfer print featuring flowers and butterflies. A green leaf border trims the outer edge. Back stamped "Canadian Pacific Railway Co.", "Canadian Pacific Ocean Services" or "Canadian Pacific Steamships Ltd."
Mfr: COP 1913–22 (CPR back stamp 1913–15, CPOS back stamp 1915–21, CPS back stamp 1922).
Ref.: Laister p. 36.

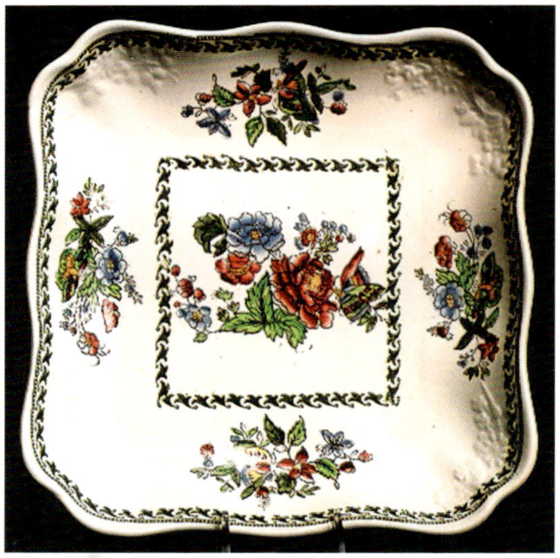

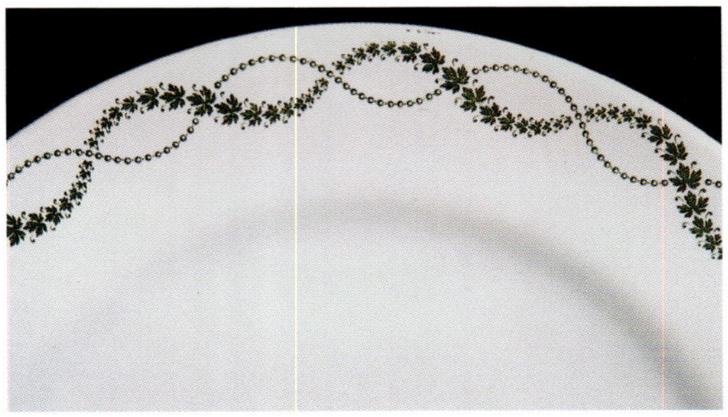

CPR-209 Beads & Leaves ★★★★★

White china decorated with intertwined border design consisting of green beads and leaves. Back stamped "Canadian Pacific Railway Co." Provenance unknown; most likely a marine pattern. Only one piece known.
Mfr: COP 1913–15.

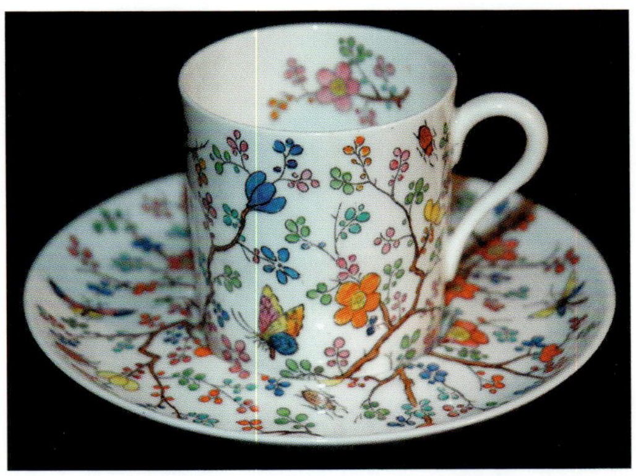

CPR-210 Shanghai ★★★★

White china with a colourful hand-painted transfer print featuring flowers, branching twigs, bugs and butterflies. Back stamped "Canadian Pacific Railway Co." The Copeland-Spode name for this pattern is Shanghai. Luckin calls this pattern Zincton.
Mfr: COP ca 1913–15.
Ref.: Luckin CP-25.

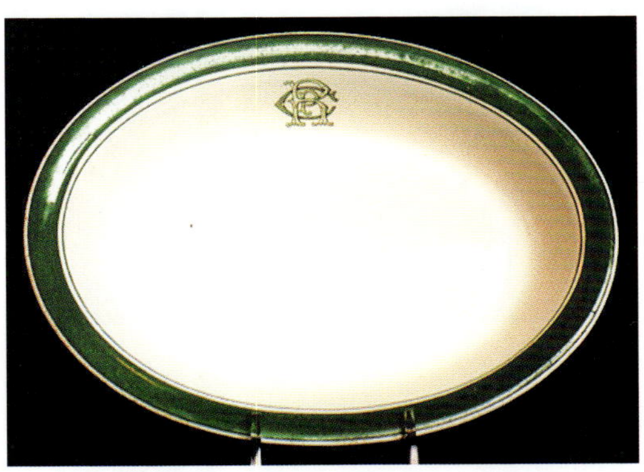

CPR-211 Green Border ★★★

White china top marked with superimposed "CPR" and trimmed with a thick, dark green band, flanked on its inner side by a green pinstripe. This pattern has been dive-recovered from numerous coastal sites. It appears to be unique to BC Coast Steamship service.
Mfr: COP 1914.
Ref.: Luckin CP-15.1.

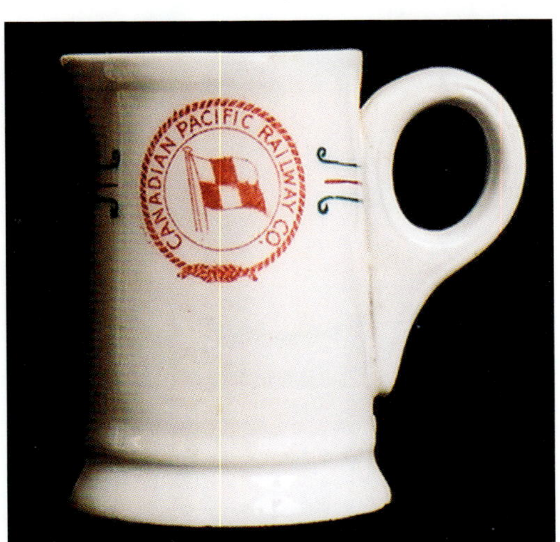

CPR-212 Sophia ★★★★

White china top marked with a rope garter logo containing the CPR house flag and trimmed with green, red and green border pinstripes. The garter text reads "Canadian Pacific Railway Co." This pattern was recovered from the wreck of the *Princess Sophia* in Lynn Canal.
Mfr: DNS 1914–18.

CPR-213 Peacock ★★★★

White china decorated with multicoloured floral transfer print and trimmed with a purple-and-yellow fleur-de-lys border design. Back stamped "Canadian Pacific Ocean Services" or "Canadian Pacific Steamships Ltd." Provenance unknown. Only seen as demi-cup and saucer sets. The transfer print complements the design of the Coloured Heron (CPR-207) and the two patterns may have been used together.
Mfr: COP 1915–22.
Ref.: Laister p. 36.

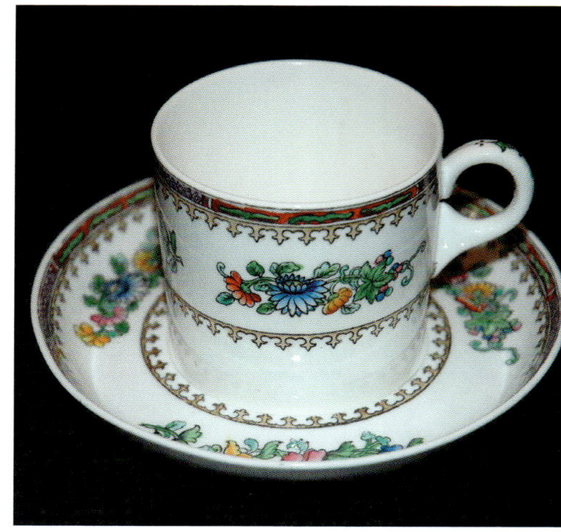

CPR-214 Savoy ★★★

White china with a looping green-and-pale-blue floral design trimmed with a green bluebell border. This is a stock pattern recovered from the wreck of the *Princess Sophia*. Savoy is the manufacturer's name for the pattern. The CPR also used this pattern at the Sicamous Hotel.
Mfr: MKN 1917.
Bruce Lanham photo.

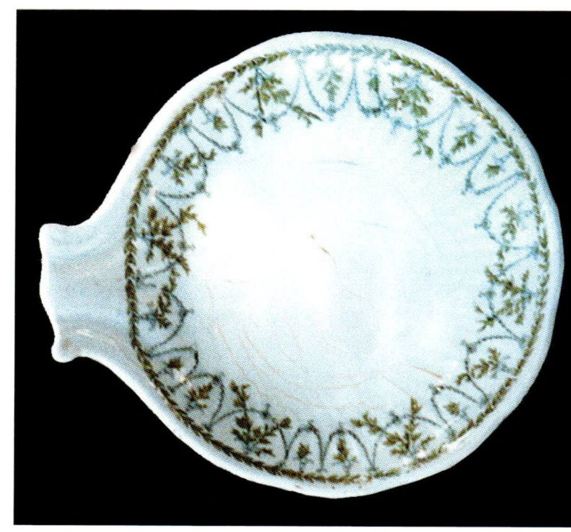

CPR-215 Green Band ★★

White china trimmed with a green band flanked on either side by green pinstripes. A stock pattern recovered from the wreck of the *Princess Sophia* and from various dock sites. Most likely companion pieces to Green Band Interscroll (CPR-204).
Mfr: MDK 1917.

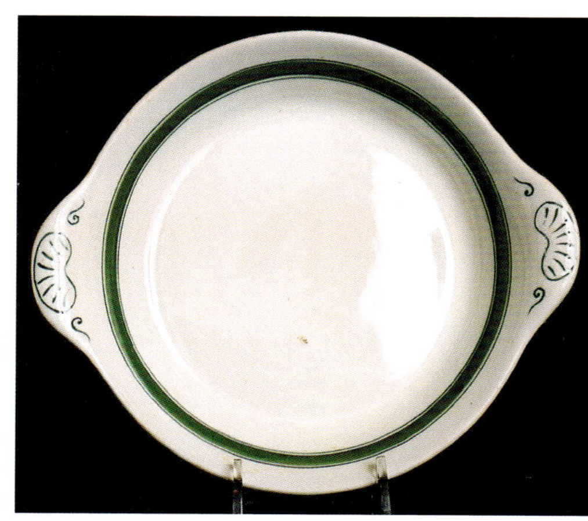

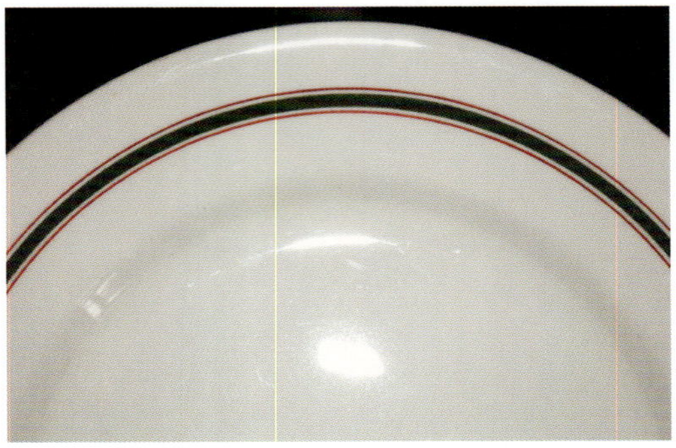

CPR-216 Green & Red Stripe ★★

White china trimmed with a green stripe, flanked on either side by red pinstripes. Stock pattern recovered from the wreck of the *Princess Sophia* and from various dock sites.
Mfr: WOD 1917

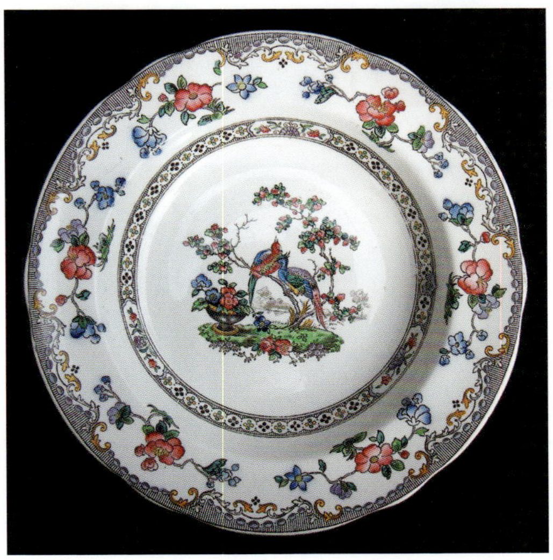

CPR-217 Brown Eden ★★★★

White china decorated with two colourful pheasant-like birds in the centre and a blue-and-red floral border transfer print. Back stamped "Canadian Pacific Steamships Ltd." This pattern, introduced on March 12, 1913, was based on an 1830s Spode pattern called French Birds. The British registration number is 615911.
Provenance unknown.
Mfr: COP 1922.

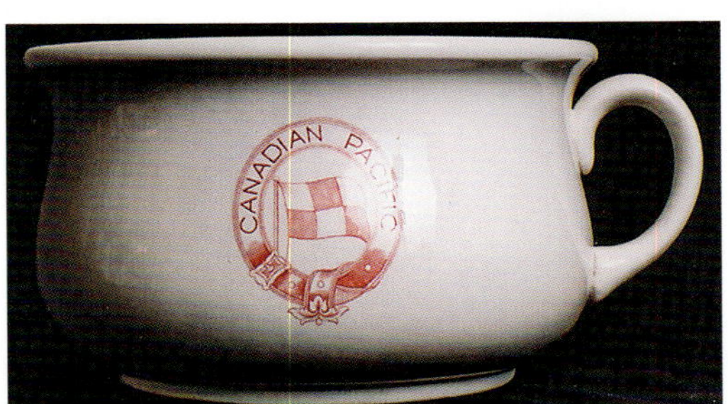

CPR-301 Oval Garter ★★★★★

White china with a red oval garter logo containing the CPR house flag. The garter text reads "Canadian Pacific". Two pieces are known, recovered from docks serviced by BC Coast Steamships.
Mfr: Poutney (England) 1924.

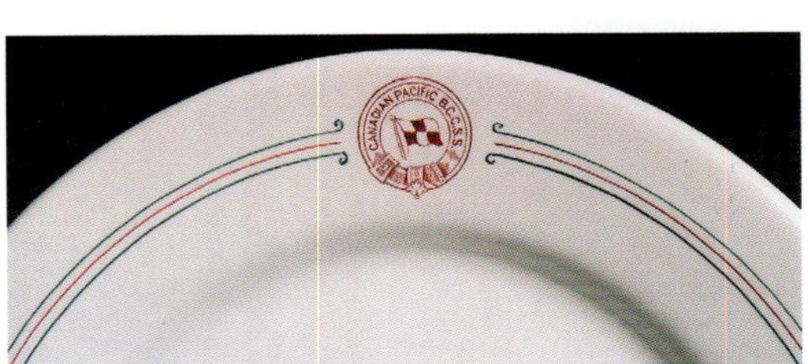

CPR-302 Tremblant BCCSS ★★

White china top marked with a round garter logo containing the CPR house flag and trimmed with green, red and green border pinstripes. The garter text reads "Canadian Pacific BCCSS".
Mfr: GLY 1927–28, 1937–45; MDK 1947–48.
Ref.: Luckin CP-21, Spence CP-4, Laister p. 38.

CPR-303 Bleu de Roi - Gold ★★★

White china top marked with a gold CPR interscroll and trimmed with a cobalt blue band, flanked on either side by gold pinstripes. This pattern has been recovered from several coastal dock sites, in particular Newcastle Island. CPR operated a resort on Newcastle from 1930 to 1940. The *Charmer, Princess Victoria, Princess May* and *Princess Norah* were at one time or another moored at the island during the summer months to provide accommodation for visiting tourists. The manufacturer, Alfred Meakin, calls this pattern Bleu de Roi; it is also known as Banff. It was most likely an afternoon tea service as it is only found as cups, saucers and luncheon plates.
Mfr: MKN ca 1914–unknown date, 1927–30, 1940s; BTH 1930–32; BAK 1927–43; RID ca 1912–unknown date; MDK 1918.
Ref.: Luckin CP-3.

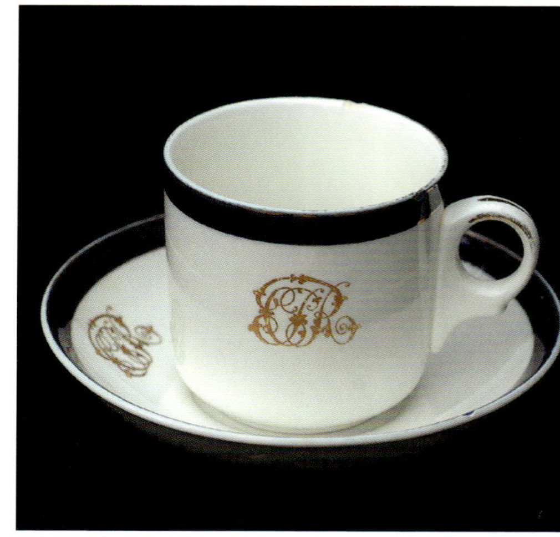

CPR-303a Bleu de Roi - Blue ★★★

Similar to CPR-303 except that it is top marked with a blue CPR interscroll. This pattern is also known as Banff.
Mfr: MKN 1927–30.
Ref.: Luckin CP-3.

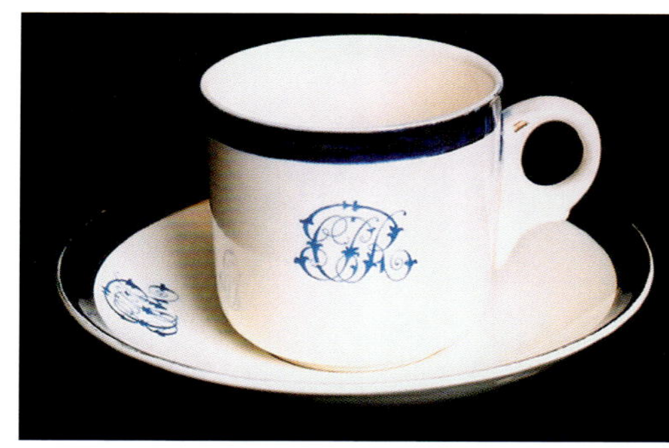

CPR-304 Tremblant Oval ★★★

White china top marked with a red oval garter logo containing the CPR house flag and trimmed with green, red and green pinstripes. The garter text reads "Canadian Pacific". This pattern was made for three years only.
Mfr: GLY 1928–30.
Ref.: Luckin CP-21.

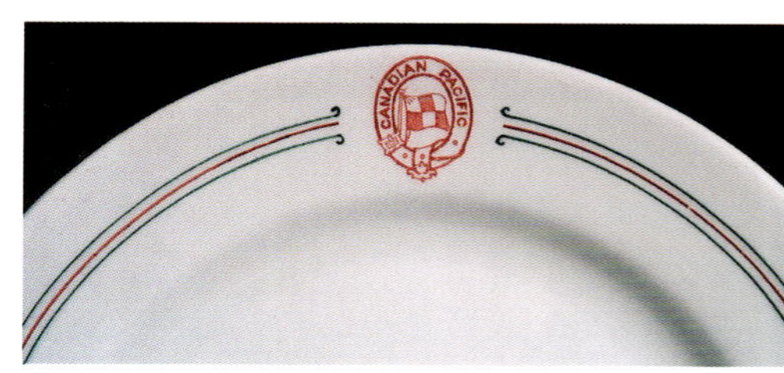

CPR-305 Gold Garter ★★★★★

White china top marked with a gold oval garter. The garter text reads "Canadian Pacific". Beneath the garter a banner contains the initials "B.C.C.S."
Mfr: WMG 1928-1932

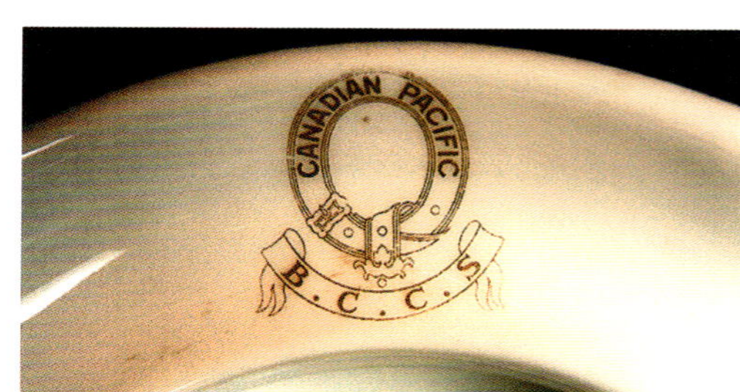

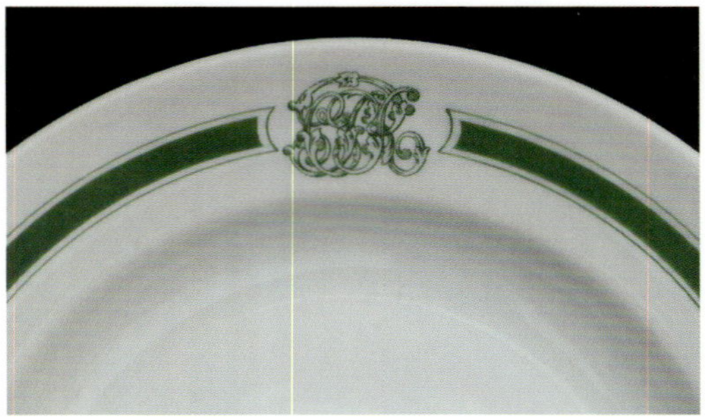

CPR-306 Green Band Crest ★★

White china top marked with an ornate green CPR interscroll and trimmed with a green band, flanked on either side by green pinstripes. This appears to have been a multipurpose pattern seeing service on ships and trains and in hotels. A crest-only version of this pattern also exists, but has not been observed on ships.

Mfr: WMG 1928–32.

Ref.: Luckin CP-15a.

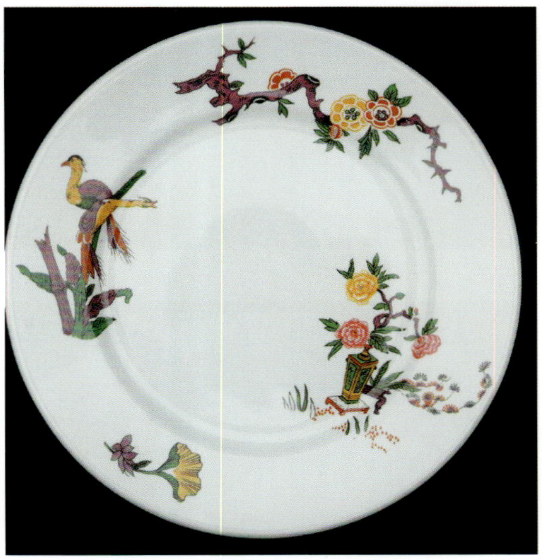

CPR-307 Empress ★★

White china decorated with a multicoloured transfer print featuring two birds of paradise, a flower vase and peony flowers. Back stamped with the CPR garter, shield and beaver, and with the British registration number 774878, which dates to 1932. This pattern was approved for service by CPR as early as 1924 but not registered until 1932. It was used on Atlantic and Pacific *Empress* liners and on BC coastal steamships. A green shield back stamp usually denotes use on a CPR steamship, while a black shield denotes a BC coastal ship. After 1958, this china is not back stamped with CPR identification. It remained in service on the *Princess Marguerite* up until 1974.

Mfr: WMG 1928–32, RLW 1932–39, MIN 1932–41, NSF 1946–48, GLY 1947–74.

Ref.: Luckin CP-11, Conroy CP-11, Krummes CPS-2, Spence CP 3, Laister p. 37.

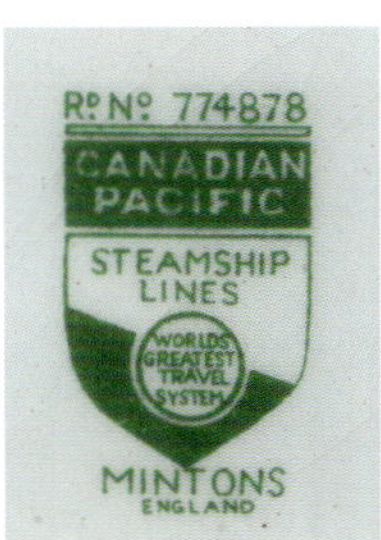

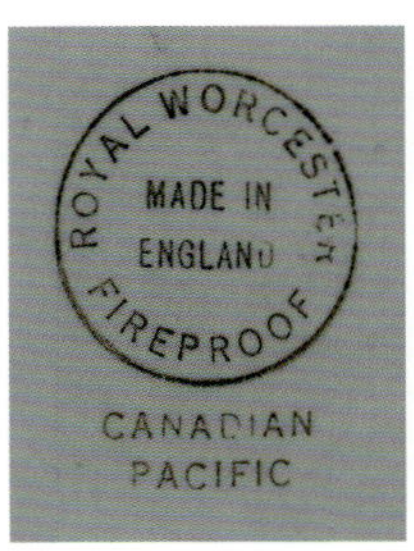

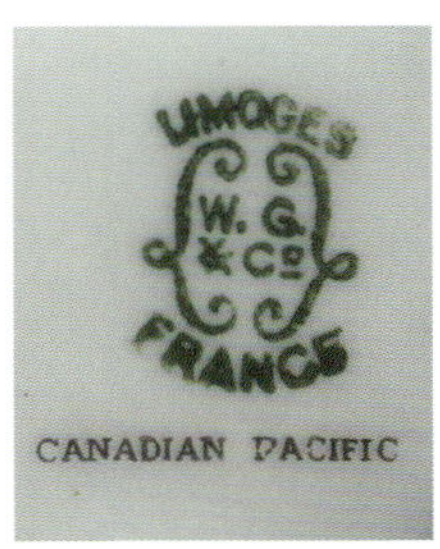

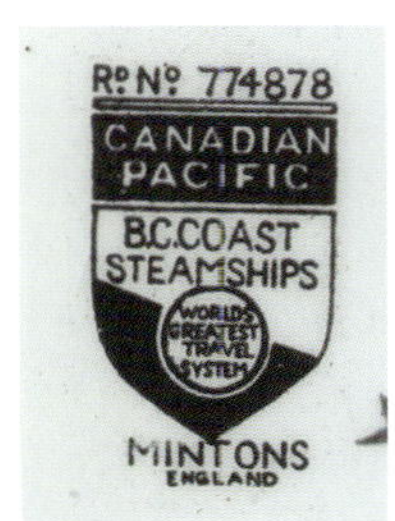

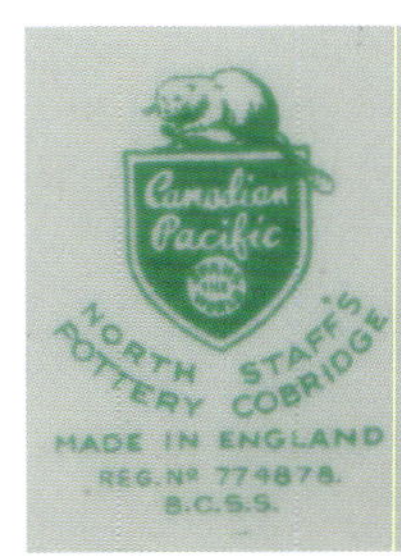

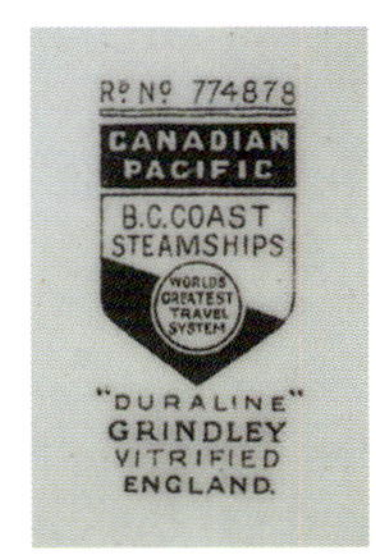

CPR-308 Blue & Brown ★★

White china with a medium blue band, flanked with brown pinstripes on either side. This is a stock pattern with no distinguishing marks. It appears to have been a tea service. Many pieces have been found at Newcastle Island and in Victoria's Inner Harbour at CPR dock sites.
Mfr: GLY 1929-33

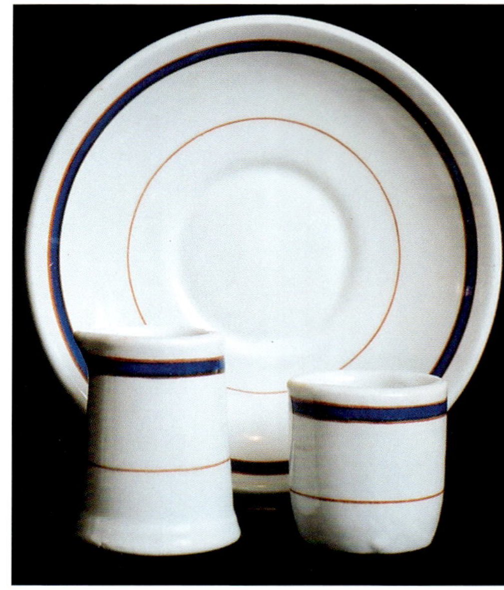

CPR-401 Brown Maple Leaf ★★

White china with a brown-and-yellow border design that features green and red maple leaves and yellow wheat stalks. Top marked with "Canadian Pacific" in a brown box that breaks the border at 12 o'clock. Several pieces, mostly tea cups, have been found in Victoria's Inner Harbour. This was most likely a fill-in pattern during the lean war years. The British registration number 745624 dates it to 1929.
Mfr: MIN 1934–44, RID 1949–54.
Ref.: Luckin CP-5a, Krummes CP-2a.

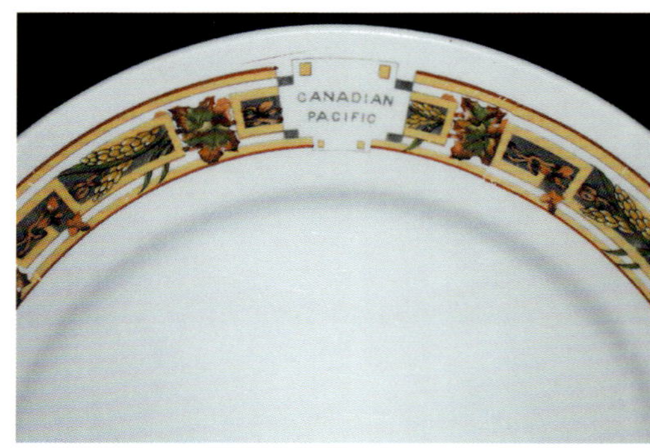

CPR-402 Blue Maple Leaf ★★

White china with a blue-and-yellow border design that features green and red maple leaves and yellow wheat stalks. Top marked with "Canadian Pacific" in a blue box that breaks the border at 12 o'clock. Several pieces of this pattern, mostly tea cups, have been found in Victoria's Inner Harbour. This was most likely a fill-in pattern during the lean war years. The British registration number 745624 dates it to 1929.
Mfr: WMG 1932, MIN 1943–44, RID 1949–54, MED 1953.
Ref.: Luckin CP-5, Krummes CP-2, Spence CP-5, Laister p. 38.

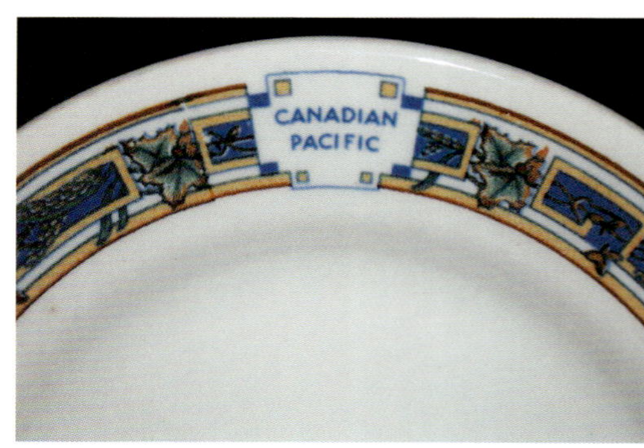

CPR-501 Gold Shield ★★

Green china decorated with a CPR BC Coast Steamship gold-foil shield. This badge occurs principally on teapots, toothpick holders and creamers.
Mfr: MED 1940–45, HAL unknown dates.
Ref.: Luckin 20.1, Laister p. 38.

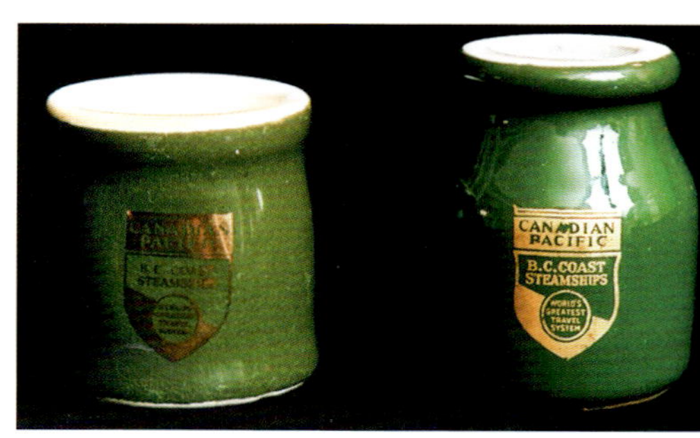

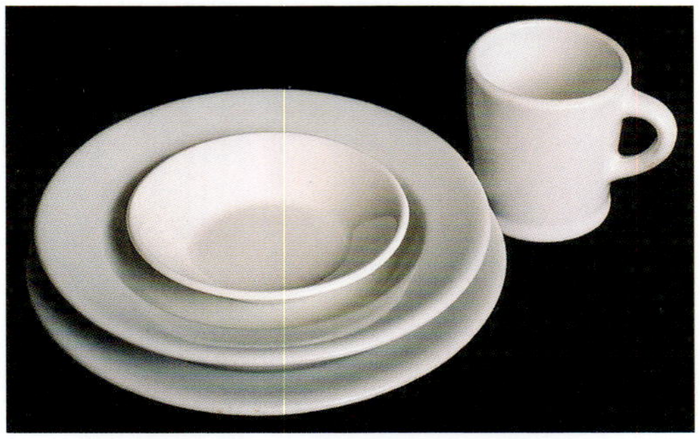

CPR-502 White Ware ★★★

White china back stamped "Canadian Pacific" in black script lettering, a North Staffordshire oval, and the CPR garter and shield. CPR records state that this pattern was put on ships during World War II when they transported troops. A later variant that Luckin calls Baillie was used in the 1960s on Atlantic *Empress* ships.
Mfr: NSF 1940–46, WOD 1943–45, MDK 1946–48, RID 1963–67.
Ref.: Luckin CP-1.2, Krummes CPS-6.

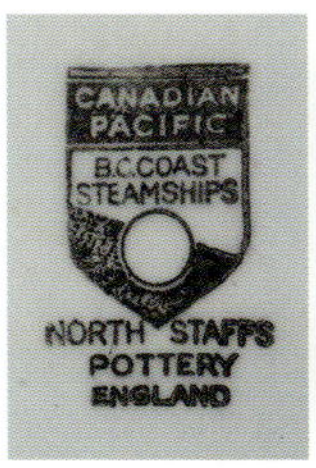

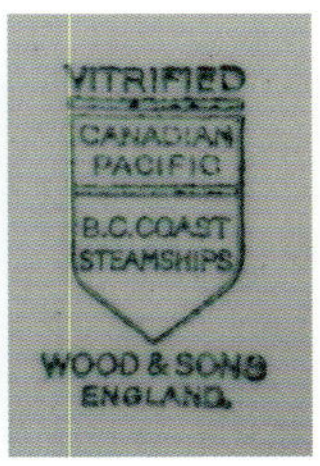

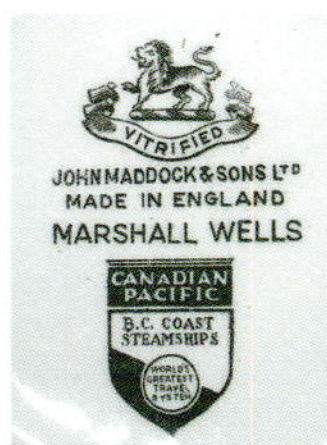

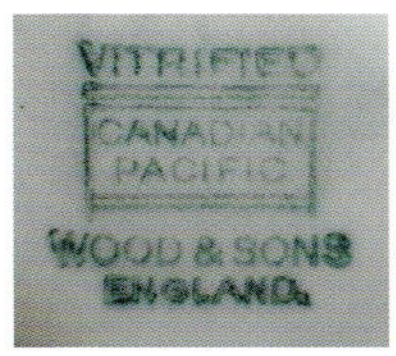

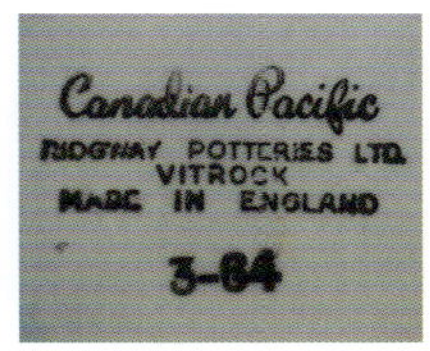

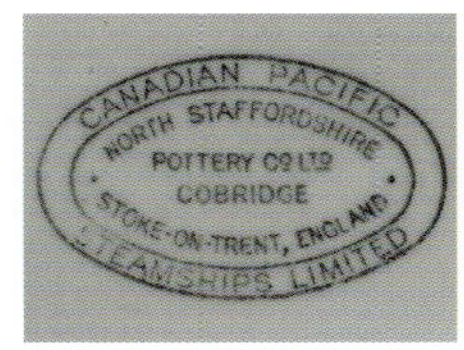

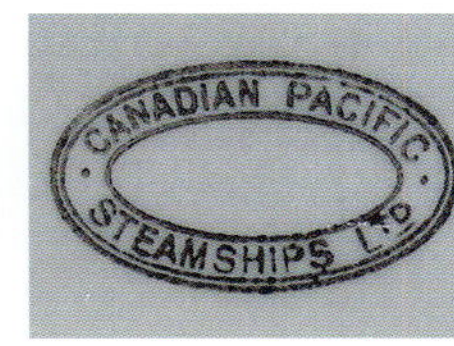

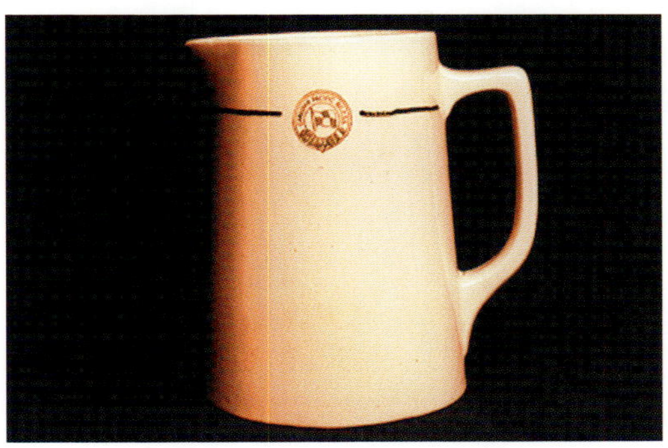

CPR-503 Green Garter ★★★★

White china top marked with a green garter logo containing the CPR house flag and trimmed with a single green border stripe. The garter text reads "Canadian Pacific BCCSS". This pattern is seen only on Medalta chamber pots, water pitchers and creamers.
Mfr: MED 1942–47.

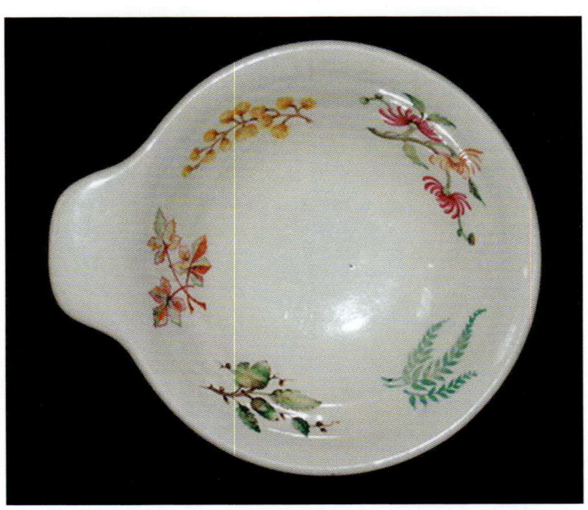

CPR-504 Foliage ★★★

Cream-coloured china decorated with a maple leaves, oak leaves, chrysanthemums, wattle flowers and ferns. This pattern was first registered in 1946 (no. 847730) and introduced on the *Empress of Canada II* in 1947. The five floral illustrations represent the countries served by CPR ships: the maple for Canada, oak for England, fern for New Zealand, chrysanthemum for China and wattle for Australia. West coast use has not been confirmed.
Mfr: NSF 1947–52, RID 1955–60.
Ref.: Luckin CP-13, Conroy CP-13, Laister p. 37.

CPR-504a Foliage ★★★
Same as CPR-504 but with an additional green border stripe.
Mfr: NSF 1947–52, RID 1955–60.
Ref.: Luckin CP-13, Conroy CP-13.

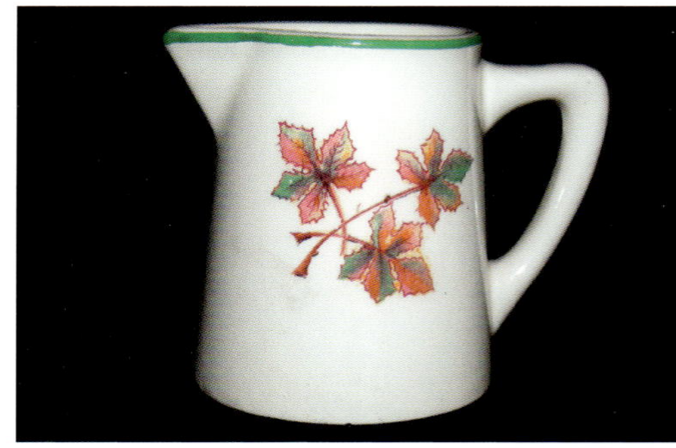

CPR-505 Yellow Flower ★★★
White china with a colourful yellow flower transfer print. Back stamped "Canadian Pacific" in a style that suggests late 1950s and early 1960s. Most likely saw service on Atlantic *Empress* ships.
Mfr: NSF 1947–52.

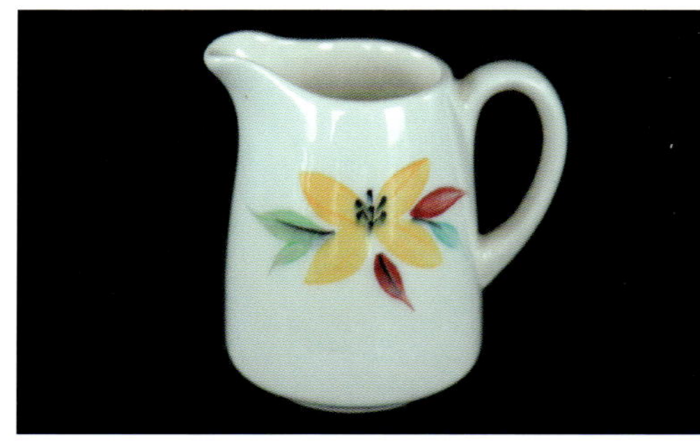

CPR-506 Abstract ★★★
Multicoloured china, back stamped "Canadian Pacific Steamships Limited". Most likely a post-war pattern that saw service on the Atlantic *Empress* ships.
Mfr: NSF 1947–52.
Phil Beaudry photo.

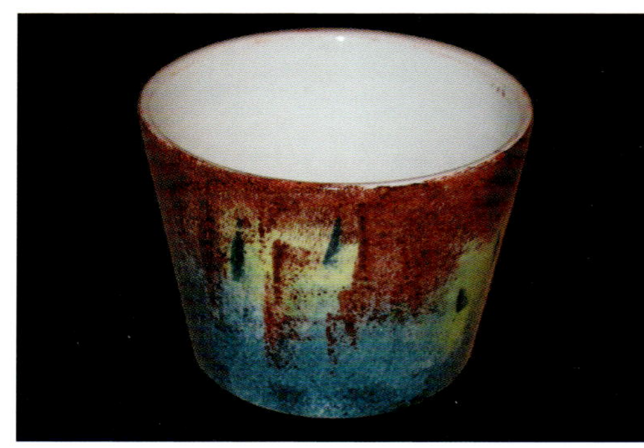

CPR-507 Kathleen ★★★
White china trimmed with a green border stripe flanked inside by a green pinstripe. Back stamped with a green CPR garter logo. Photographs show this pattern in use in the *Princess Kathleen* coffee shop during late 1940s. This pattern has also been dive-recovered from wreck of the *Kathleen*.
Mfr: GLY 1948.
Ref.: Luckin CP-16.1.

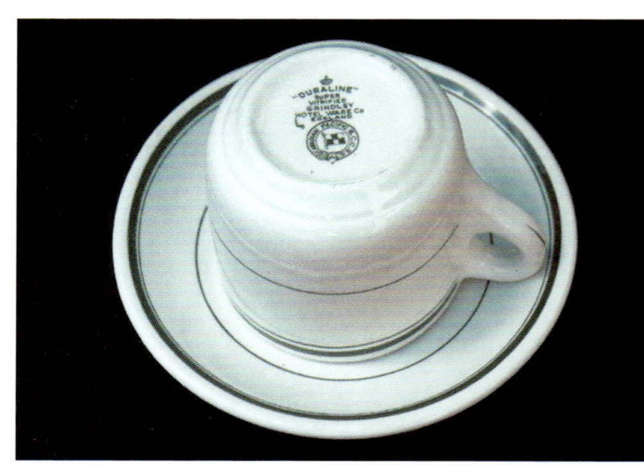

CPR-601 Brown Lustre ★★
Brown china used for baking. Back stamped "Canadian Pacific" in a style that suggests late 1950s and early 1960s. This pattern probably saw service in all aspects of CPR operations. It also appears as cream-coloured (601a) and green (601b).
Mfr: RLW ca 1956.

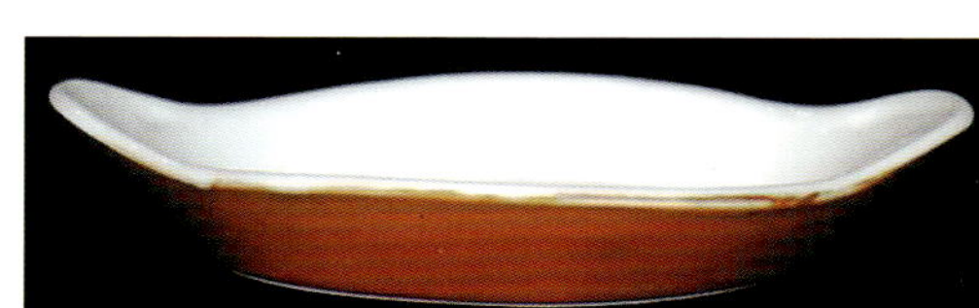

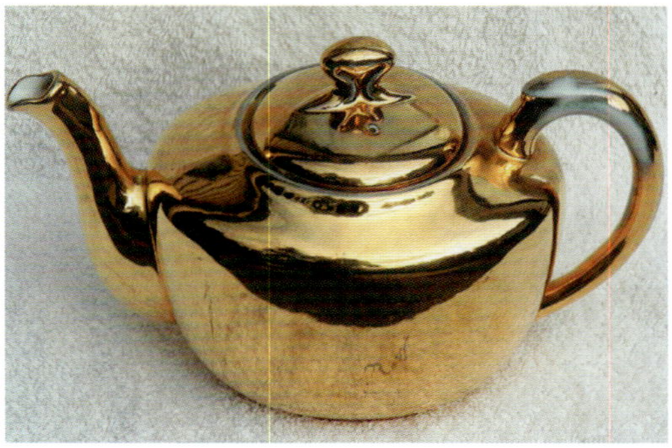

CPR-602 Gold Lustre ★★★
Gold china back stamped "Canadian Pacific" in a style that suggests late 1950s and early 1960s. Provenance unknown.
Mfr: RLW 1950s.

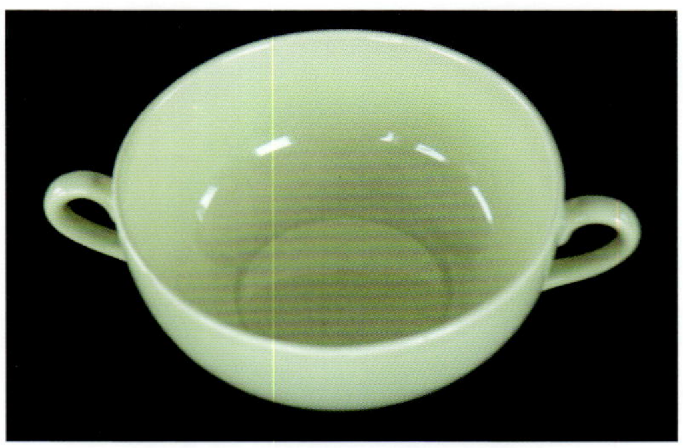

CPR-603 Deco Ware ★★★
Green china back stamped "Canadian Pacific" in a style that suggests late 1950s and early 1960s. Provenance not known. It most likely saw service on Atlantic *Empress* ships.
Mfr: MKN 1962–68.

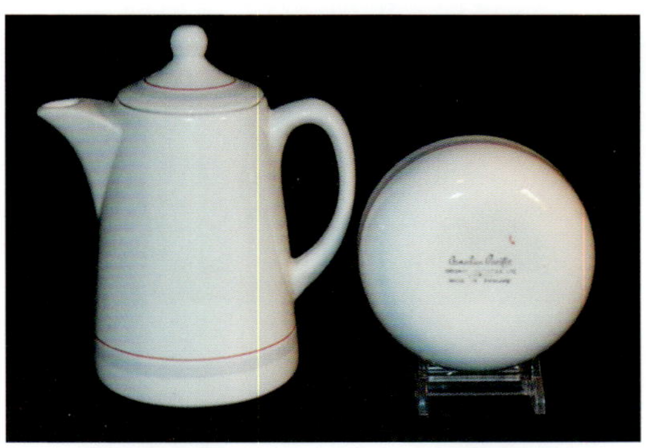

CPR-701 Halifax ★★
White china with a grey border band flanked on the inside by a maroon pinstripe. Back stamped "Canadian Pacific", written in script. The British registration number 878088 dates this pattern to 1956. It was used on Atlantic *Empress* ships and on west-coast railcar ships.
Mfr: RID 1962–69.
Ref.: Luckin CP-17a, Spence CP-5, Laister p. 39.

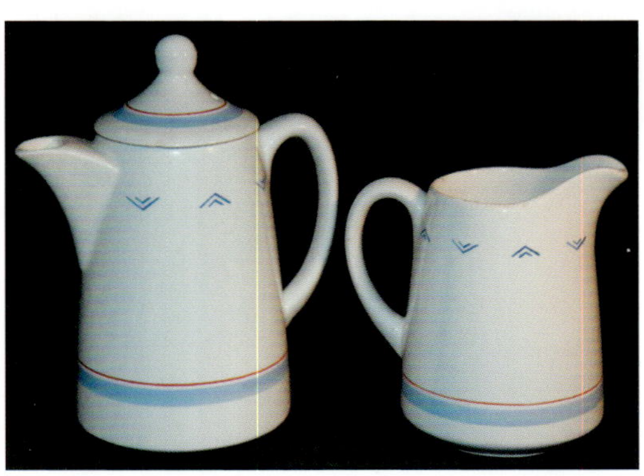

CPR-702 Halifax Aztec ★★
White china with Aztec chevron decorations and a pale-blue border band flanked on the inside by a maroon pinstripe. Back stamped "Canadian Pacific", written in script. The British registration number 878088 dates this pattern to 1956. It was used on Atlantic *Empress* ships and on west-coast railcar ships.
Mfr: RID 1962–69.
Ref.: Luckin CP-17, Laister p. 39.

China Mail Steamship Company

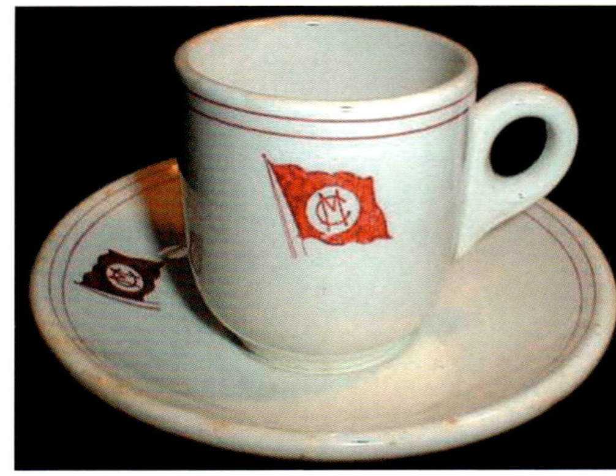

CMS-1 China Mail ★★★★

White china top marked with the China Mail house flag and trimmed with two thin side-by-side pinstripes. The flag is red with a white ball in the middle containing the black initials "CM". On china the initials are red.
Mfr: MDK 1915–23.
Ref.: Spence CML-1.

Coeur d'Alene and St Joe Transportation Company

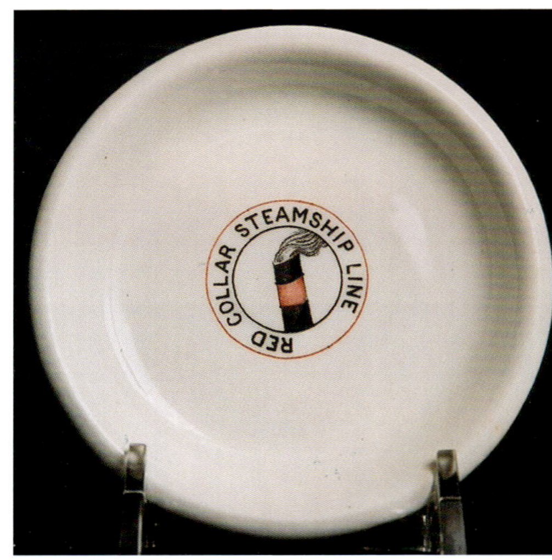

CDA-1 Red Collar ★★★★★

White china decorated with a red-and-black banded smoke stack within a ring. The words "Red Collar Steamship Line" encircle the smoke stack inside the ring. The outer line of the ring is red and the inner line black.
Mfr: MDK ca 1908–10.
Ref.: Luckin RCL-1, Spence RCL-1.

Dalles, Portland and Astoria Navigation Company

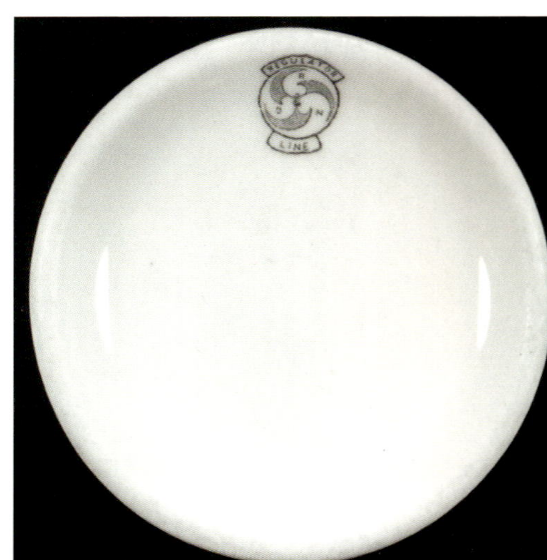

DPA-1 Regulator ★★★★★

White china with circular reddish-brown badge. The logo looks like a three-bladed spinning propeller within a circle and contains the letters "C", "R" and "N" between the propeller blades. Above the circle is the name "Regulator" and beneath the circle is the word "Line". The Columbia River and Northern Railway Company (CRN) bought out DPA in 1902, then was absorbed by the Spokane, Portland and Seattle Railway in 1908.
Mfr: OPC ca 1902–08.
Ref.: Luckin CR&N-1, Spence REG-1.
Richard Luckin photo.

Dollar Steamship Company

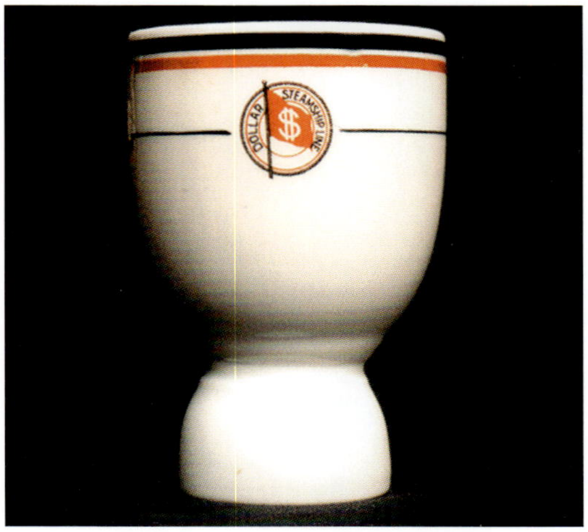

DSC-1 President Hoover ★★★

White china top marked with the DSC logo and trimmed with black and red border stripes. The logo consists of a ring with a rope trim on the outside; the ring contains the DSC house flag and "Dollar Steamship Line".
Mfr: OPC 1923–37, NPT 1926.
Ref.: Conroy DOL-1, Spence DO-1.

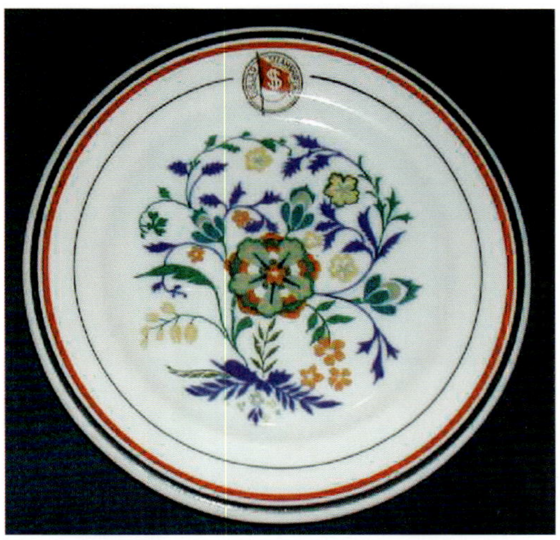

DSC-2 Sorrento ★★★

White china top marked with the DSC logo and trimmed with black and red border stripes. The centre of the plate is decorated with multicoloured floral design called Sorrento by the Onondaga Pottery Company. Great Northern Railroad ordered a similar version of this pattern for dining car use between 1923 and 1939. The floral design occurs on 5.5-inch (14-cm) plates only.
Mfr: OPC 1927.

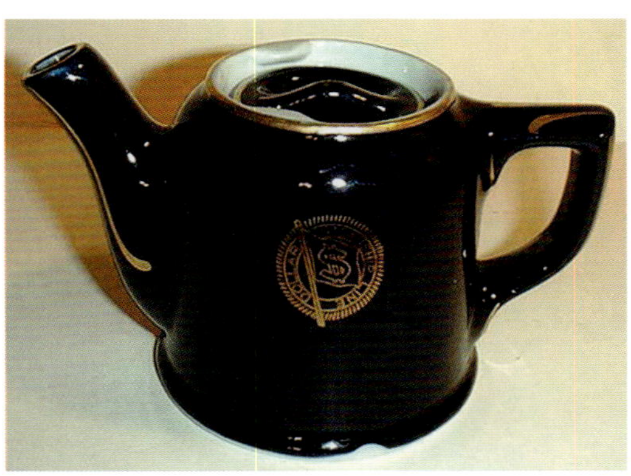

DSC-3 Cobalt Teapot ★★★★★

Blue china top marked with a gold DSC logo. The logo consists of a ring with rope trim on the outside that contains the Dollar Line house flag; the ring contains the DSC house flag and "Dollar Steamship Line".
Mfr: HAL ca 1927.
Photographer unknown.

Grand Trunk Pacific Coast Steamship Company

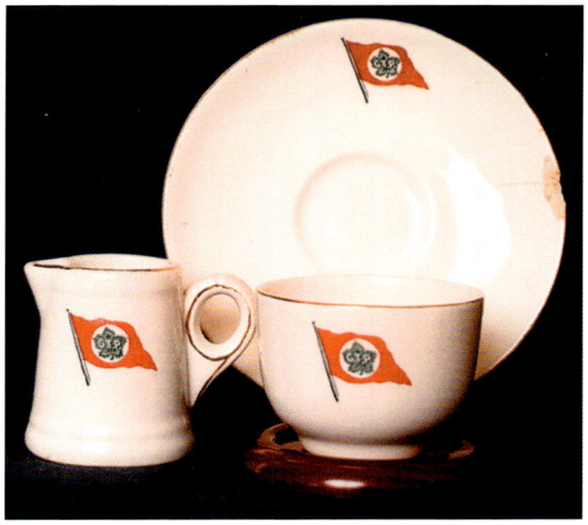

GTP-1 Flag & Leaf ★★★★

White china top marked with the GTP house flag and trimmed along the edges with gold. The flag is orange with a green maple leaf superimposed on a white circle. Across the maple leaf are the letters "GTP".
Mfr: MDK 1914–21.
Ref.: Luckin GTP-3.

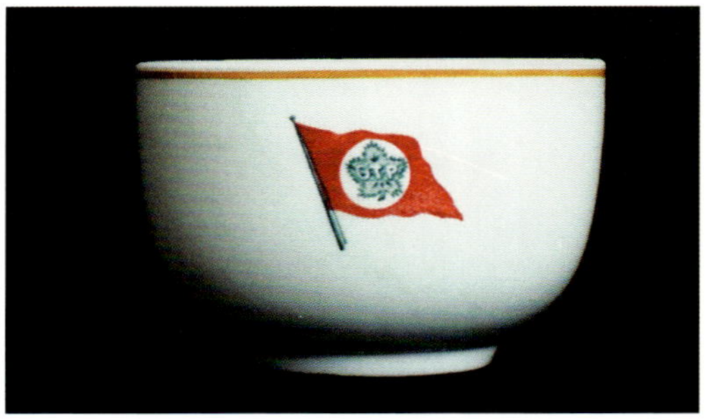

GTP-1a Flag & Leaf ★★★★

Same as GTP-1, except that the edge trim is mustard yellow instead of gold.
Mfr: MDK 1914–21.
Ref.: Luckin GTP-3.

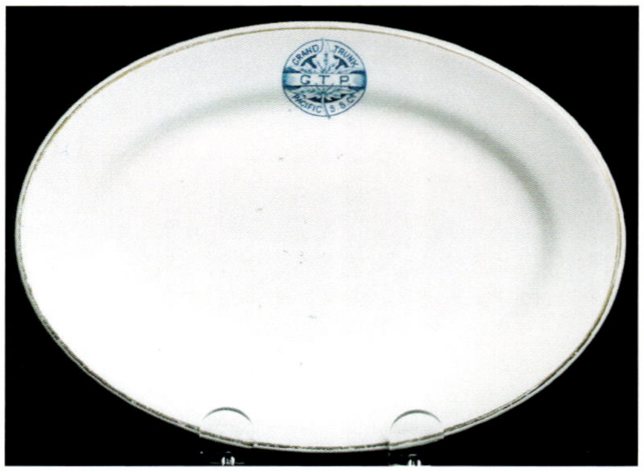

GTP-2 Prince Rupert ★★★★

White china top marked with a blue GTP ring logo and trimmed with gold along the edges. The ring text reads "Grand Trunk Pacific SS Co." and within the ring is a maple leaf with a banner across it reading "G.T.P."
Mfr: MDK ca 1920.
Ref.: Luckin & Spence GTP-6.2.

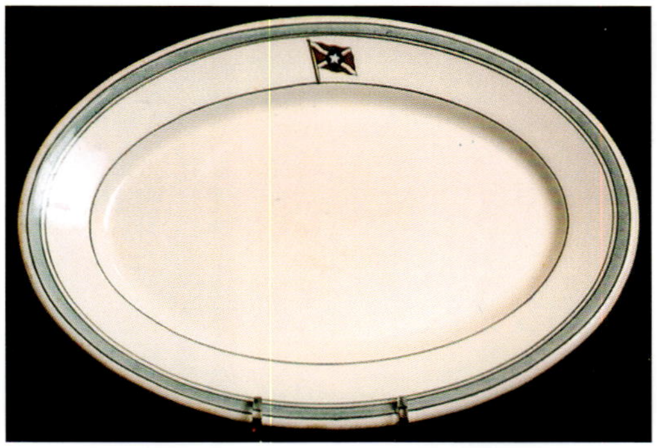

Great Northern Pacific Steamship Company

GNP-1 **Pacific Coast** ★★★

White china top marked with the GNP house flag and trimmed with a pale blue band flanked on each side with teal pinstripes.
Mfr: OPC 1913–16.
Ref.: Luckin & Conroy GN-10, Spence GN 10.

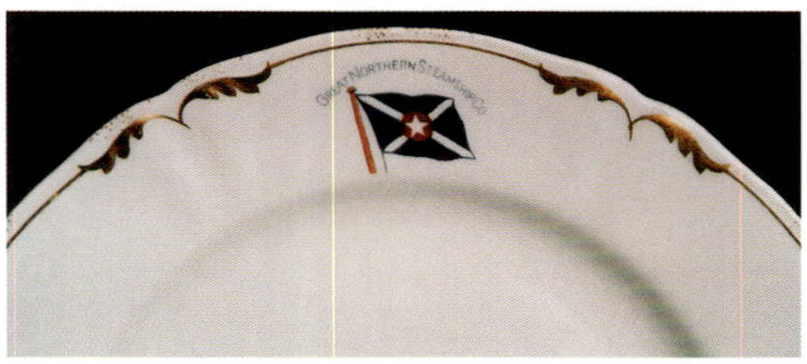

Great Northern Steamship Company

GNS-1 **Dakota** ★★★★

White china top marked with the GNS house flag and trimmed with an elegant gold border. "Great Northern Steamship Co." arches over the flag. This pattern was named for the SS *Dakota*.
Mfr: OPC 1905.

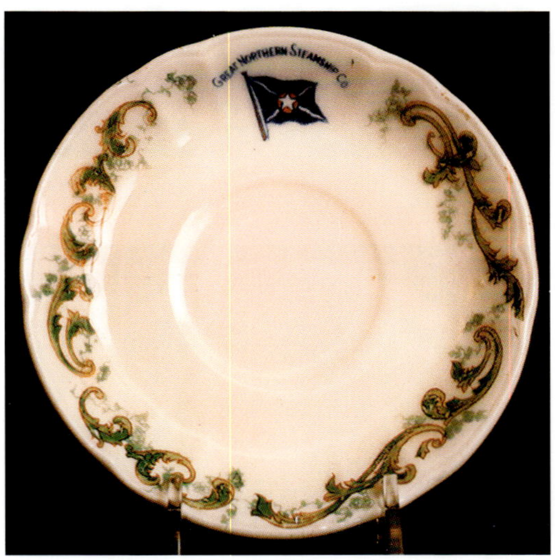

GNS-2 **Floral** ★★★

White china featuring the GNS house flag with "Great Northern Steamship Co." arched over it. The flag breaks a green floral border design at 12 o'clock. This pattern appears to have been used over the life of this line.
Mfr: OPC 1904–13.
Ref.: Luckin GN-10.1, Spence GN-1.

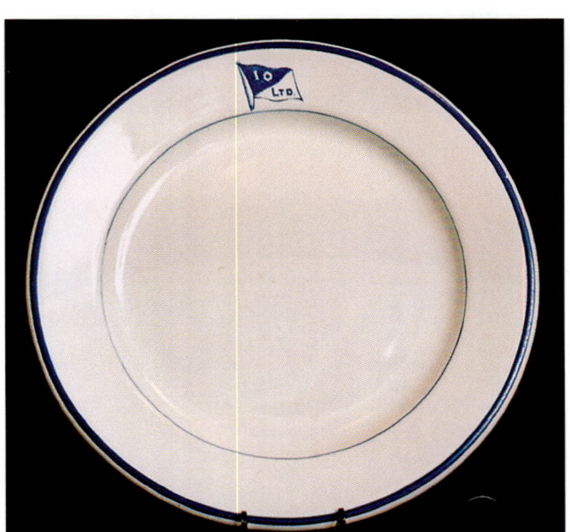

Imperial Oil Company

IOC-1 **Imperial Windsor** ★★★

White china top marked with the IOC house flag and trimmed with a blue border stripe. A blue pinstripe trims the well of plates.
Mfr: RDN ca 1930s, ADM ca 1950, Unknown ca 1936; Parsons-Steiner of Toronto distributor.
Ref.: Krummes IMP-2, Spence IMO-1.

IOC-2 ESSO Houseflag ★★★

White china top marked with the ESSO house flag and trimmed with a red border stripe.
Mfr: GLY 1980.
Ref.: Krummes IMP-3, Spence ESSO-1.

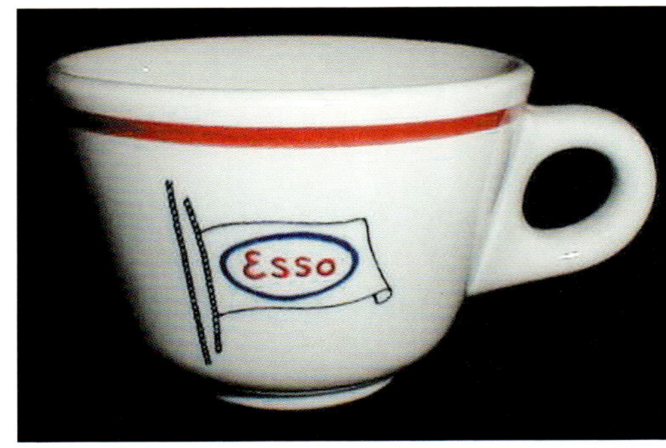

Inland Navigation Company

INC-1 Inland Chief ★

Tan china top marked with a stylized First Nations chief's head and trimmed with rope border.
Mfr: OPCo. 1951–60.

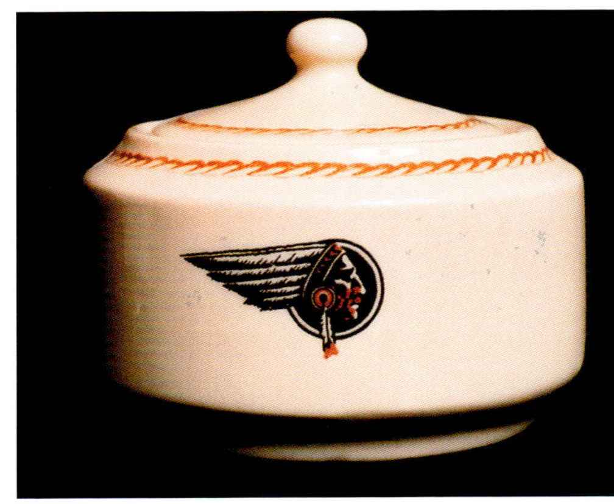

Inter-Island Steam Navigation Company

ISN-1 House flag ★★★

White china top marked with the ISN house flag and trimmed with a lime green border stripe flanked on either side with red pinstripes.
Mfr: GLY 1923.
Alberto Bisagno photo.

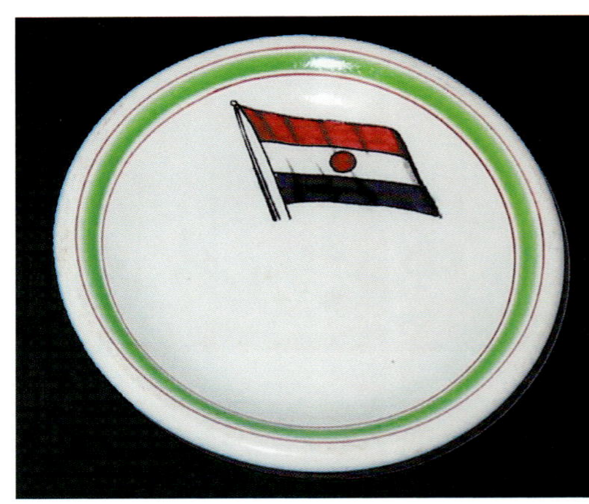

ISN-2 Haleakala ★★★

White china with the ISN garter logo and two blue border pinstripes. The logo consists of a double ring containing the company name and in the centre the house flag superimposed on Diamond Head. Pieces of this pattern have been found back stamped with "SS *Haleakala*". The *Haleakala* was launched in 1923 for ISN's Honolulu-Hilo service.
Mfr: Royal Bavarian ca 1923, WAR 1941–42.
Ref.: Spence IIS-1.

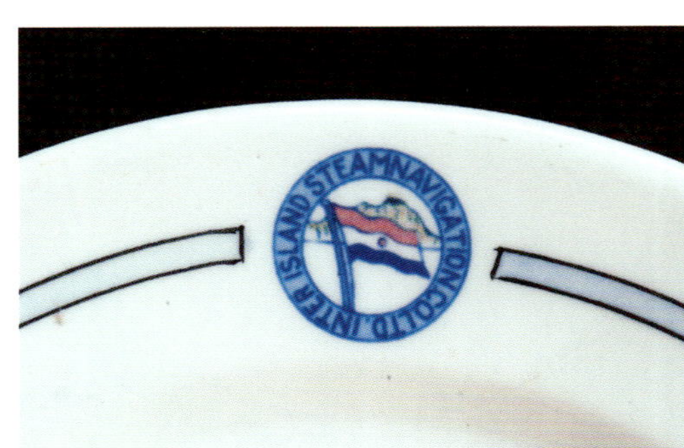

Los Angeles Steamship Company

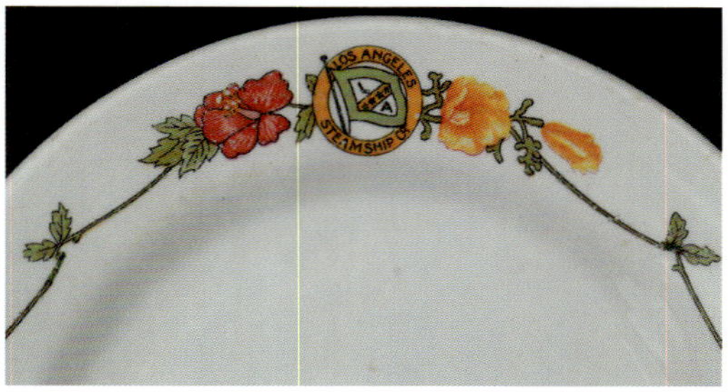

LAS-1 California Poppy ★★★

White china with the LAS logo breaking the border decoration of red and yellow California poppies and green leaves. The logo consists of a yellow life ring containing the company name and house flag. This was a copyrighted pattern, no.73086, dating to 1927.
Mfr: NPT 1926–29.
Ref.: Spence LAS-1.

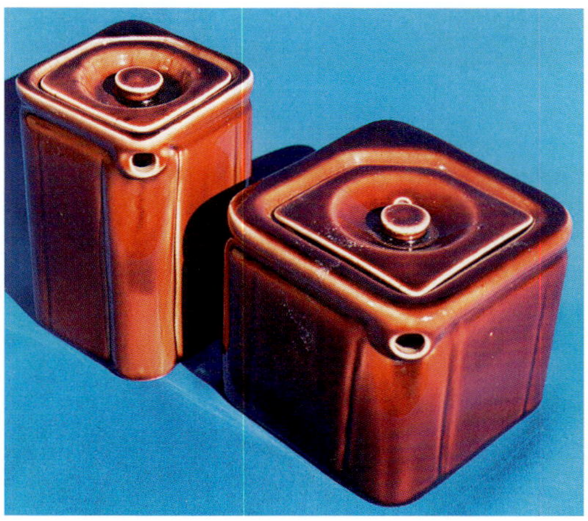

LAS-2 Brown Cube ★★★

Brown rectangular art-deco china – cream, sugar and individual teapots – with no top mark. Back stamped "LASSCo."
Mfr: NPT 1926–29, GCC 1920s.
Daniel Krummes photo.

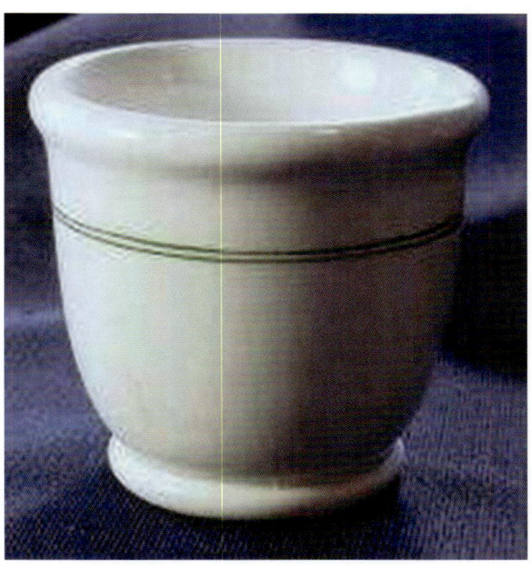

LAS-3 Green Stripes ★★

White china with two green border pinstripes. Back stamped "Nathan Straus". Made exclusively for LAS.
Mfr: NPT 1926–29.

Matson Navigation Company

MNC-1 Matsonia ★★

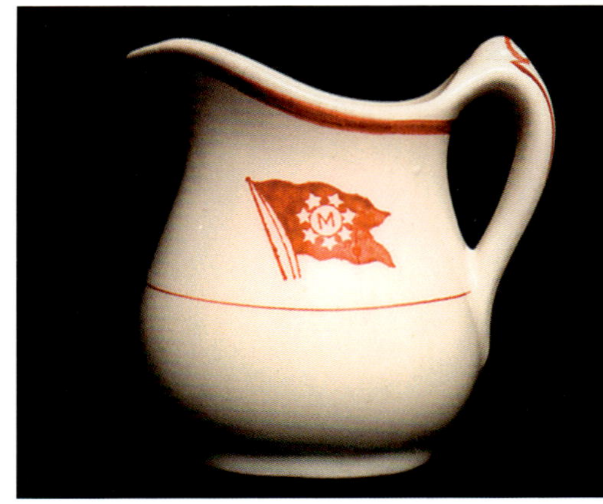

White china top marked with the Matson house flag and trimmed with a medium-weight red border stripe. Back stamped with the name of one of three ships that operated the San Francisco-Honolulu route: *Matsonia* (1919–37), *Maui* (1920–38) or *Wilhelmina* (1910–32).
Mfr: WOD 1922, MDK 1927.
Ref.: Conroy MAT-2, Spence MA-2.

MNC-2 Lurline ★★

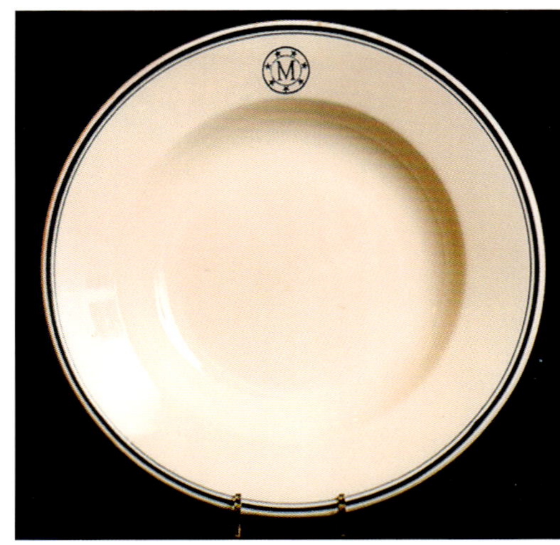

White china with a blue Matson ring logo and trimmed with a medium-weight blue border stripe flanked inside by a blue pinstripe. The ring logo has a blue "M" in the centre with seven blue stars encircling it in the outer ring. This pattern was reportedly used on the SS *Lurline* in the late 1920s.
Mfr: MDK 1927, OPC 1928–37.
Ref.: Conroy MAT-1

MNC-3 Bombay ★★

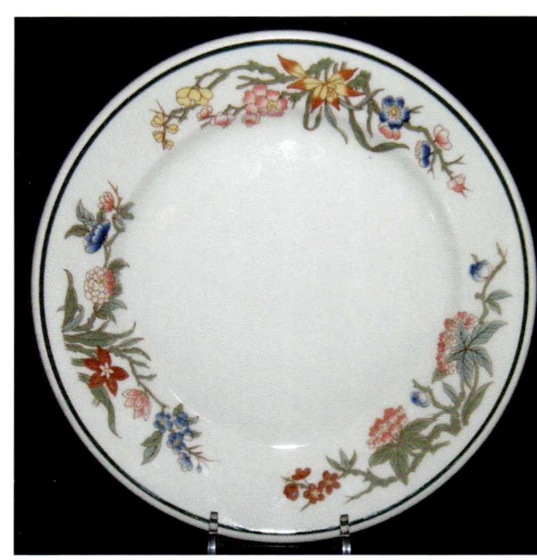

White china with a multicoloured floral design around the well and trimmed with a dark green border stripe. This was an OPC stock pattern customized for Matson. The OPC back stamp reads "Made especially for Matson Navigation Co."; the MYR back stamp reads "Made especially for Matson Lines". A 1946 magazine ad identifies OPC as the supplier to Matson Line. According to Fred Stindt, a retired Matson chief steward, all pieces of Bombay with the green stripe were created solely for Matson. The Bombay pattern was used in first class.
Mfr: OPC 1930–52, MYR 1952–57.
Ref.: Spence MA-1.

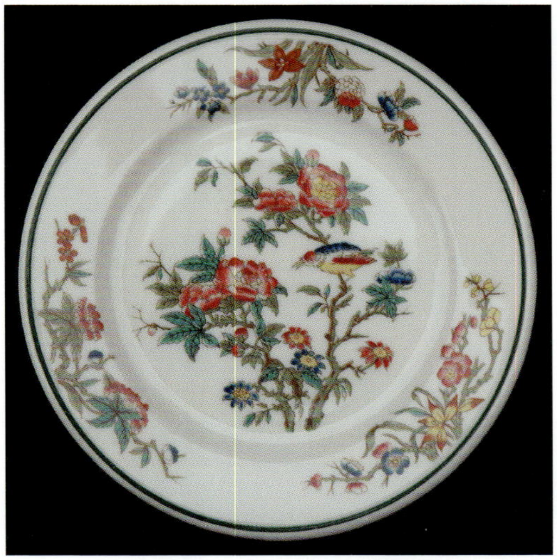

MNC-3a Bombay ★★
Same as MNC-3, but with an added floral transfer print in the well of the plate, which includes a small, colourful bird perching on a branch. This well design only occurs on 6.5-inch (16.5-cm) plates.
Mfr: OPC 1930–52, MYR 1952–57.

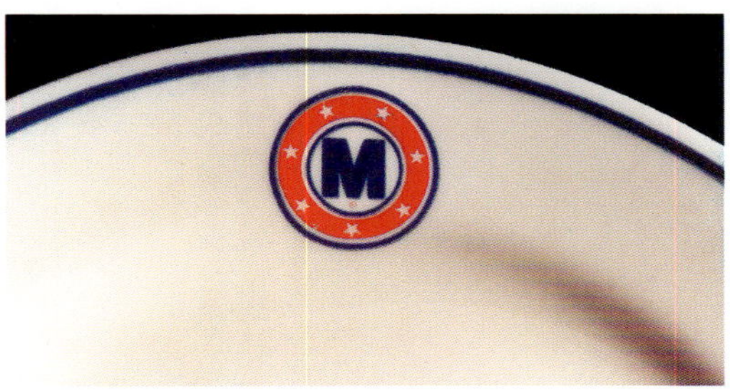

MNC-4 Pfeiffer ★★★★
White china top marked with a red, white and blue MNC ring logo and trimmed with a medium-weight blue border stripe. The ring logo is a red ring containing seven white stars, with a blue "M" in the centre. The Matson flagship, *R.J. Pfeiffer*, launched in 1992.
Mfr: HRL 1992.
Daniel Krummes photo.

McCormick Steamship Company

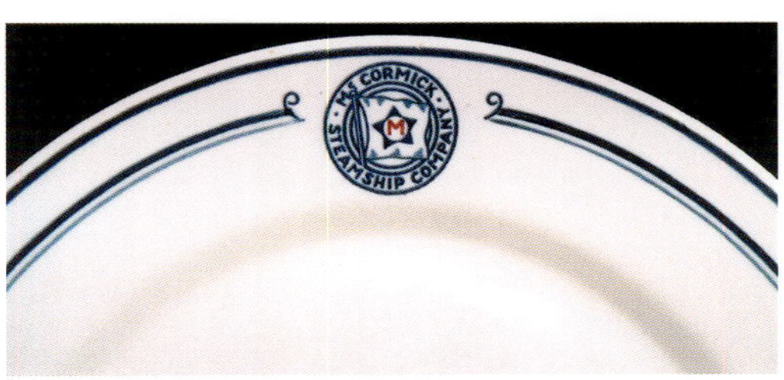

MCS-1 McCormick ★★★★
White china with the MCS badge and a blue medium-weight pinstripe on the rim. The badge consists of the house flag encircled by a ring containing "McCormick Steamship Company". The flag is white with a red "M" inside a white circle on a blue star.
Mfr: WAR 1930–32, WAL 1931–40.
Ref.: Spence MCS-1.

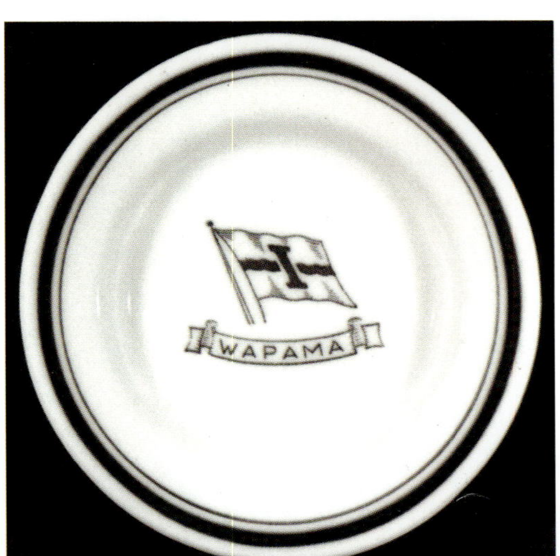

MCS-2 Wapama ★★★★★
White china top marked with an unidentified shipping company house flag and trimmed with cobalt blue outer stripe flanked on the inside by a red pinstripe. The SS *Wapama* was initially operated by MCS, until Alaska Transport bought her in 1937 (she was retired in 1947). The *Wapama* is now a museum ship at the San Francisco National Marine Historic Park.
Mfr: WKR 1942–47.
Ref.: Luckin MCR-1.
Richard Luckin photo.

Monticello Steamship Company

MSC-1 Napa Valley ★★★

White china top marked with red superimposed "MSSCo" and trimmed with two red pinstripes. The earliest date on this china coincides with the launch of the MSC steamer *Napa Valley*.
Mfr: WOD 1910–17.

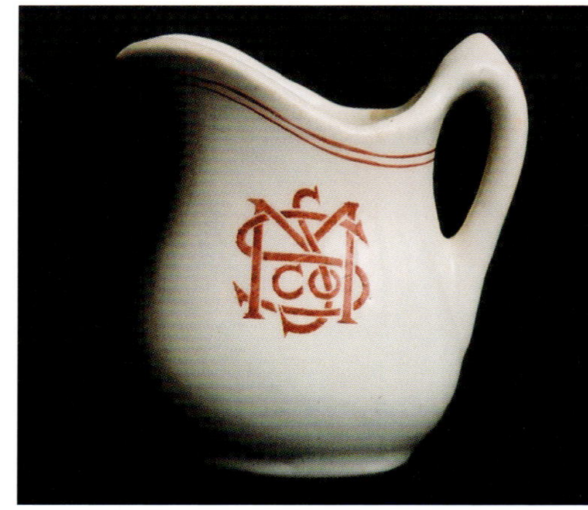

MSC-2 Vallejo ★★★★

White china top marked with a double-ring MSC logo and trimmed with a medium-weight blue-green border stripe. The logo reads "Monticello S.S. Co." arched upwards and "Mare Island San Francisco" arched downwards within the ring. A thick yellow-orange band runs diagonally through the ring and contains the word "Vallejo". Behind the band is a state building tower motif.
Mfr: OPC 1920–26.
Ref.: Luckin MOT-1.

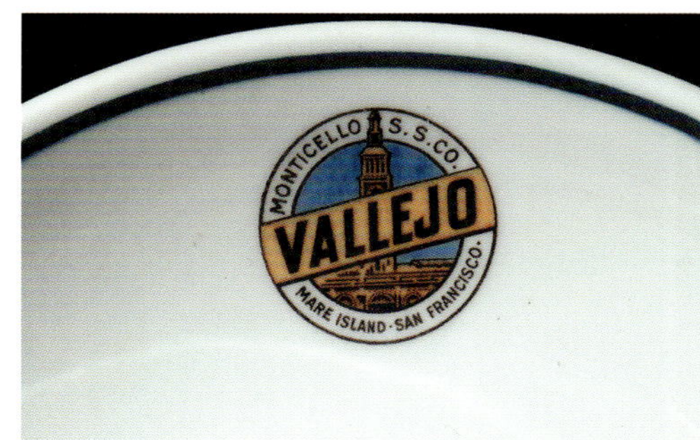

North American Transportation and Trading Company

NAT-1 Healy ★★★★★

White china top marked with the NAT house flag. The *John J. Healy* was placed in service on the Yukon River in 1897.
Mfr: KTK 1897–1900.
Daniel Krummes photo.

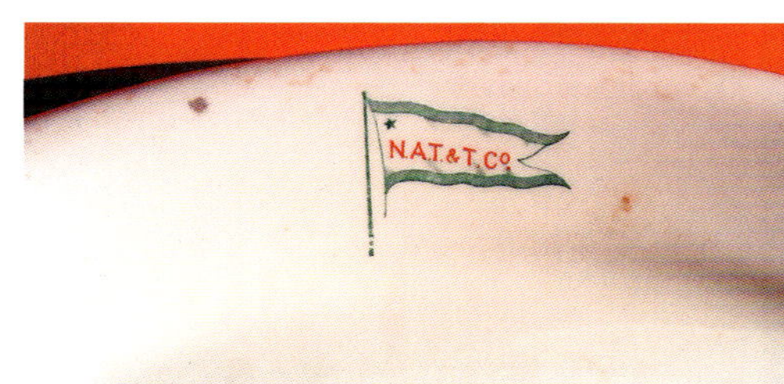

Northland Navigation Company

NLN-1 Northland ★★★

Clear drinking glass top marked with the NLN house flag. The flag is reddish orange with a black "N" on a white horizontal diamond.
Mfr: Dominion Glass 1951–68.

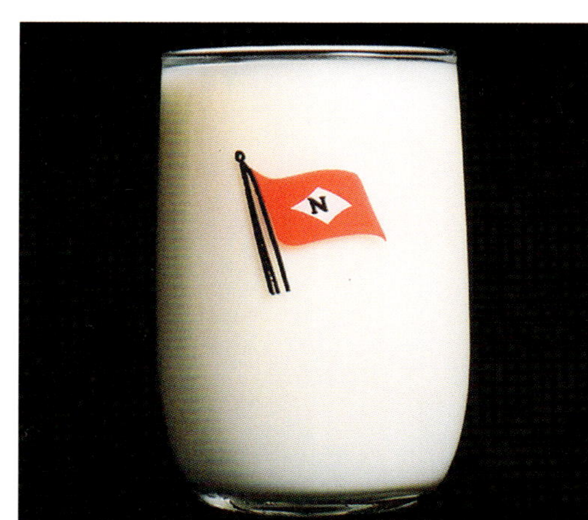

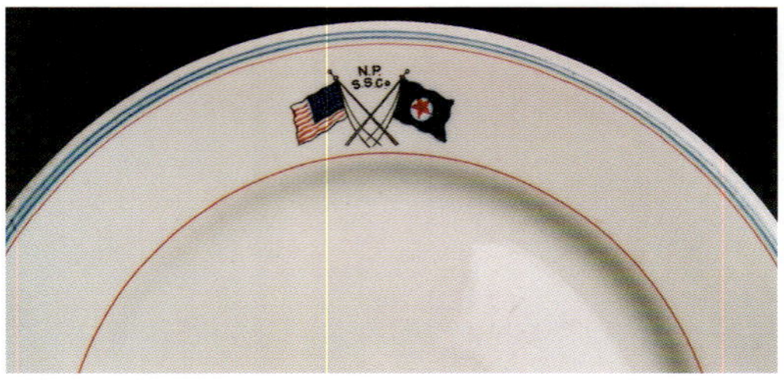

North Pacific Steamship Company

NPS-1 **Crossed Flags** ★★★

White china top marked with the NPS logo – the USA flag crossed with the NPS house flag – and "NPSSCo." between the flags. Three pinstripes trim the outer edge of service pieces: two outer blue stripes and one inner red. Large plates have an extra red pinstripe around the well.
Mfr: WOD 1911.
Ref.: Spence NPS-1.

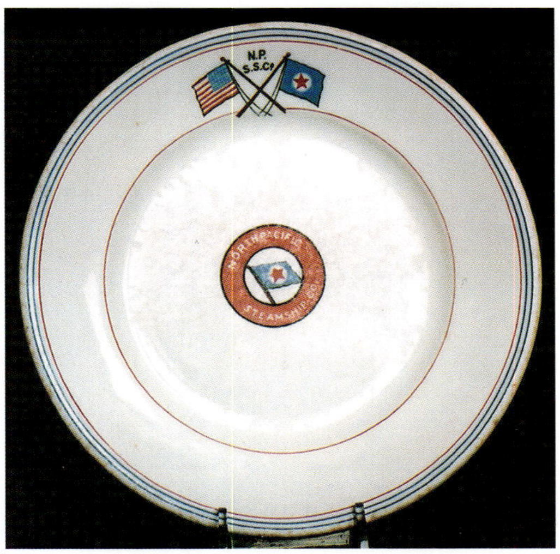

NPS-2 **George W Elder** ★★★

White china top marked with the NPS crossed flag logo (see NPS-1) on the rim and a red ring logo in the well of plates. Three pinstripes trim the outer edge of service pieces: two outer blue stripes and one inner red. Large plates have an extra red pinstripe around the well.
Mfr: RID 1909.

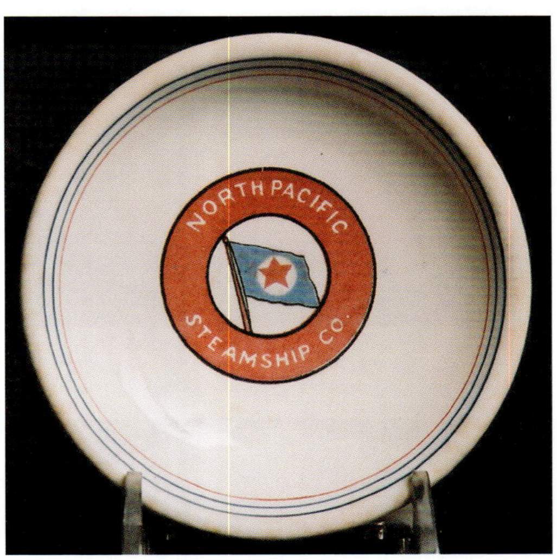

NPS-3 **Alliance** ★★★★

White china top marked with a red ring logo containing the NPS house flag. Three pinstripes trim the outer edge of service pieces: two outer blue stripes and one inner red.
Mfr: WOD 1911, RID 1912–17.
Ref.: Luckin NPS-1, Spence NPS-2.
NPS-3a contains an additional red pinstripe around the logo.

Northwestern Steamship Company

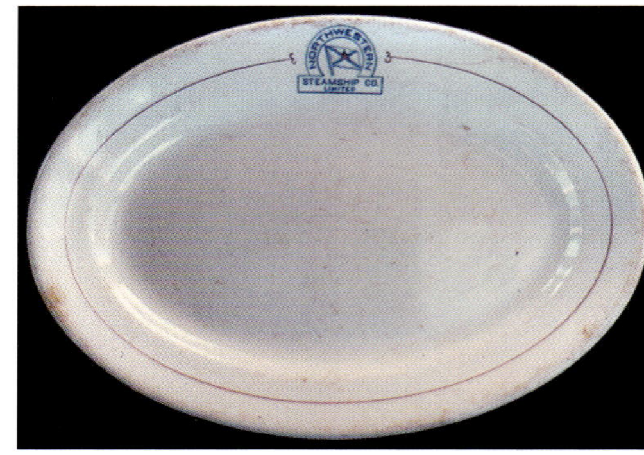

NWS-1 **Ohio** ★★★★

White china top marked with a blue-and-red NWS logo and trimmed with a single red pinstripe. The logo is a half circle containing the NWS house flag. Arching upwards around the flag is the word "Northwestern" and a box beneath the circle contains "Steamship Co. Limited". This pattern has been recovered from the wreck of the SS *Ohio*, which sank in 1909.
Mfr: MDK 1904–08, KTK 1904–08.

Oceanic Steamship Company

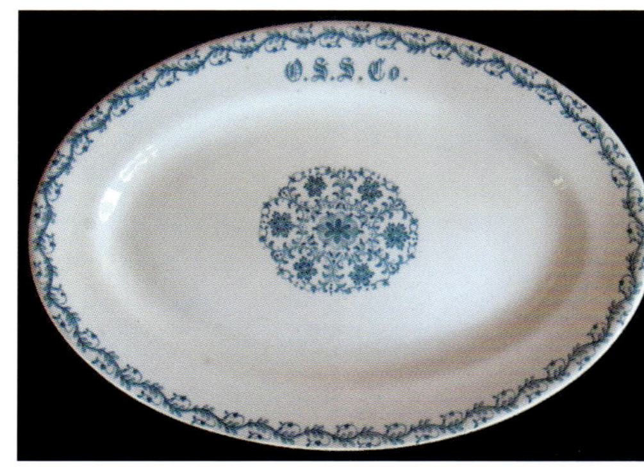

OSC-1 **Brunswick** ★★★★★

White china top marked with old-English blue-grey lettering, "O.S.S. Co." and trimmed with a blue-leaf border design. Only three pieces of this pattern are known to exist. It was distributed by Nathan Dohrman of San Francisco.
Mfr: MDK 1896–1905.
Ref.: Spence OC-1.

Oregon Railroad and Navigation Company

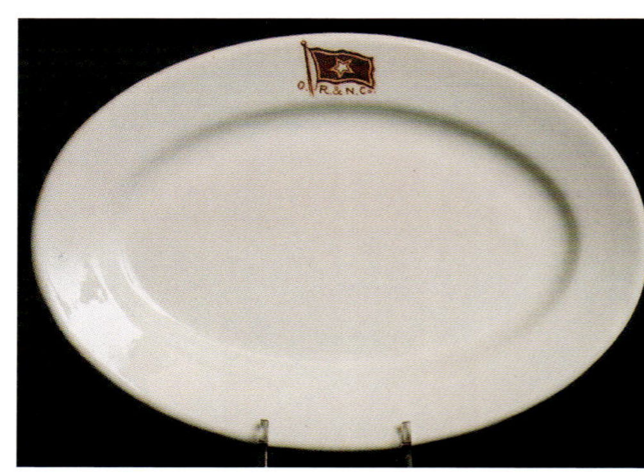

ORN-1 **Columbia** ★★★★★

White china top marked with the ORN house flag, which is brown with white border stripe and a white star in the middle, and "O R & N Co." in brown beneath the flag.
Mfr: OPC 1901.
Ref.: Luckin OR&N-1.

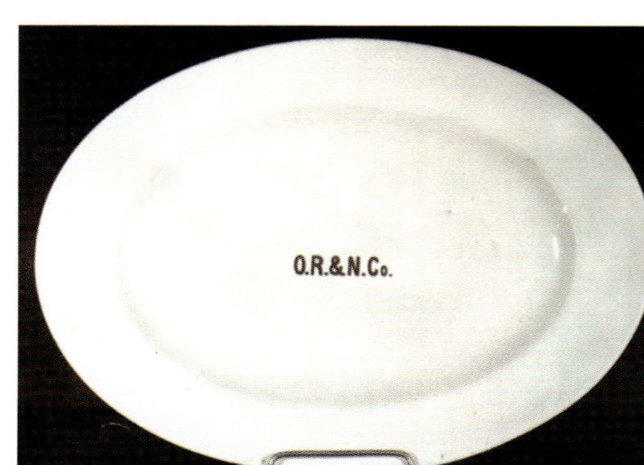

ORN-2 **Walla Walla** ★★★★

White china with black "O.R.& N. Co." in the middle of dinnerware.
Mfr: GWD 1879–1910.
Ref.: Luckin OR&N-2.
Richard Luckin photo.

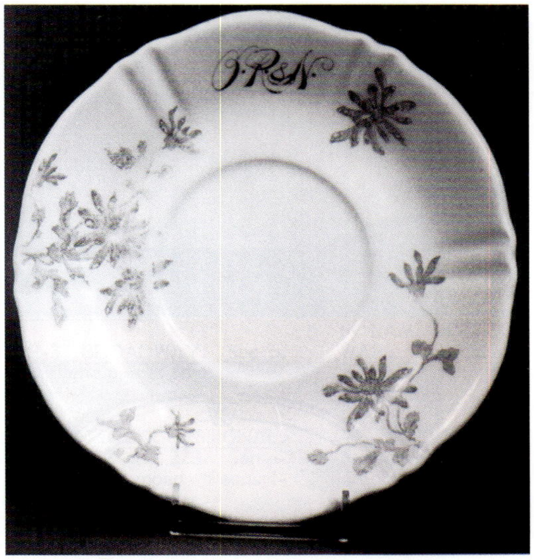

ORN-3 **Floral Script** ★★★★★

White china with tan flowers and green leaves and embossed with purple script reading "OR & N". Not confirmed as a steamship pattern.
Mfr: HAV 1892–1910.
Ref.: Luckin OR&N-1.1.
Richard Luckin photo.

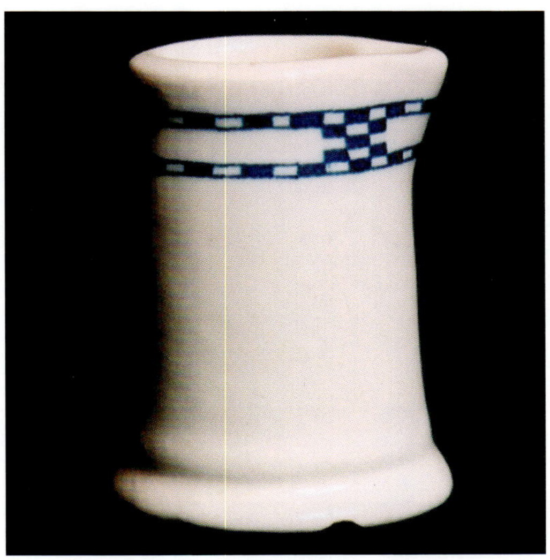

ORN-4 **Oregon** ★★

White china with a cobalt-blue-and-white checkered border pattern. Diver James Seely-White found this pattern at various ORN dock sites along the Willamette River. It's likely a stock ORN pattern, but definitely not from the SS *Oregon*. She was laid up in 1894 well before the china was produced. Oregon is the manufacturer's name for this pattern.
Mfr: WOD 1907–10.
Ref.: Luckin OR&N-1.2, Spence OR&N-1.

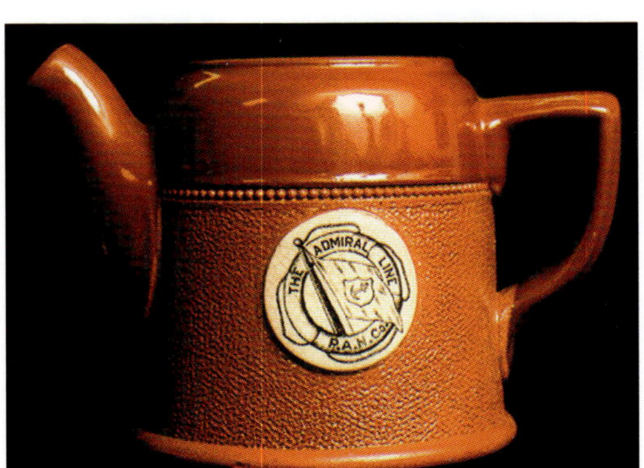

Pacific Alaska Navigation Company

PAN-1 **Lovatt** ★★★★★

Brown Chicago teapot with a white PAN badge on the side. The badge is a life ring with the PAN house flag inside. The flag features an anchor within a shield. Only one piece of this pattern is known to exist.
Mfr: LVT 1914-16.

PAN-2 **Sampson** ★★★

White china with a pale blue floral border design. Numerous pieces of this stock pattern were recovered from the wreck of the *Admiral Sampson*, which was rammed and sunk by CPR's *Princess Victoria* on August 26, 1914. It is not top marked or back stamped with the line name.
Mfr: GWD pre-1914.
Brian Fardoe photo.

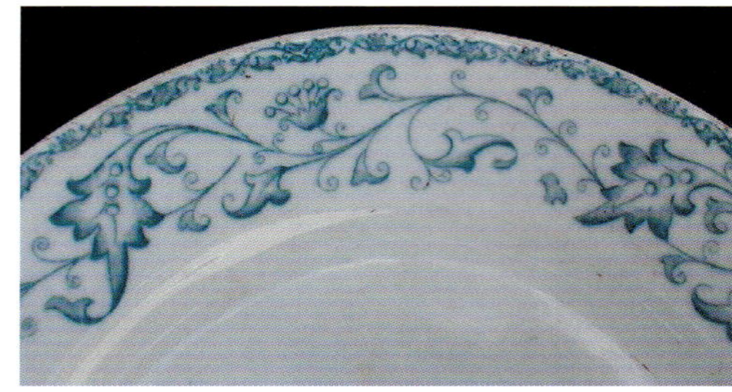

Pacific Coast Steamship Company

PCS-1 **House Flag** ★★★★

White china top marked with the PCS house flag. The flag is blue with red Maltese cross inside a horizontal white diamond. "P.C.S.S. Co." appears in blue under the flag.
Mfr: OPC ca 1890s.
Ref.: Spence PCS-3.

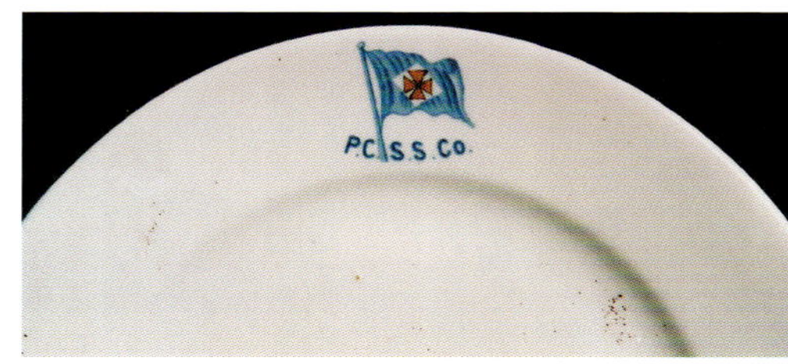

PCS-1a **House Flag** ★★★★★

White china top marked with the PCS house flag. This pattern differs from PCS-1 in that the Maltese cross in the flag is blue and there are no letters beneath the flag.
Mfr: OPC ca1890s.

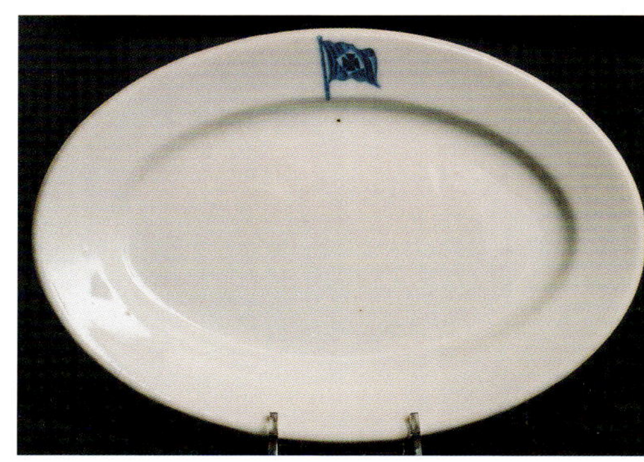

PCS-2 **Coos Bay** ★★★★

White china top marked with the PCS ring logo. The ring has an outer red circle and an inner blue circle. The ring text, in blue, reads "Pacific Coast Steamship Co." The blue PCS house flag appears inside the ring; it has a red Maltese cross inside a white diamond.
Mfr: RID 1904–05, MDK 1905, BUF 1905.
Ref.: Luckin PCS-1, Spence PCS-2.

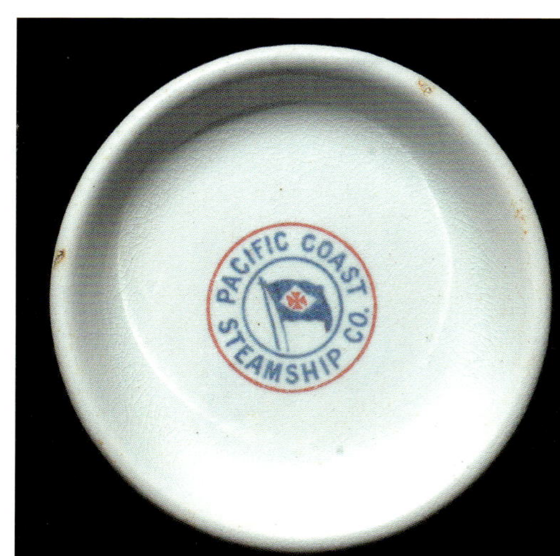

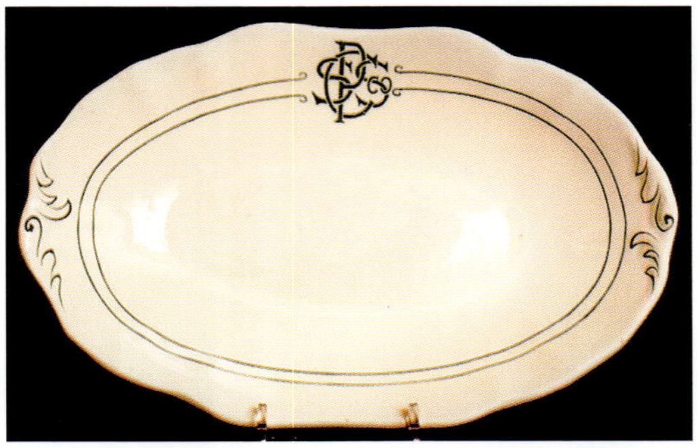

PCS-3 **Santa Rosa** ★★★

White china top marked with superimposed green letters, "PCSS Co", and trimmed with two green border pinstripes. This pattern was recovered from the wreck of the *State of California*, which sank in Gambier Bay, Alaska, in 1913.
Mfr: MDK 1909, 1912–14, WOD 1910–11.
Ref.: Luckin PCS-2, Spence PCS-1.

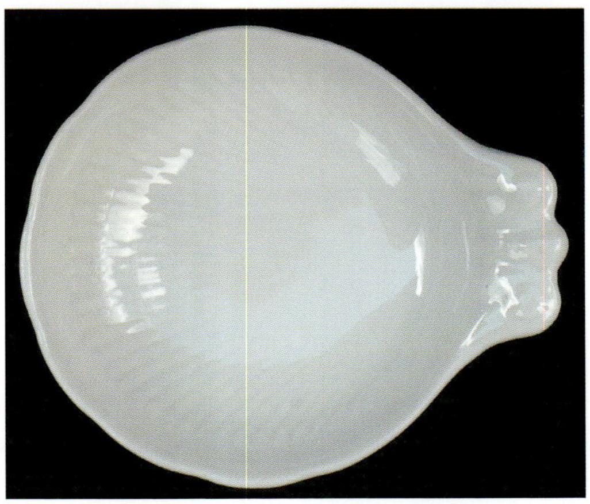

PCS-4 **White Ware** ★

White china with no identification. A stock pattern recovered by divers from the *State of California*, which sank in Gambier Bay, Alaska, in 1913.
Mfr: OPC 1912–13.

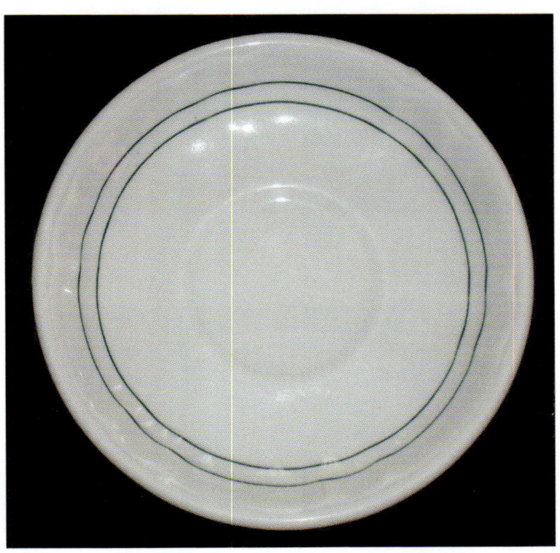

PCS-5 **Green Pinstripes** ★

White china trimmed with two green border pinstripes. A stock pattern recovered by divers from the *State of California*, which sank in Gambier Bay, Alaska, in 1913. This is most likely a companion to the Santa Rosa pattern.
Mfr: WOD 1911–13, MDK 1910.

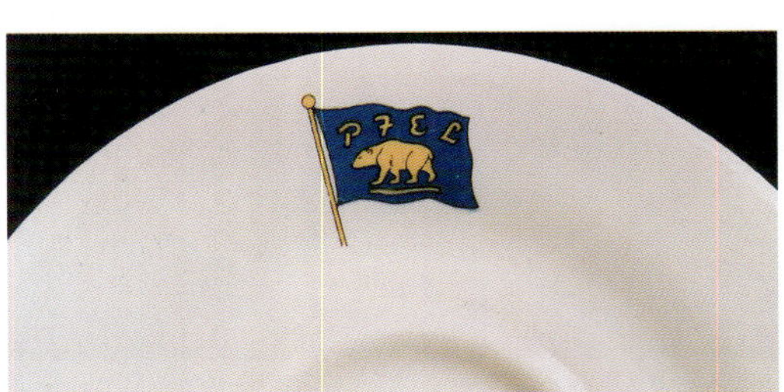

Pacific Far East Line

PFE-1 **Philippine Bear** ★★

White china top marked with the PFEL house flag. The flag is bright blue with a yellow bear and "PFEL" in yellow above the bear.
Mfr: OPC 1954–55, WEL 1960s, STR 1968–71.
Ref.: Conroy PFEL-1 Spence PF-1.

Pacific Mail Steamship Company

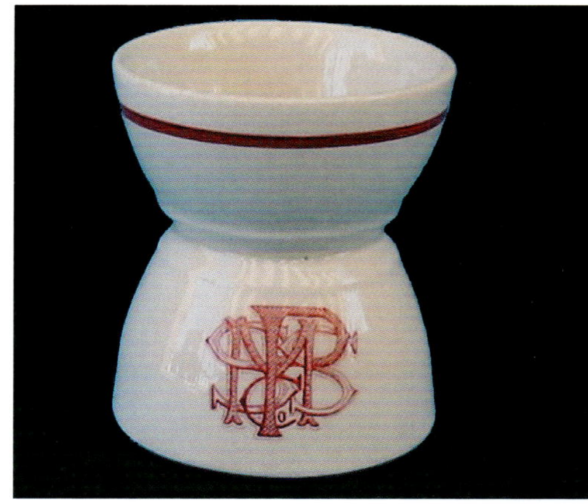

PMS-1 Superimposed ★★★★

White china top marked with red interlocking letters "PMSS" and trimmed with a medium red border stripe.
Mfr: MDK 1896–1900.
Ref.: Spence PMS-1.
Alberto Bisagno photo.

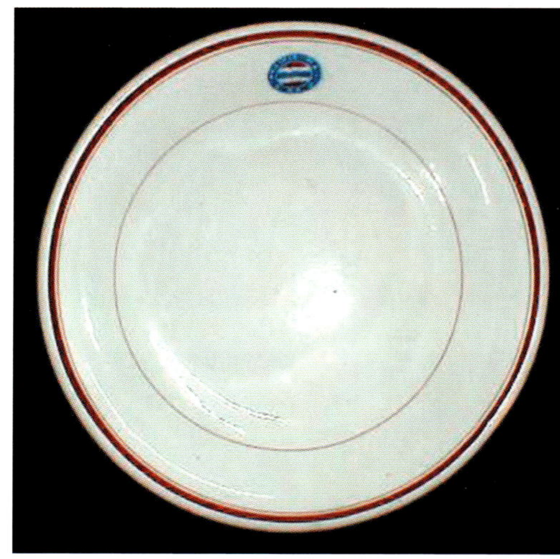

PMS-2 Korea ★★★★

White china top marked with the PMS ring logo and trimmed with a red border stripe, flanked on the inside with a red pinstripe. The Pacific Mail house flag is inside the ring. White lettering in the blue ring reads "Pacific Mail SS Co." PMS used this logo on advertising and postcards in about 1909.
Mfr: OPC 1900–1902, RID 1908.
Ref.: Spence PMS-2.

Pacific Navigation Company

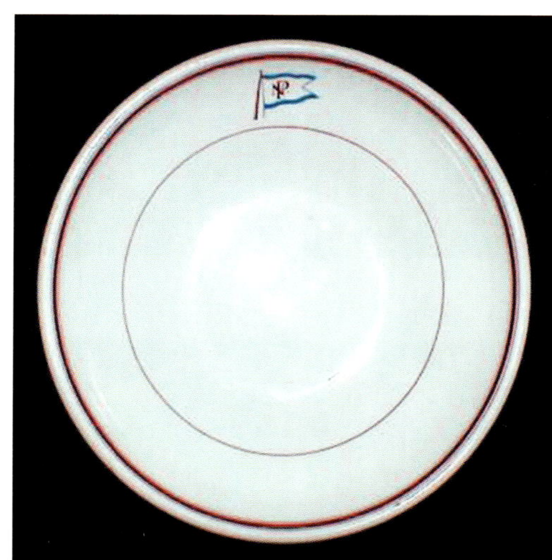

PNC-1 Yale ★★★★★

White china top marked with the PNC house flag and trimmed with a medium-weight red border stripe. Plates have a red pinstripe around the well. The house flag is a white burgee with blue bands top and bottom and the red letters "PN" superimposed in the middle of the flag. Only two pieces of this pattern are known to exist.
Mfr: MDK 1910-16.

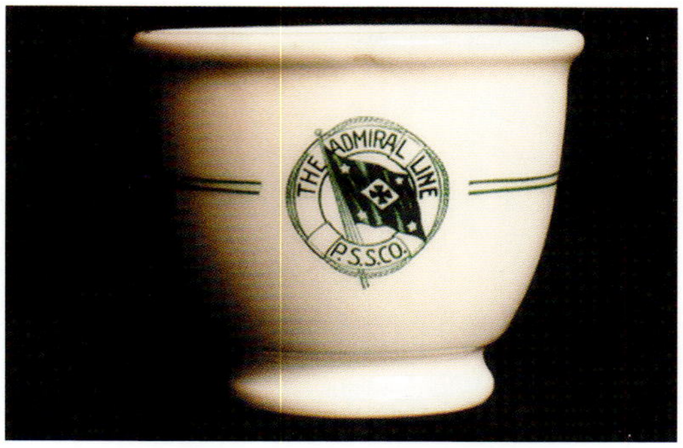

Pacific Steamship Company

PSC-1 H.F. Alexander ★★

White china top marked with a green Admiral Line life-ring logo and trimmed with two green pinstripes. The life ring contains "The Admiral Line P.S.S. Co." encircling the PSC house flag. This is the first top marked PSC pattern.
Mfr: BSH 1922, BUF 1923, JAC 1924.
Ref.: Luckin ADM-1.

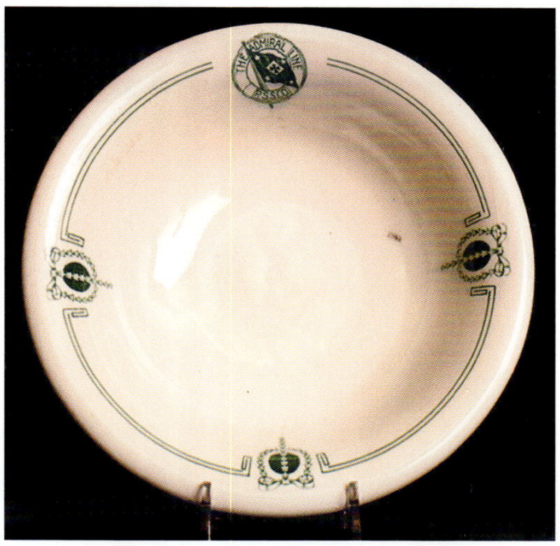

PSC-2 Emma Alexander ★★

White china top marked with a green Admiral Line life-ring logo and trimmed with two green pinstripes interspersed with decorative medallions. This is a transitional Admiral Line pattern.
Mfr: GLY 1926, NPT 1927–28.
Ref.: Spence PSS-1.

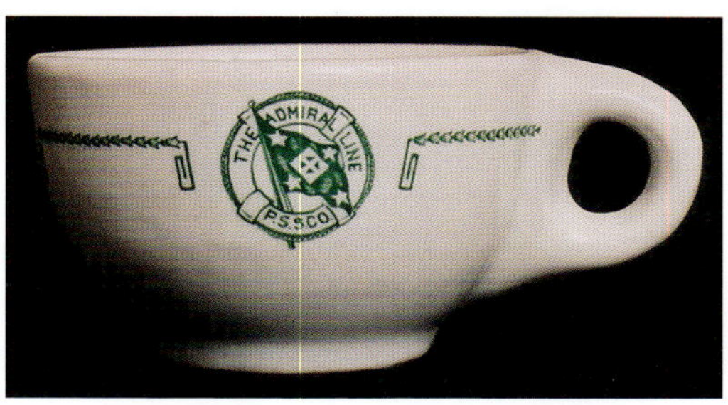

PSC-3 Admiral Benson ★★

White china top marked with a green Admiral Line life-ring logo and trimmed with a chain of green bluebell flowers interspersed with decorative medallions. The *Admiral Benson* was one of the last of the passenger steamers acquired by the Admiral Line. This was also the last of the Admiral Line patterns.
Mfr: NPT 1927–28, BUF 1929–36.
Ref.: Jennings PSS-1a.

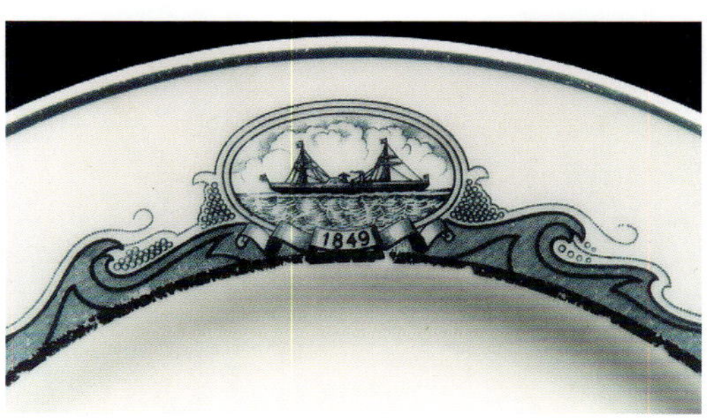

Panama Mail Steamship Company

PMC-1 Columbia ★★★★

White china top marked with the image of a side-wheel steamer in profile and "1849" beneath it. The rim is decorated with a blue-grey art deco wave pattern. The 1849 date corresponds with the arrival of the first Pacific Mail steamer, *California*, at San Francisco – Panama Mail was obviously proud of its lineage. Back stamped "P.M.S.S." This pattern was recovered from the wreck of PML's steamer *Columbia*, lost off Point Costa, Baja California, on September 12, 1931.
Mfr: BSH 1927–28, OPC 1928–30.
Ref.: Luckin SP-10.
Daniel Krummes photo.

Pope & Talbot Lines

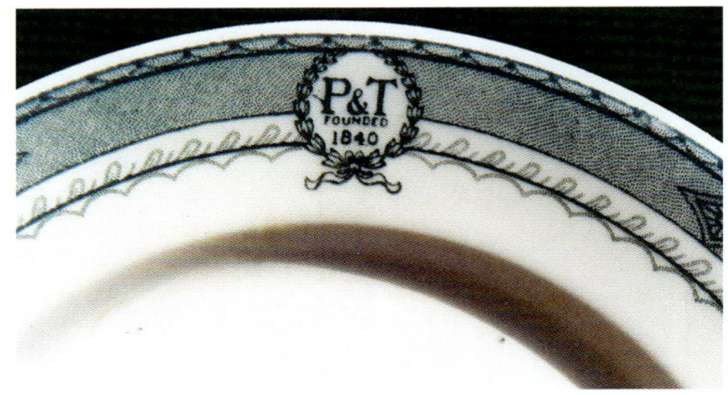

P&T-1 **Seafarer** ★★★★

White china top marked with a dark green P&T logo and green border design featuring geometric shapes. The logo is a wreath of green laurel leaves tied at the bottom in a bow. The text inside the wreath reads "P&T founded 1840" in green lettering. P&T re-entered the shipping business in 1947 with seven new C-3 class vessels. The first of these vessels was the *P&T Seafarer.*
Mfr: SCM 1947–54.
Daniel Krummes photo.

Puget Sound and Alaska Steamship Company

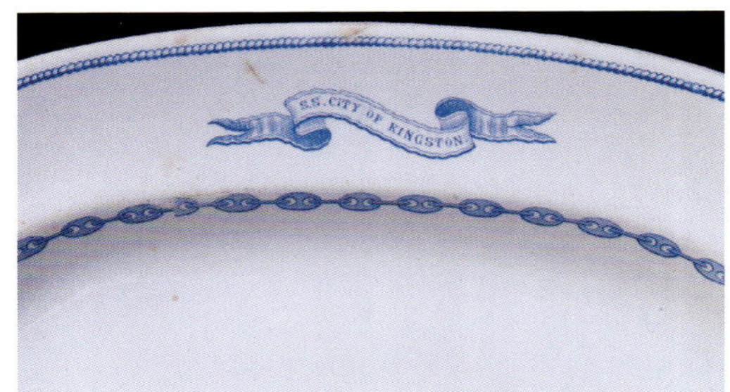

PSA-1 **City of Kingston** ★★★★★

White china top marked with a blue banner containing the words "SS City of Kingston". The outer rim has a rope border and the well of the platter is trimmed with a blue-studded anchor-chain motif. The SS *City of Kingston* was launched in 1884 and sank in a collision on the Tacoma waterfront in 1899.
Mfr: COP ca 1892.
Phil Beaudry photo.

Puget Sound Navigation Company

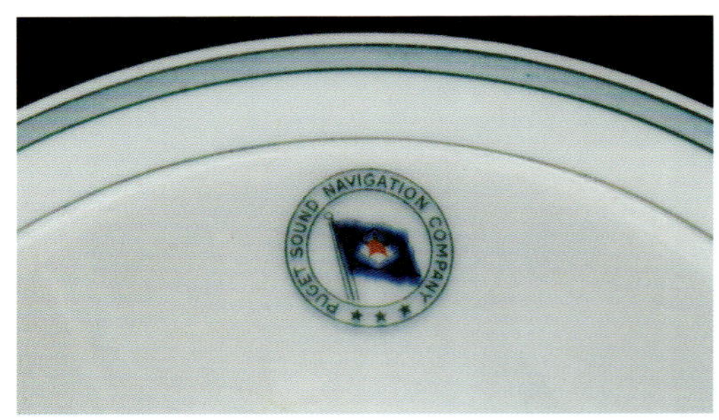

PSN-1 **Chippewa** ★★★

White china top marked with the PSN ring logo and trimmed with a pale blue band flanked on each side by teal pinstripes. The house flag appears inside the ring. PSN changed its house flag to the black ball flag in 1927. The *Chippewa* was renovated to become a vehicle ferry in 1926.
Mfr: OPC 1924–27.
Ref.: Spence PSN-1.

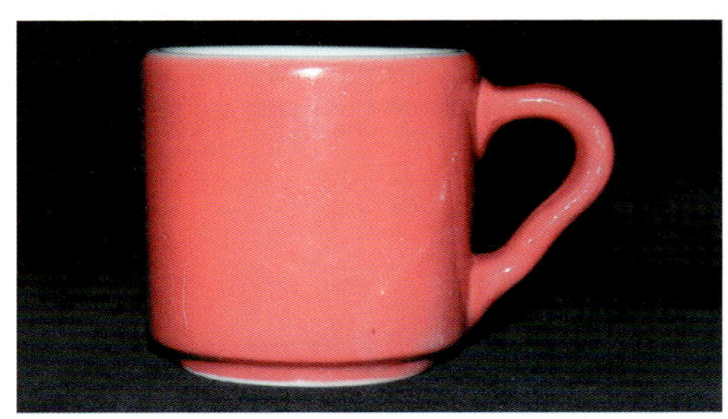

PSN-2 **Kalakala** ★

Rose-coloured china with no distinguishing marks. This pattern was reportedly used in the dining room of the *Kalakala.* A former PSN employee donated pieces of this pattern to the Kitsap County Museum in Bremerton, Washington.
Mfr: TPC ca 1935–45.

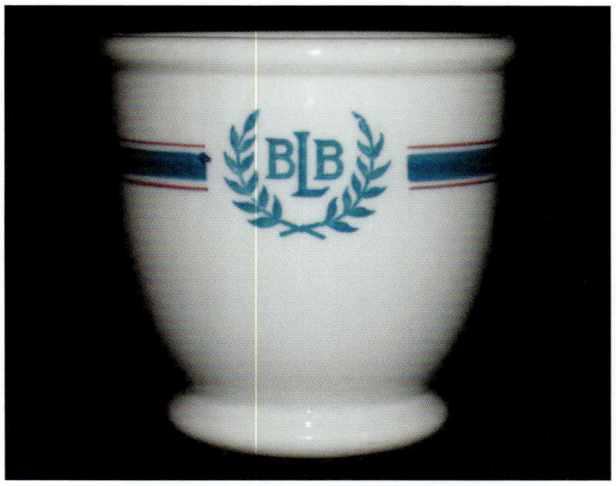

PSN-3 **Chinook** ★★★★

White china top marked with the Black Ball Line logo and trimmed with a medium blue band (5 mm) flanked on each side by burgundy pinstripes. The logo consists of "BLB" in blue, enclosed between two blue fern branches. This pattern was introduced aboard the MV *Chinook* in 1947 and continued to be used by the Black Ball Line of Canada when the *Chinook* was transferred to Canadian registry. The pattern is clearly visible in promotional photographs of the *Chinook* dining room.

Mfr: SGO 1947–51.

San Francisco and Portland Steamship Company

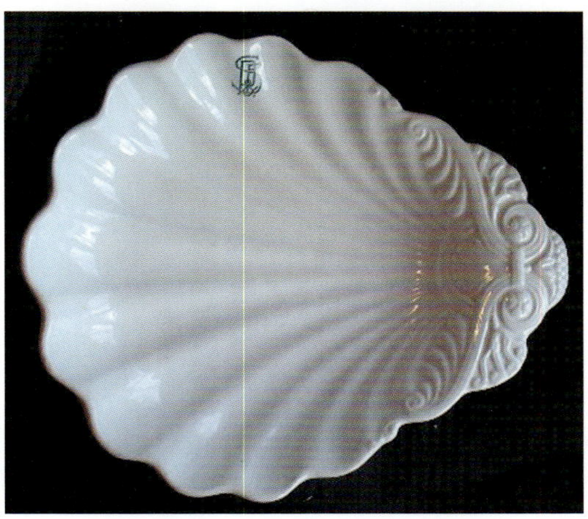

SFP-1 **Rose City** ★★★★★

White china top marked with "SF&P" superimposed in green.

Mfr: WOD 1904–07.

Dean Willis photo.

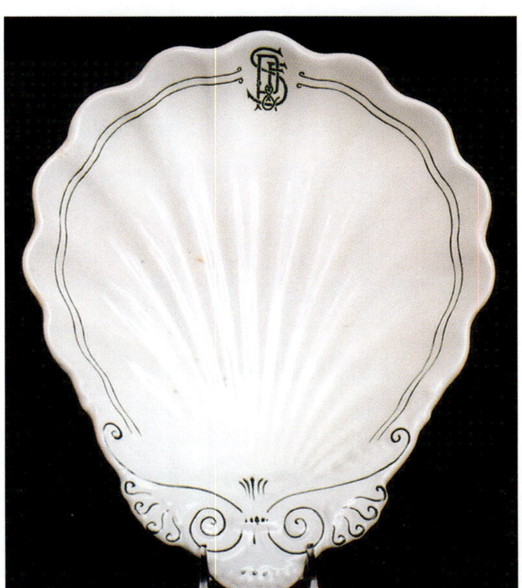

SFP-1a **Beaver** ★★★★

White china top marked with "SF&P" superimposed in green. The addition of two pinstripes make this different from SFP-1. SFP commissioned the two large steamers *Bear* and *Beaver* in 1910.

Mfr: MDK 1910–13.

Daniel Krummes photo.

Seaboard Shipping Company

SSC-1 Seaboard ★★★

White china top marked with the SSC house flag and trimmed with a blue border stripe. The flag is a blue swallowtail pennant with a red company logo and "Seaboard" in white letters. SSC has used this flag since 1936 and it is still in use.
Mfr: unknown ca 1970s.

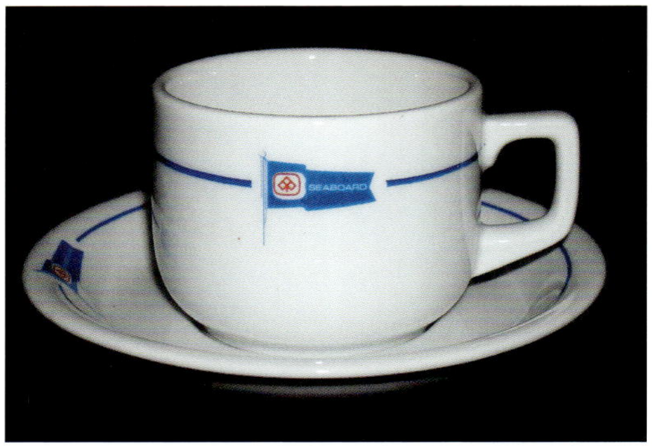

Southern Pacific Company

SPC-1 Harriman Blue ★★★

White china top marked with the SPC badge and trimmed with a scalloped blue border design. Pieces that are not top marked are back stamped "S.P.Co."
Mfr: MDK ca 1905.
Ref.: Luckin SP-4.
Richard Luckin photo.

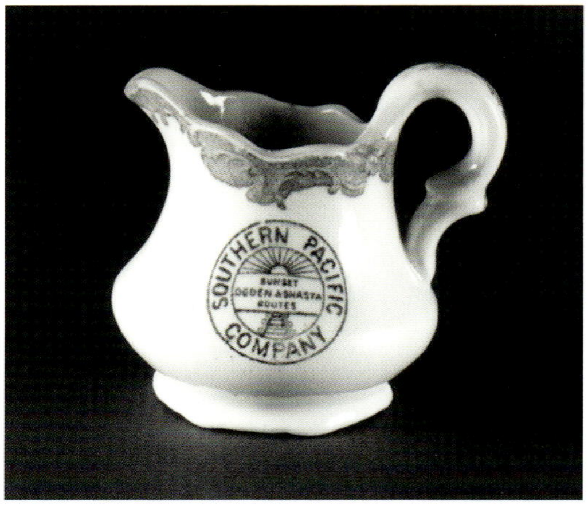

Standard Oil Company of California

SOC-1 San Francisco ★★

White china top marked with the corporate Chevron logo and trimmed with two blue pinstripes. SOC used this pattern on tankers and in company cafeterias.
Mfr: MOS 1992, BUF ca1980s.
Ref.: Conroy CHV-1.

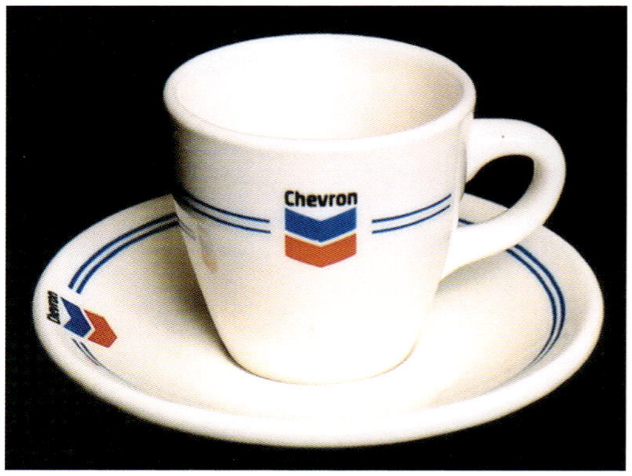

States Steamship Company

SSS-1 Swastika ★★★

Silver-plated pieces top marked with a double-ring logo containing the SSS name and house flag. So far this logo has only been found on silver-plated items.
Mfr: International Silver Co. 1932–37.

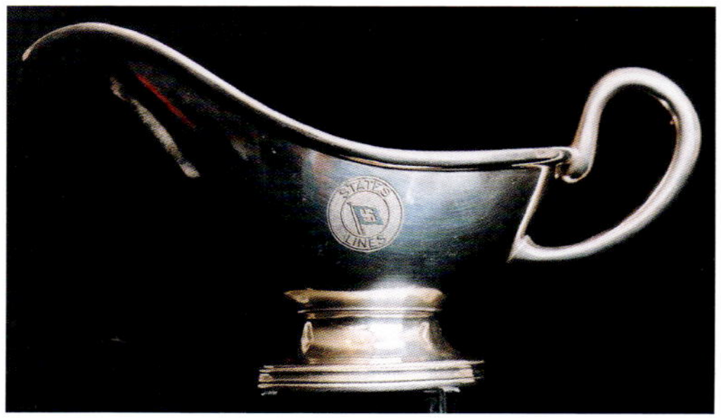

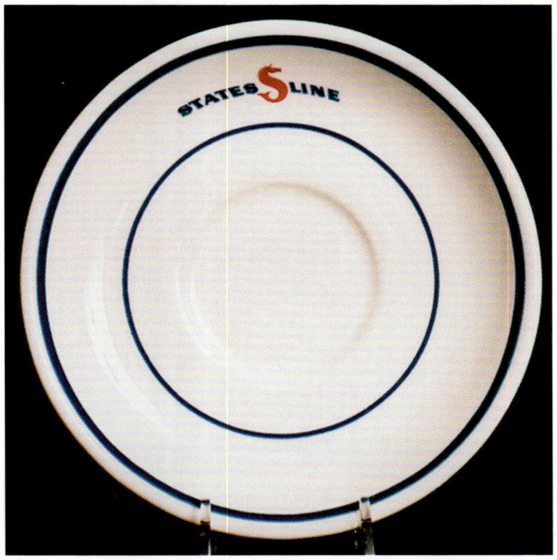

SSS-2 **Seahorse** ★★

White china top marked with the SSS seahorse logo and trimmed with an outer dark blue border stripe. Another blue stripe decorates the edge of the plate well. The logo reads "States Line" with a red seahorse forming an "S" between the words. SSS adopted the seahorse as its corporate logo in 1954.
Mfr: CRB 1967, STR 1967–75, JAC 1976–78.
Ref.: Conroy STL-1.

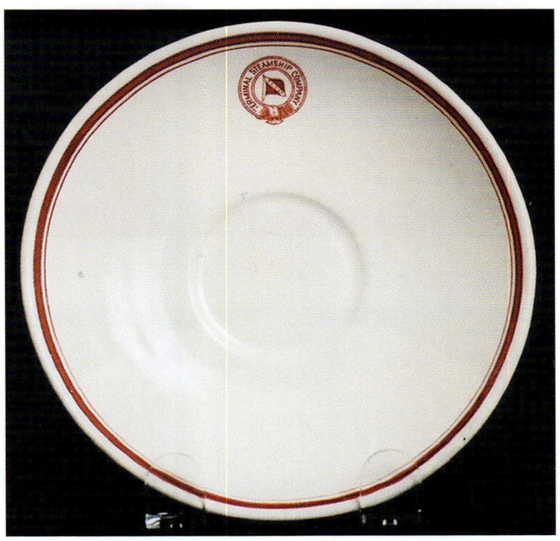

Terminal Steamship Company

TSS-1 **Terminal** ★★★★

White china top marked with a TSS garter logo and trimmed with a red border stripe flanked on the inside by a red pinstripe. The garter logo features the TSS house flag in the centre with "Terminal Steamship Company" encircling it.
Mfr: MDK 1907.

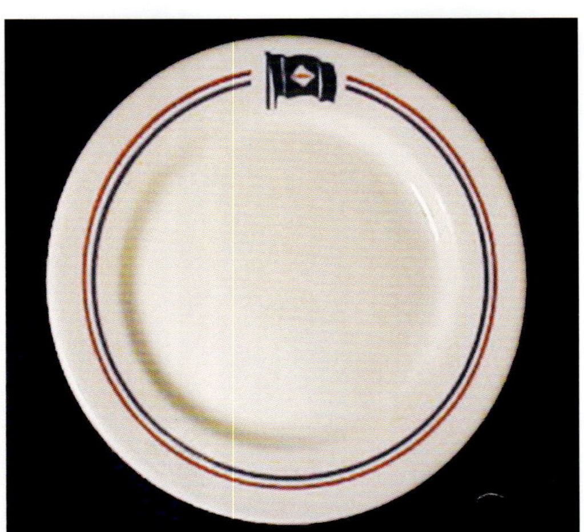

Union Oil Company of California

UOC-1 **House Flag** ★★★

White china top marked with a blue UOC house flag and trimmed with two border stripes, an outer red and inner blue.
Mfr: BUF ca 1946.

UOC-2 Los Angeles ★★★

White china with a blue Union 76 logo, which was introduced in 1947.
Mfr: OPC 1977–81.
Ref.: Conroy UOC-1, Spence UN-1.

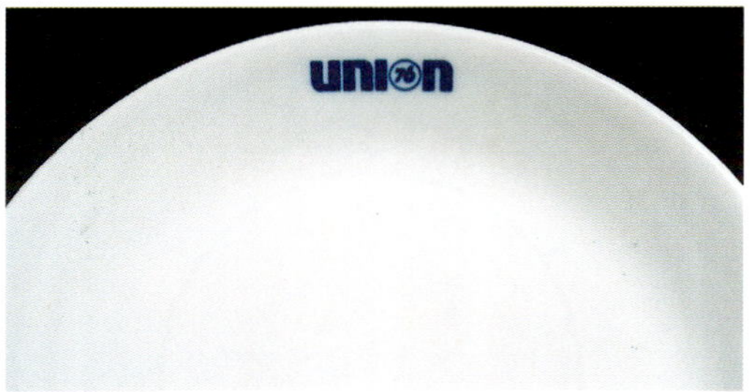

Union Steamship Company of British Columbia

USC-1 St Andrews Cross ★★★★

White china with an early USC badge: an open ring with "Union" in the upper arch and "Steamship" in the lower arch; inside the ring is a St Andrews cross with "BC" in the centre. This pattern has been found on the wreck of USC's *Capilano 1*, which sank in 1915.
Mfr: unknown pre-1914.

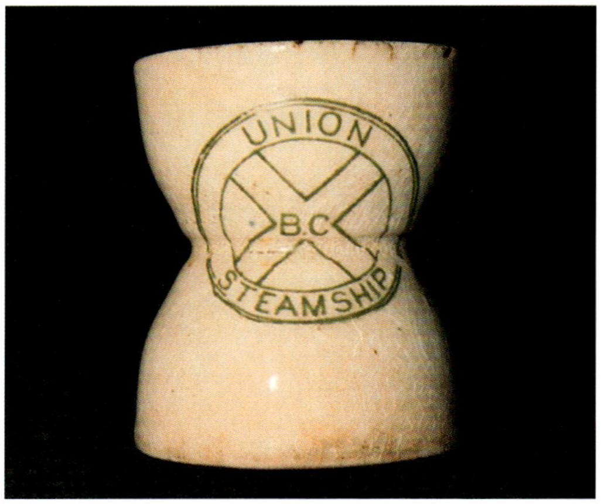

USC-2 Colonial ★★

White china decorated with a dark green bows-and-beads border design. This is a stock pattern found at numerous USC dock sites and on the wreck of the *Capilano 1*, which sank in 1915. "Colonial", the manufacturer's name for the pattern, is often part of the back stamp.
Mfr: MDK 1910–22.

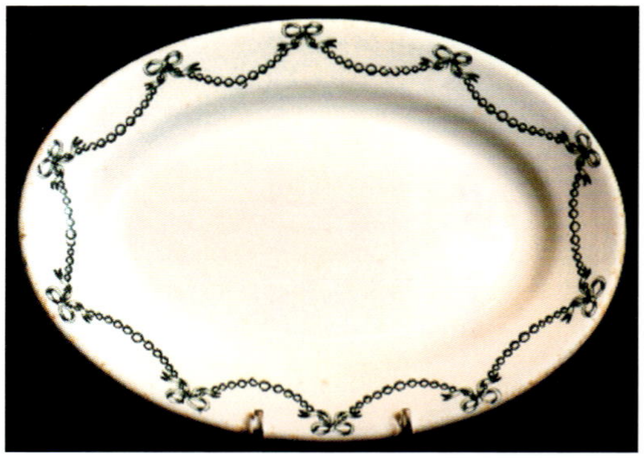

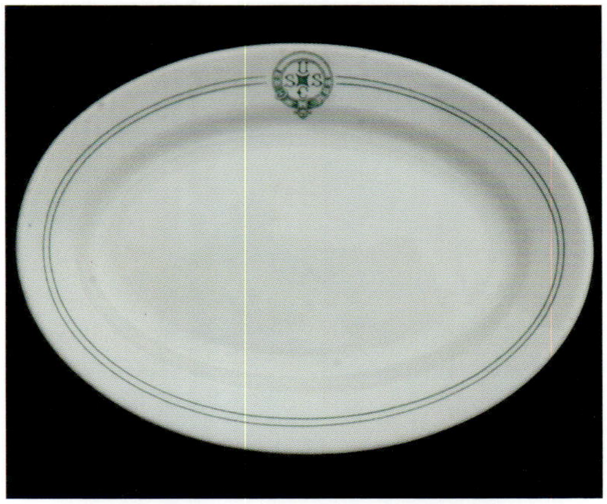

USC-3 Banner ★★★

White china top marked with a green banner containing "USSCo of BC" and trimmed with two side-by-side green pinstripes.
Mfr: MDK 1920–25, GLY 1925.
Ref.: Luckin USC-1.

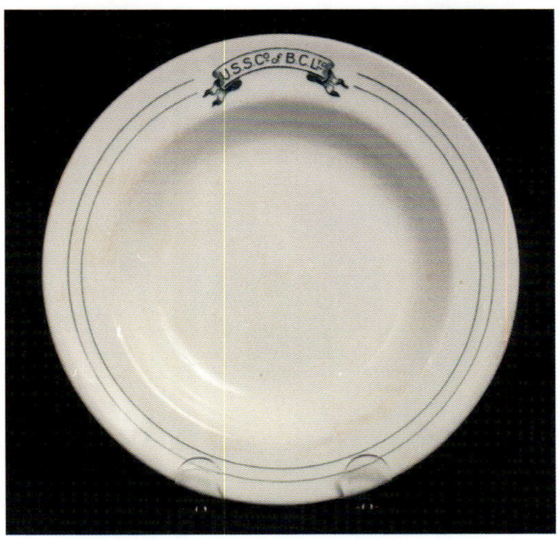

USC-4 Garter ★★★★

White china top marked with a green garter logo and trimmed with two side-by-side border pinstripes. The garter contains "U.S.S.C." in green, arranged in the shape of a cross. This pattern was short lived, as Union bought out several estates in 1926 and changed their logo to reflect these acquisitions in 1927.
Mfr: MDK 1924–26.

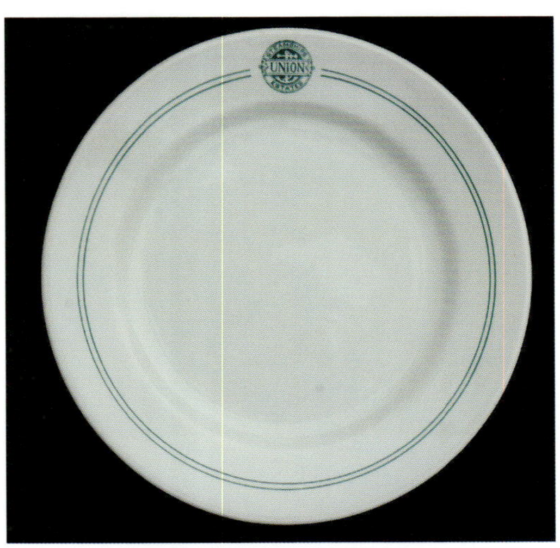

USC-5 Estates ★★★

White china top marked with a dark green USC ring logo and trimmed with two side-by-side dark green border pinstripes. In the centre of the ring there is a stylized anchor with a banner across it that reads "Union". "Steamship Estates" appears inside the ring encircling the anchor. This pattern reflects the fact that, by 1927, USC was operating steamships as well as a number of vacation estate properties.
Mfr: MDK 1928; GLY 1927–30, 1938–39; GLB 1932–40; MED 1942–47.

USC-5a **Estates** ★★★★

This pattern, seen only on coffee mugs, is the same as USC-5 except that the badge is larger and there are no border pinstripes.
Mfr: GLY 1930–36.

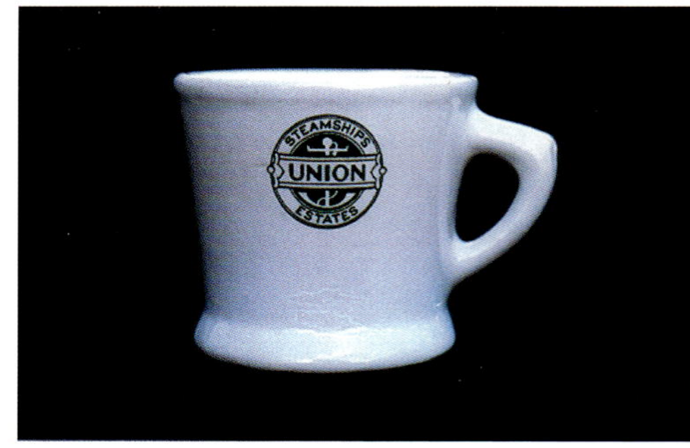

Western Canada Steamship Company

WCS-1 **Lake Atlin** ★★

White china top marked with the WCS house flag and trimmed with two side-by-side lime green border pinstripes. The flag is a white swallowtail with a red St George's cross and a green maple leaf in the top left quadrant. All pieces of this china appear to be dated 1948.
Mfr: Globe 1948.

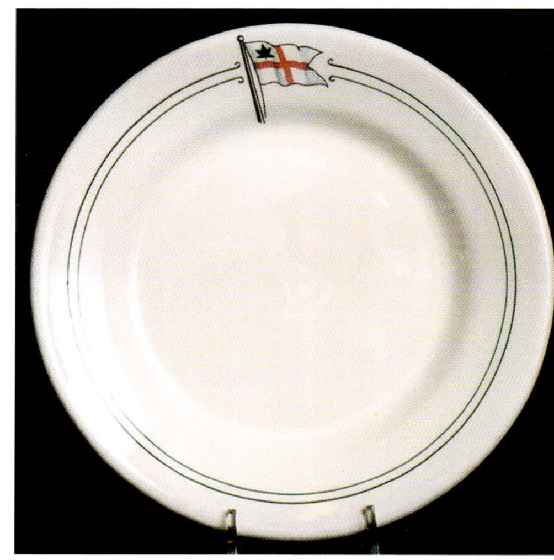

Western Steam Navigation Company

WSN-1 **Mainlander** ★★★★★

White china top marked with the WSN house flag. The flag is blue on top and red below with a diagonal white stripe running from the top left to bottom right with red initials "V.L." in it. The WSN ship *Mainlander* was launched on July 31, 1900.
Mfr: OPC 1900–1907.

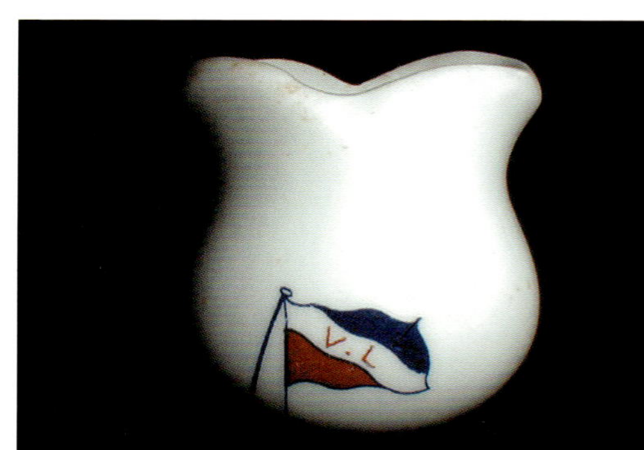

Canadian Government Services Patterns

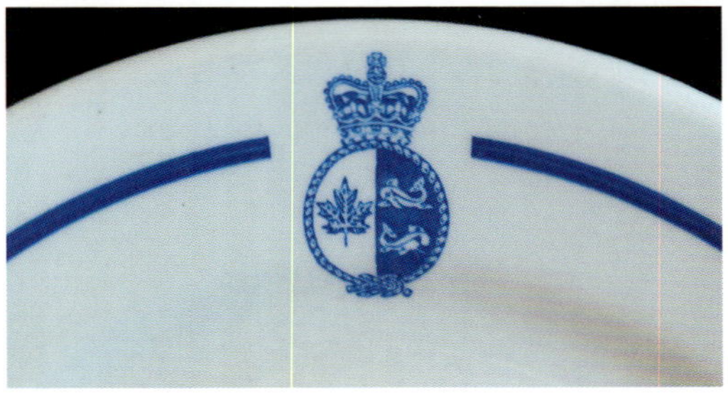

Canadian Coast Guard

CCG-1 Coast Guard ★

White china top marked with the CCG badge and trimmed with medium-weight blue border stripe. This pattern was introduced in 1964 and decommissioned in 1996.

Mfr: MAD 1963–65, GLY 1966–76, SYR 1980–92.

Ref.: Krummes CCG-1.

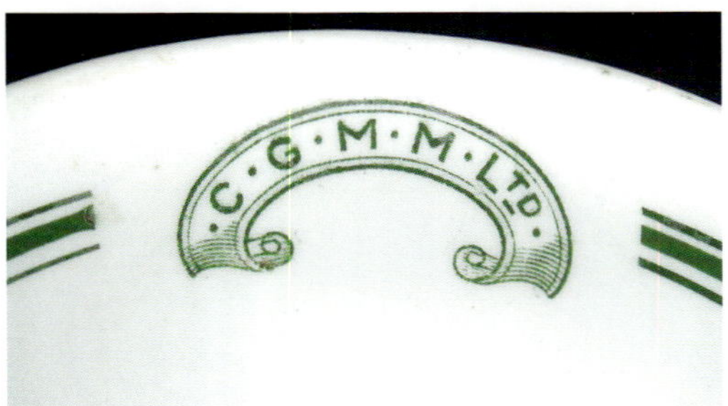

Canadian Government Merchant Marine

CMM-1 Merchant Marine ★★★★★

White china with a green banner containing "CGMM Ltd" and trimmed with green border stripe flanked on each side by green pinstripes. Only two pieces are known to exist.

Mfr: unknown.

Bill Kellett photo.

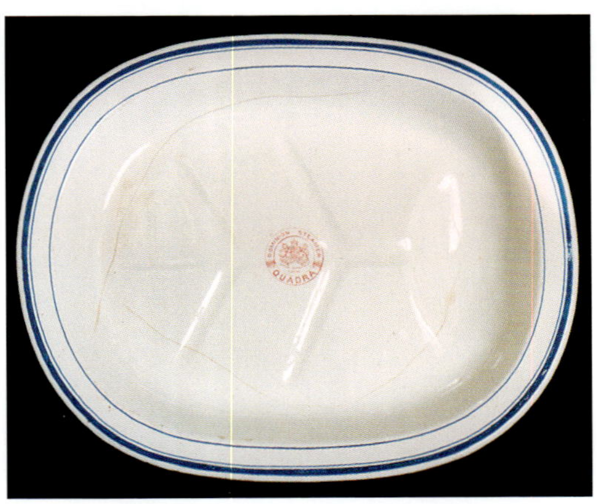

Canadian Department of Marine and Fisheries

DMF-1 Quadra ★★★★★

White china top marked with a red Dominion of Canada badge in the centre and trimmed with a blue border stripe flanked inside by a blue pinstripe. The CGS *Quadra* served on the BC coast from 1892 to 1917.

Mfr: AWB 1892–1917.

DMF-2 **DMF Crest** ★★★★★

White china decorated with a multitude of DMF badges, thought to be a test piece to demonstrate what the badge would look like on china.
Mfr: RID 1910.

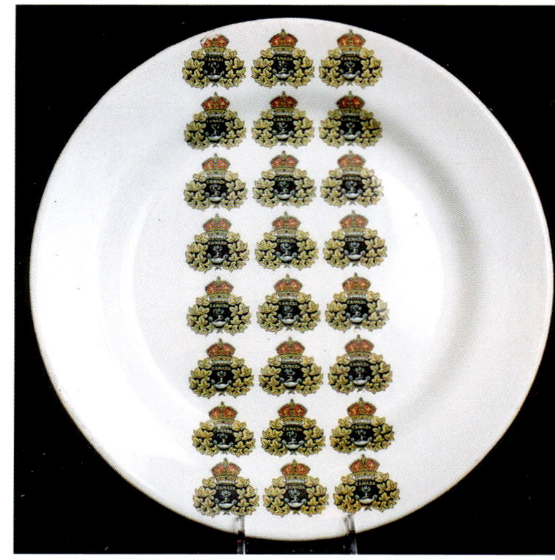

DMF-3 **Red Badge** ★★★★

White china top marked with a red DMF garter and trimmed with a red border stripe flanked on each side by a green pinstripe.
Mfr: GLY 1916.

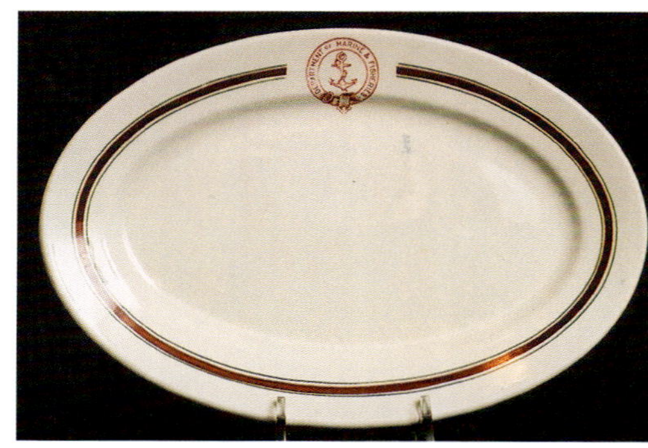

DMF-4 **Green Badge** ★★★★

White china top marked with a green DMF garter and trimmed with a green border stripe flanked on each side by a red pinstripe.
Mfr: RID 1912–20.
Greg Cochkanoff photo.

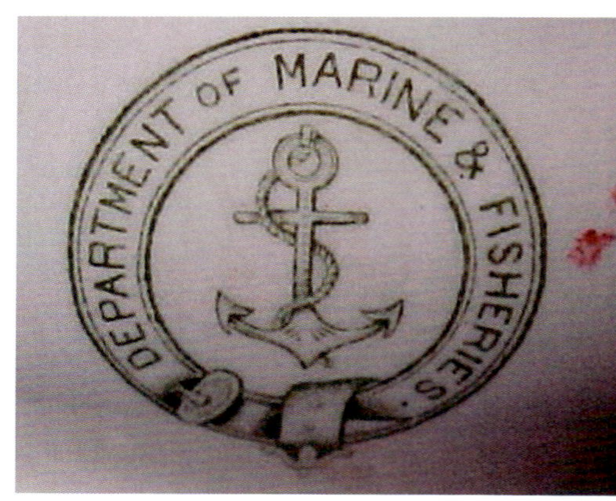

Royal Canadian Navy

RCN-1 **WR Gold** ★★★

White china top marked with gold superimposed letters "WR". This is an RCN wardroom pattern that saw service aboard early ships and in on-shore mess facilities. It was introduced in the early 1900s and discontinued in 1924, replaced with the Admiralty Gold pattern (RCN-4).
Mfr: COP ca 1921.

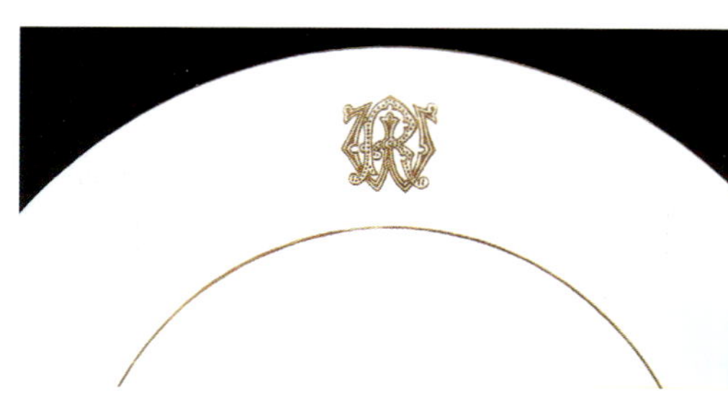

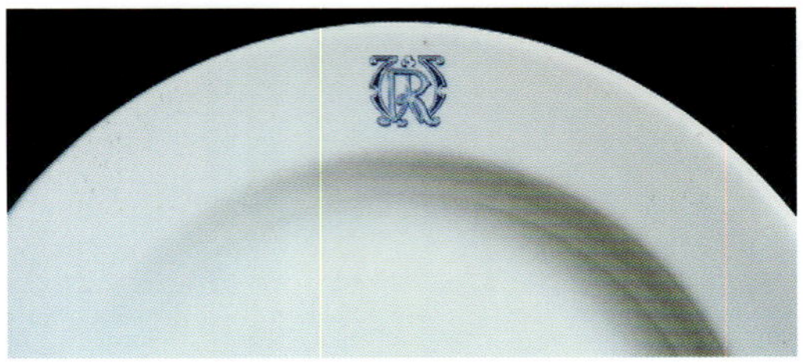

RCN-1a **WR Blue** ★★★

Same pattern as RCN-1, except that all the decoration is blue instead of gold. This is a wardroom pattern that saw service aboard early RCN ships. It was introduced in 1921 and discontinued in 1937. This pattern has been dive-recovered from the wreck of HMCS *Thiepval*, lost in 1930.

Mfr: COP unknown dates, CRN 1930, MIN 1930, CLD 1933.

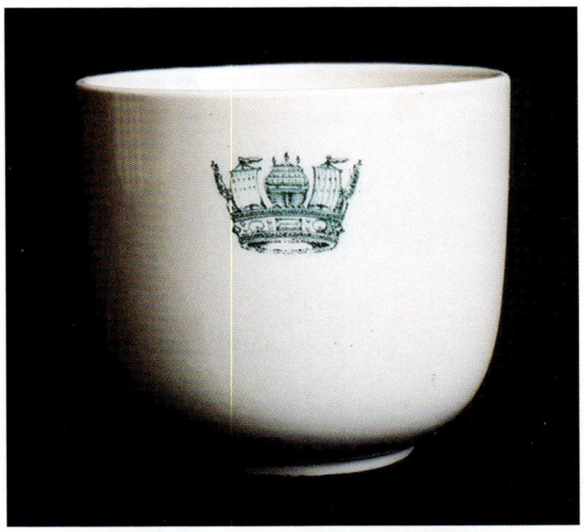

RCN-2 **Admiralty Crown** ★★★★

White china top marked with a green Admiralty crown. This pattern was introduced into the Royal Navy's Chief and Petty Officers Mess in 1923 and discontinued in 1941. It has been found in and around the Esquimalt dockyard, west of Victoria, and on the wreck of HMCS *Thiepval*, lost in 1930.

Mfr: COP 1928 and 1936, BST 1936.

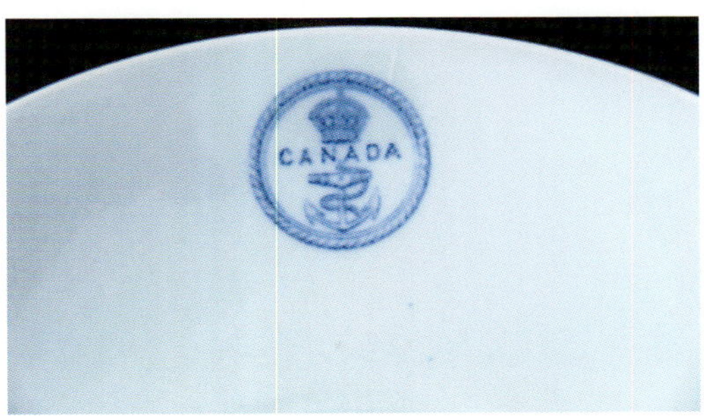

RCN-3 **Rainbow** ★★★★

White china top marked with a blue RCN button badge. The Imperial crown dates the badge to the period of 1902 to 1953. This pattern may be off Canada's first cruiser HMCS *Rainbow*, commissioned on May 4, 1910. This piece was dive-recovered at Swanson Bay.

Mfr: unknown.

Vivienne Grant photo.

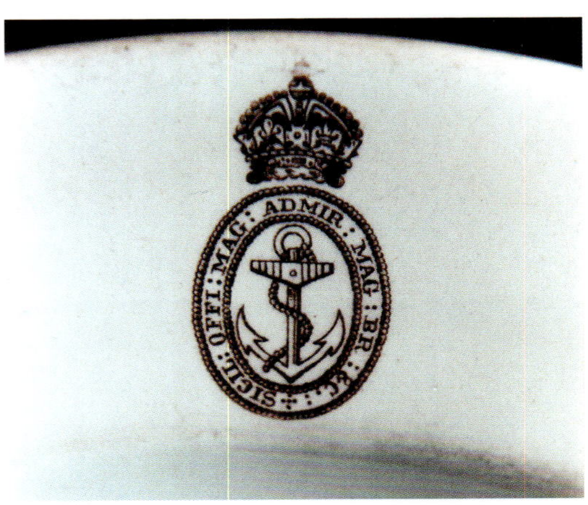

RCN-4 **Admiralty Gold** ★★★★

White china top marked with a gold Admiralty crest. The inscription in the ring of the crest, "SIGIL: OFFI: MAG: ADMIR: MAG: BR & Co." means "the seal of the Office of the Lord High Admiral of Great Britain". This pattern was introduced into the Royal Navy in 1924, replacing WR Gold (RCN-1). The RCN version is often back stamped with "Canada". Replaced in 1946 with the RCN badge.

Mfr: COP 1938, FOL 1945, Coalport 1954, BTH 1955.

RCN-4a **Admiralty Blue** ★★

Same as RCN-4 except that the decorations are blue instead of gold. This pattern was introduced into the Royal Navy in 1937, replacing WR Blue (RCN-1a). It was used in wardrooms and in non-commissioned officers' messes. The RCN variation is often back stamped with "Canada". It was discontinued in 1946 with the introduction of the RCN badge.
Mfr: SOV 1937–46, CRD ca 1934.

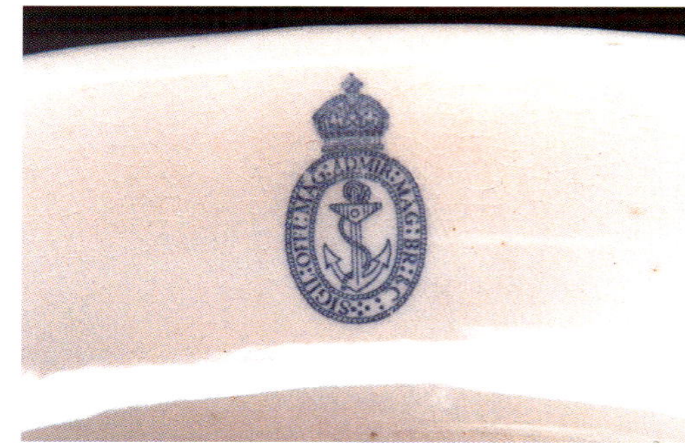

RCN-5 **Flag Officer 1** ★★★

White china top marked with a gold RCN crest containing nine maple leaves and the Imperial crown. This pattern was introduced into the RCN in 1946, replacing Admiralty Gold (RCN-4). Fluted china with this badge was used in the ship captain's cabin, by flag officers, and in an admiral's residence. Non-fluted china was used in wardrooms for visiting dignitaries.
Mfr: CRN 1946–52.

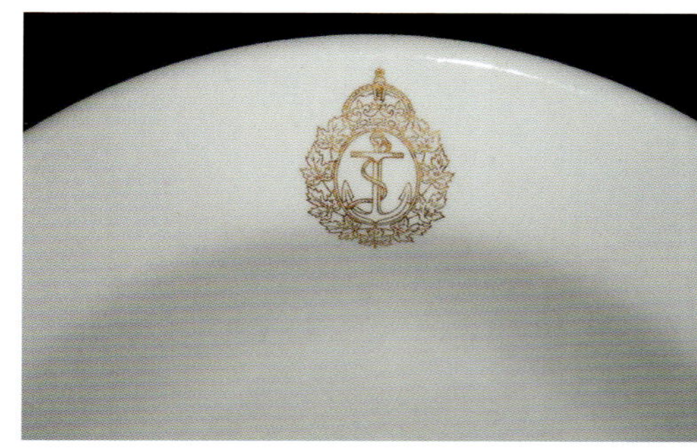

RCN-5a **Wardroom 1** ★★

The same pattern as RCN-5, except the crest is blue instead of gold. This pattern was introduced into the RCN in 1948, replacing Admiralty Blue (RCN-4a).
Mfr: CRN 1948–52.

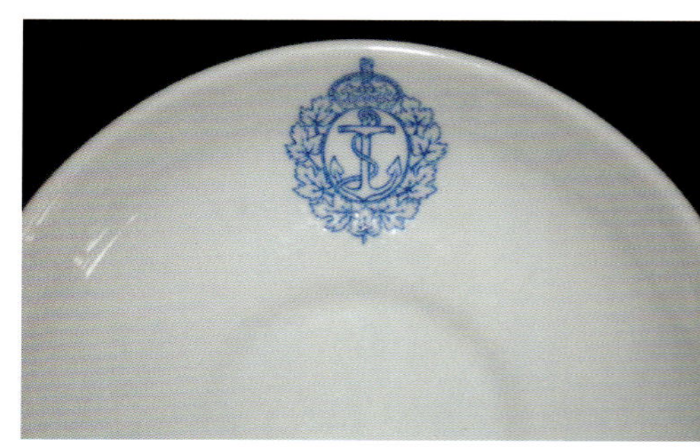

RCN-6 **Flag Officer 2** ★★

White china top marked with a gold RCN crest containing ten maple leaves and the Imperial crown. This pattern replaced Flag Officer 1 (RCN-5) in 1952, three years after Newfoundland joined Canada. No one has been able to find this pattern and it is only speculated to exist.
Mfr: unknown.
No photo available.

RCN-6a **Wardroom 2** ★★

The same pattern as RCN-6, except the crest is blue instead of gold. This pattern replaced Wardroom 1 (RCN-5a) in 1952, three years after Newfoundland joined Canada.
Mfr: MED 1952-53

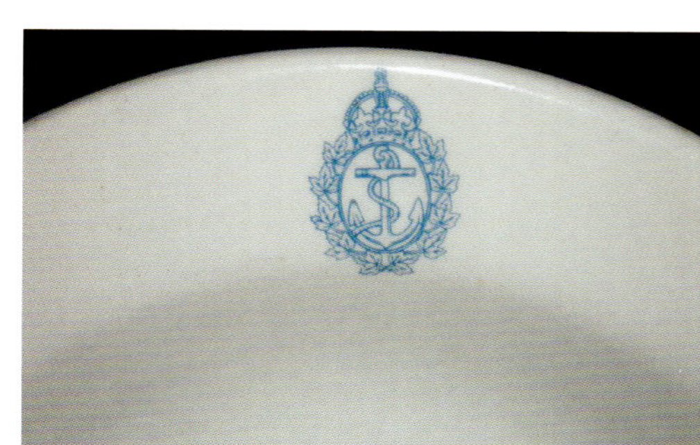

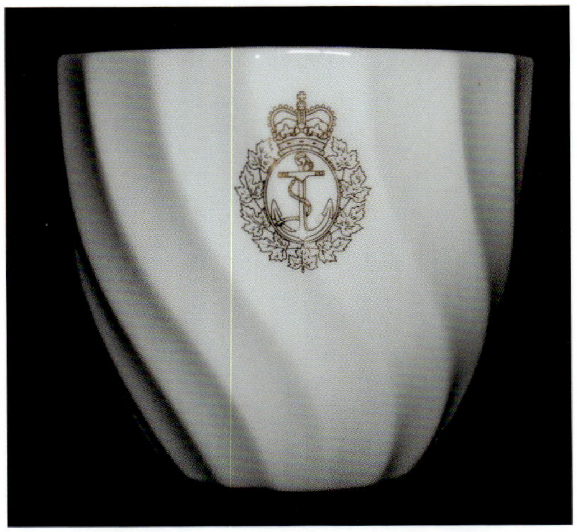

RCN-7 Flag Officer 3 ★★

White china top marked with a gold RCN crest containing nine maple leaves and the St Edward's crown. This pattern was produced in 1955 and has the wrong number of maple leaves. Fluted china with this badge was used in the captain's cabin, by flag officers and in an admiral's residence. Plain china with the gold badge was used at formal mess dinners and for diplomatic functions.
Mfr: GJS 1955.

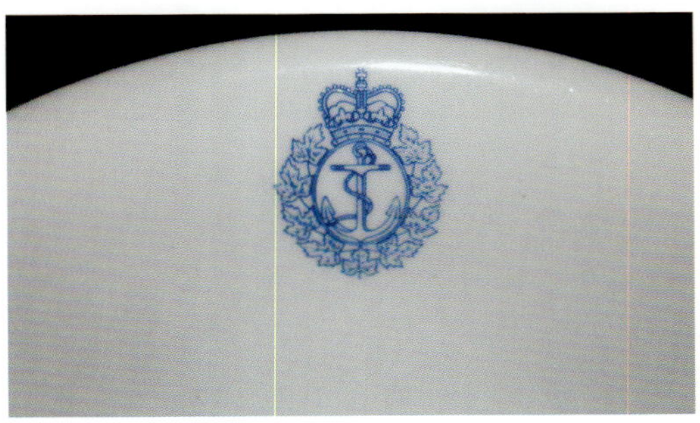

RCN-7a Wardroom 3 ★★

White china top marked with a blue RCN crest containing nine maple leaves and the St Edward's crown. This pattern was produced in 1955 and has the wrong number of maple leaves. It was used in the wardroom on ships and on shore.
Mfr: FOL 1955–56, RID 1955.

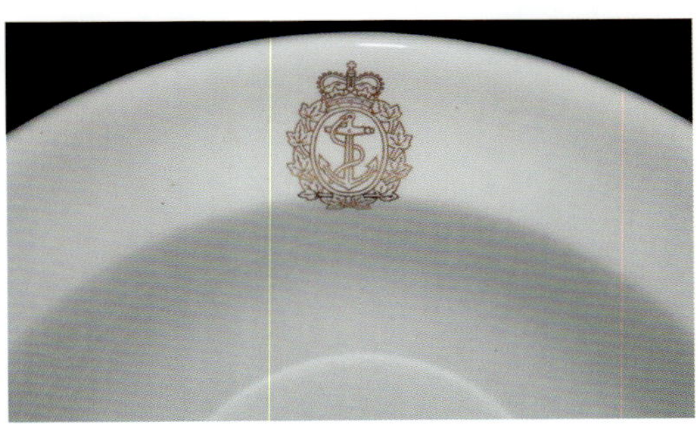

RCN-8 Flag Officer 4 ★★

White china top marked with a gold RCN crest containing ten maple leaves and the St Edward's crown. This pattern replaced Flag Officer 3 (RCN-7) in 1956 to recognize Queen Elizabeth's accession to the throne in 1953. Fluted china was used by commanding officers, base commanders and admirals. Plain china with the gold badge was used at formal mess dinners and for diplomatic functions.
Mfr: GJS 1956–57, CRN 1958–60 and 1963, CLD 1961, TSK 1960 and 1965–66.

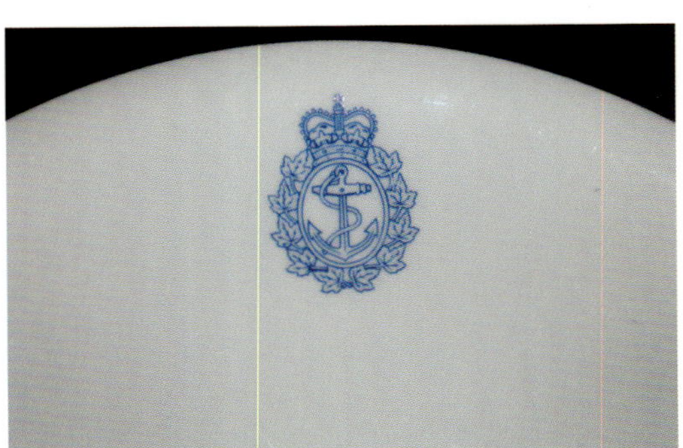

RCN-8a Wardroom Blue 4 ★★

White china top marked with a blue RCN crest containing ten maple leaves and the St Edward's crown. This pattern replaced Wardroom 3 (RCN-7a) in 1955 to recognize Queen Elizabeth's accession to the throne in 1953.
Mfr: GRM 1955–58, FOL 1956, GJS 1957, WEG 1957, CRN 1958–59, GLD 1959, CRD 1959 and 1966, GLD 1959–60, RID 1960 and 1963, STH 1963–67.

RCN-9 CF Gold ★★★

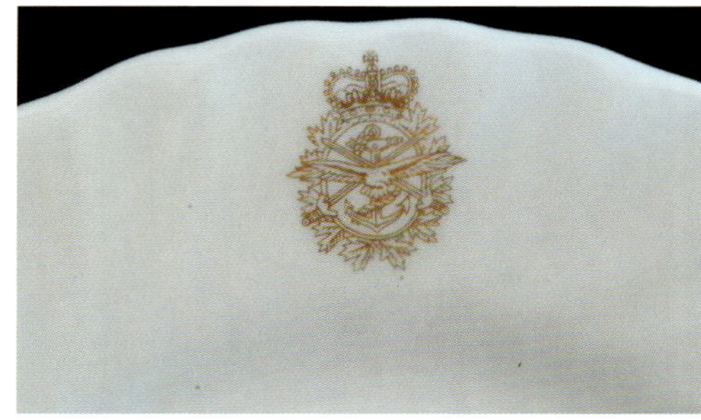

White china top marked with a gold Canadian Forces logo. This pattern was created in 1968. Fluted china was used by commanding officers, base commanders and admirals. Non-fluted china with the gold badge was used at formal mess dinners and for diplomatic functions.
Mfr: MIN 1968, WEG unknown dates.

RCN-9a CF Mustard ★

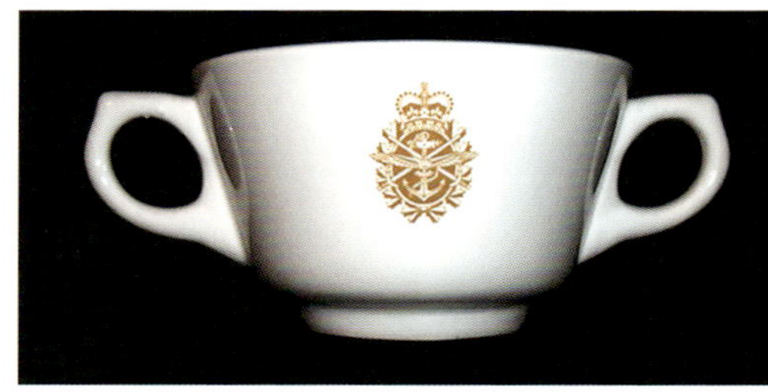

White china top marked with a mustard yellow Canadian Forces badge and trimmed with a yellow border stripe. This pattern was created in 1968 and is used today by all Canadian Forces departments.
Mfr: SYC 1972–95.

RCN-10 Pacific Admiral ★★★★★

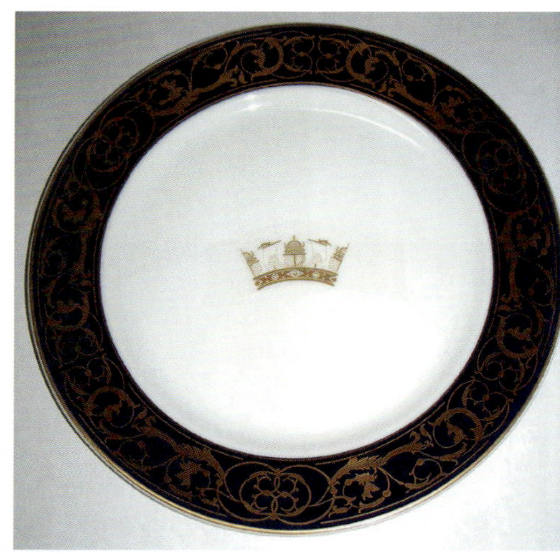

White china top marked with a gold Admiralty crown and trimmed with a dark blue border that is decorated with a gold filigree design. This pattern was commissioned for the admiral in charge of Maritime Forces Pacific fleet in 2007.
Mfr: DUD 2008.
Jamie Webb photo.

United States Government Services Patterns

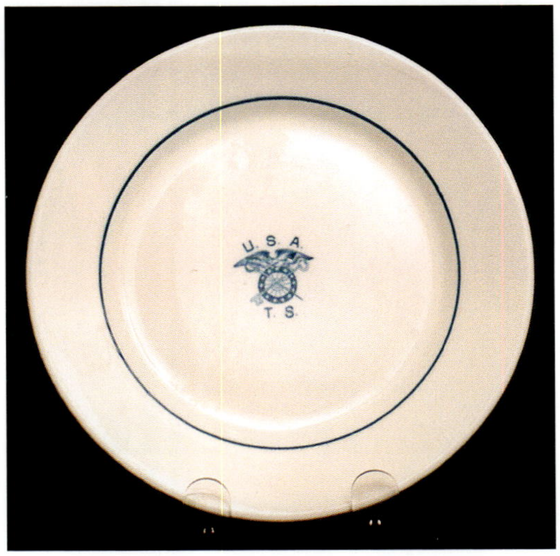

US Army Transport Service

USAT-1 Eagle & Wheel ★★★

White china top marked with a blue US Army Quartermaster Corps logo and trimmed with a blue border pinstripe. The logo consists of a spoked wheel with a sword running through it from left to right and a key running through it from right to left. An eagle is perched on top of the wheel; "T.S." (Transport Service) appears below the wheel. This design was used by the USAT on uniform buttons from 1930 to 1942.
Mfr: TRN 1904, GWD 1930-42.

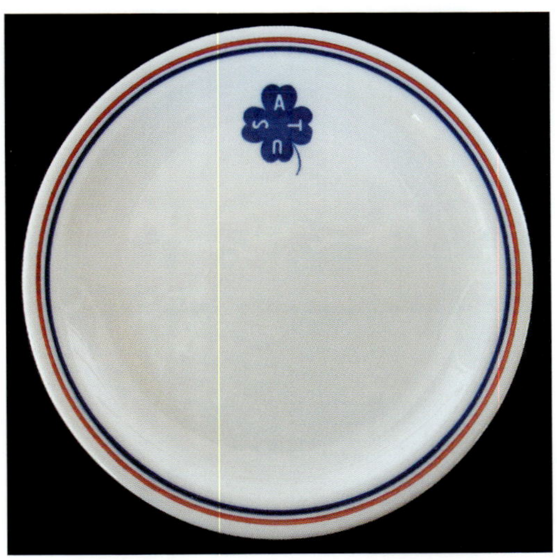

USAT-2 Clover Leaf ★★

Ivory-coloured china with a blue USAT four-leaf clover logo and trimmed with a red border stripe that is flanked on the inside by a blue stripe.
Mfr: SYR 1942–43.
Ref.: Conroy UAST-1.

US Bureau of Fisheries

USBF-1 Albatross ★★★

White china top marked with the USBF house flag and trimmed with a red pinstripe flanked inside by a dark blue pinstripe. The *Albatross* was a USBF research vessel that operated on the west coast from 1915 to 1921. This is the only USBF pattern known to exist. The use of a house flag suggests that it was used on ships as well as at headquarters and on trains.
Mfr: BUF 1927.
Ref.: Conroy UABF-1, Spence UBF-1.

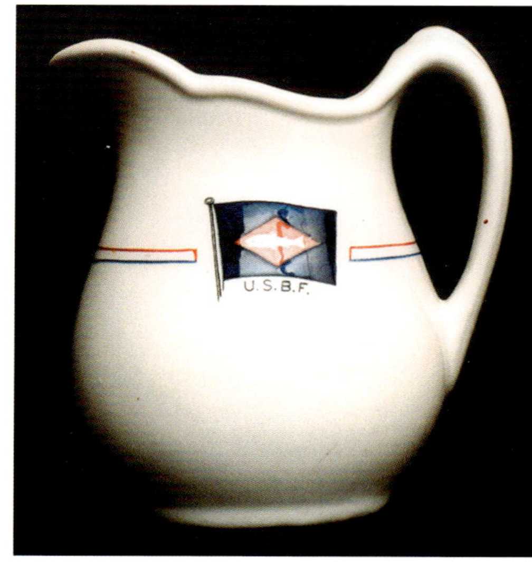

US Coast and Geodetic Survey

USCS-1 House Flag ★★★

White china top marked with the USCS house flag and trimmed with a gold border stripe.
Mfr: BUF 1915–40s.

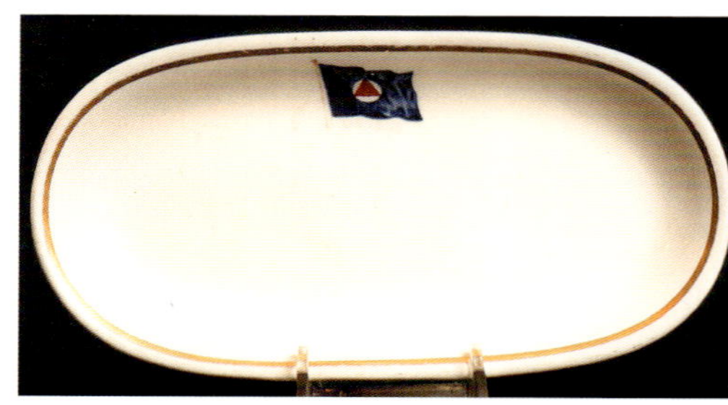

USCS-2 Blue Badge ★★★

White china top marked with a blue USCS crest and trimmed with a blue border stripe.
Mfr: SGO 1950.
Ref.: Conroy vol. 2, p. 429.
Photo courtesy of Barbara Conroy.

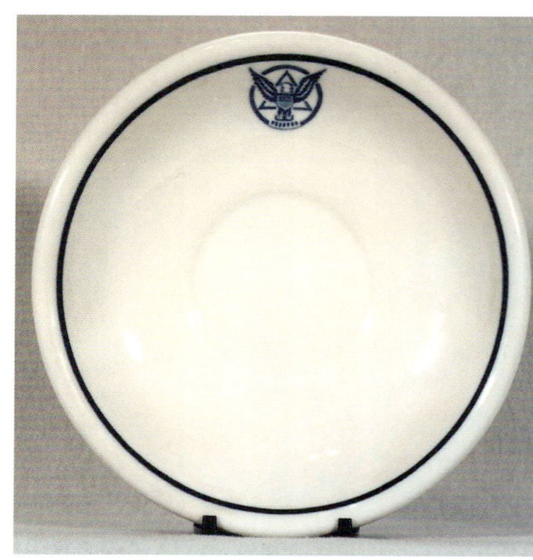

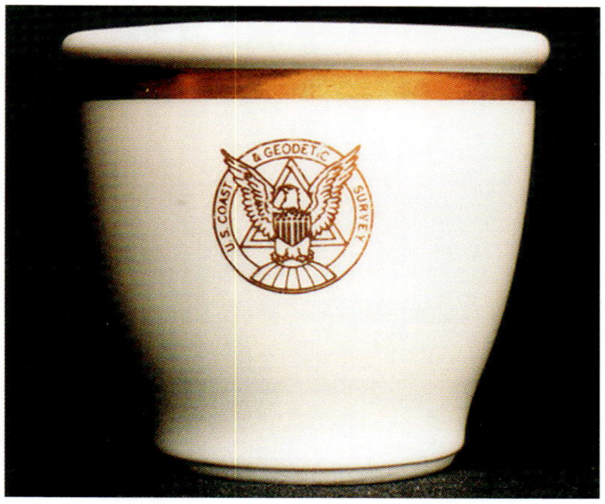

USCS-3 Gold Badge ★★

White china top marked with a gold USCS crest and trimmed with a gold border stripe. This pattern was in use just before the USCS was incorporated into the NOAA in 1970.
Mfr: MYR 1962.

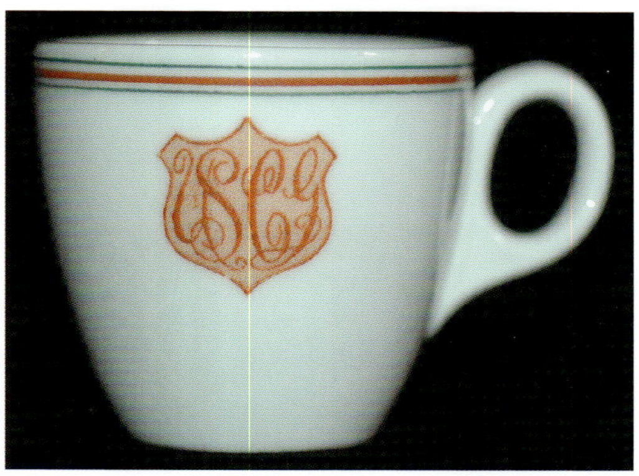

US Coast Guard

USCG-1 Orange Shield ★★★★★

White china with an orange chevron badge containing "USCG" in script and trimmed with an orange border stripe flanked on each side by green pinstripes.
Mfr: LAM 1915–24.

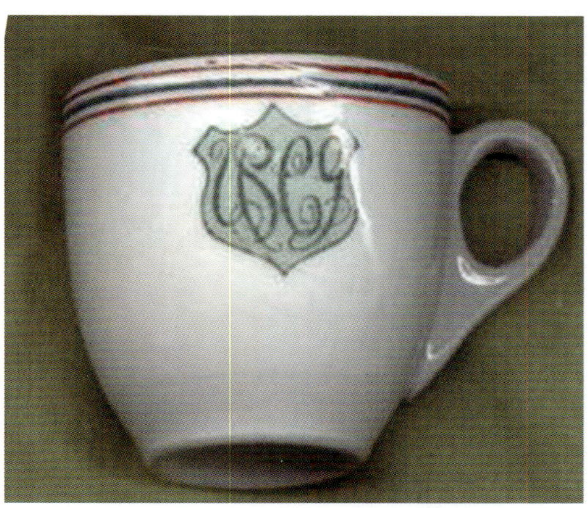

USCG-1a Green Shield ★★★★★

Same as USCG-1 except that the badge is green and the colours of the trim stripes are reversed (green border stripe flanked by orange pinstripes). This pattern was produced for the captain's mess.
Mfr: LAM 1915–24.
Photographer unknown.

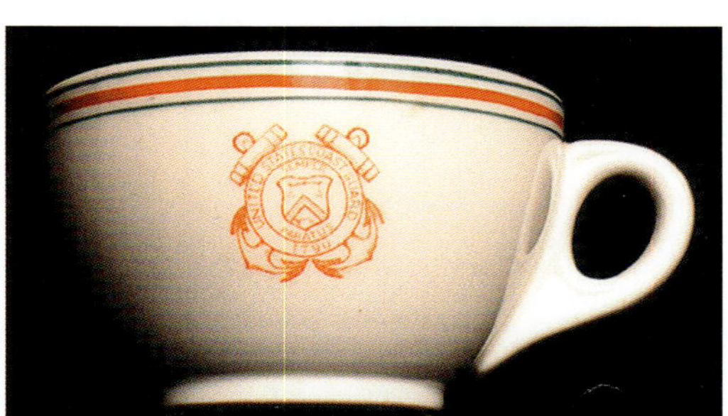

USCG-2 Treasury ★★★

White china top marked with an orange USCG crest and trimmed with an orange border stripe flanked on each side by a green pinstripe. The crest consists of a double ring superimposed on two crossed anchors. The ring text reads "United States Coast Guard 1790" and the badge of the US Treasury Department is inside the ring.
Mfr: JAC 1923.

USCG-3 Blue Crest ★★★

White china top marked with a blue USCG crest, a double ring superimposed on two crossed anchors. The ring text reads "United States Coast Guard 1790" and the badge of the US Treasury Department is inside the ring. This pattern was produced during World War II.
Mfr: JAC 1939–45.

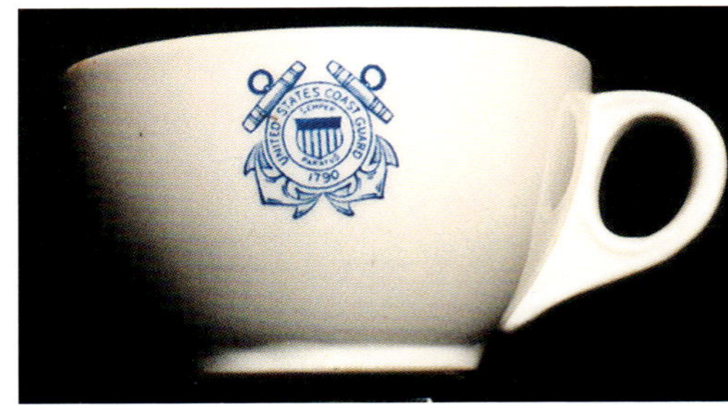

USCG-4 Wardroom ★★

White china top marked with an orange USCG crest and trimmed with an orange border stripe flanked on each side by a green pinstripe. The crest consists of a double ring superimposed on two crossed anchors. The ring text reads "United States Coast Guard 1790" and the badge of the US Treasury Department is inside the ring.
Mfr: JAC 1942, 1961, 1976; WKR 1955, 1967–69; MYR 1962–65, 1974–75, 1979–82.
Ref.: Conroy USCG-2.

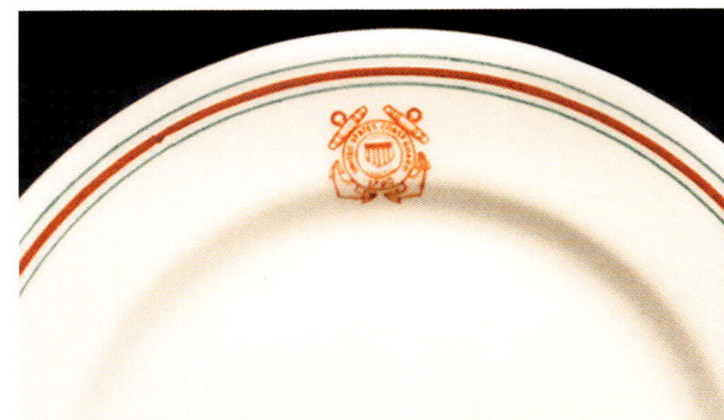

USCG-4a Commander ★★★

Same pattern as USCG-4, except that the badge is green and the colours of the trim stripes are reversed (green border stripe flanked by orange pinstripes). This pattern was used for captain's service.
Mfr: JAC 1948.

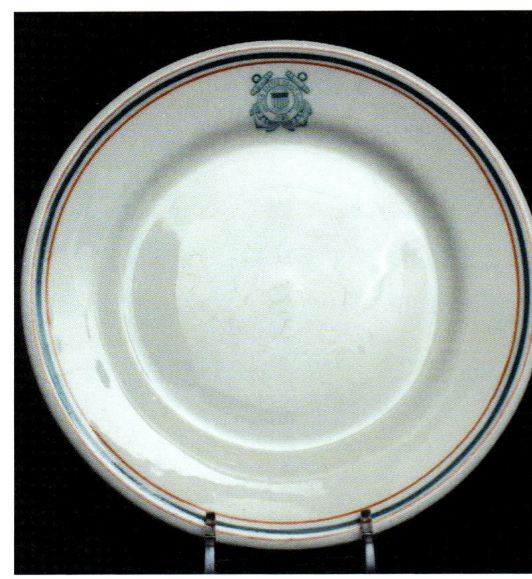

USCG-5 Gold Badge ★★★★

White china top marked with a gold USCG badge and trimmed with gold border trim and stripe around well.
Mfr: COR ca 1962–70s.

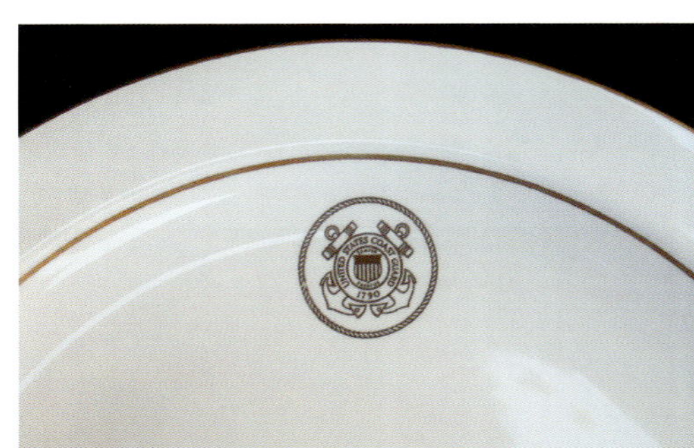

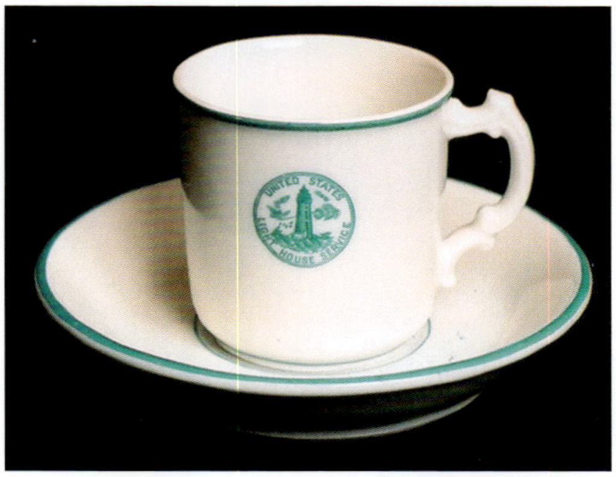

US Lighthouse Service

USLS-1 Lighthouse ★★★★

White china top marked with a pale blue USLS logo and trimmed with a light blue border stripe. The logo depicts a lighthouse on a rock surrounded by clouds and birds. "United States" appears above the lighthouse and "Lighthouse Service" below.
Mfr: HAV 1905–10, 1918; BSH 1911; WAR 1921–30.
Ref.: Spence USLS-1, Krummes USL-2.

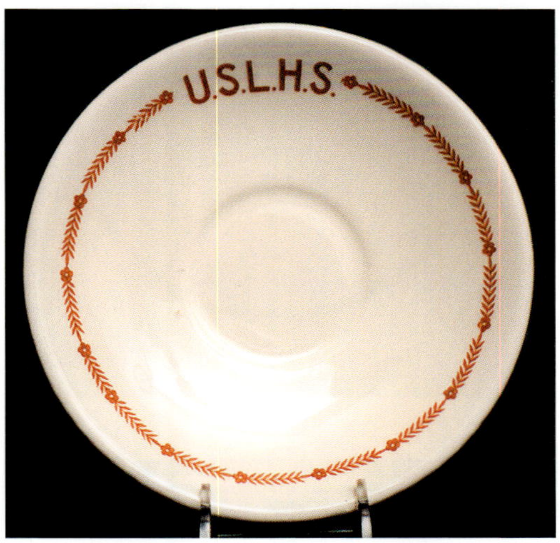

USLS-2 Minott Ledge ★★★★

White china top marked with an orange leaf and daisy border design broken by "U.S.L.H.S." in orange.
Mfr: JAC 1930–37.
Ref.: Luckin USLH-1, Spence USLS-2, Krummes USL-1.

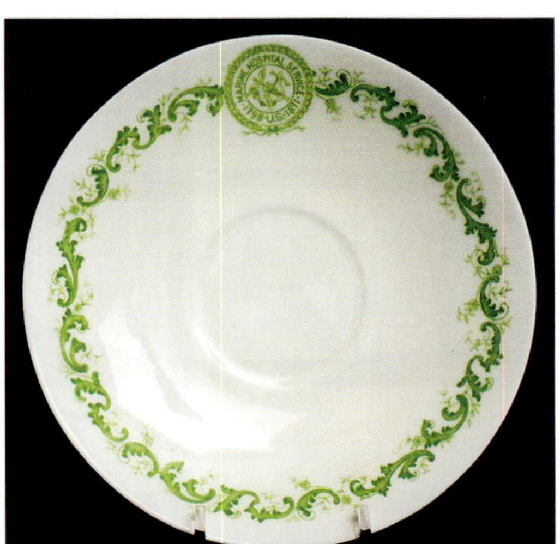

US Marine Hospital Service

USMH-1 Astoria ★★★★

White china top marked with a green USMHS badge and trimmed with green fern-leaf border design. The badge consists of a wreath with a Medical Corp symbol and anchor crossed in the centre. "Marine Hospital Service 1798-US-1871" appears in the wreath encircling the anchor. This pattern is from the Knappton, Washington, quarantine hospital that examined all new immigrants and sick sailors entering the USA.
Mfr: LAM 1897–1902.

USMH-1a **Astoria** ★★★★

Same pattern as USNH-1, except that the decorations are orange.
Mfr: LAM 1897–1902.

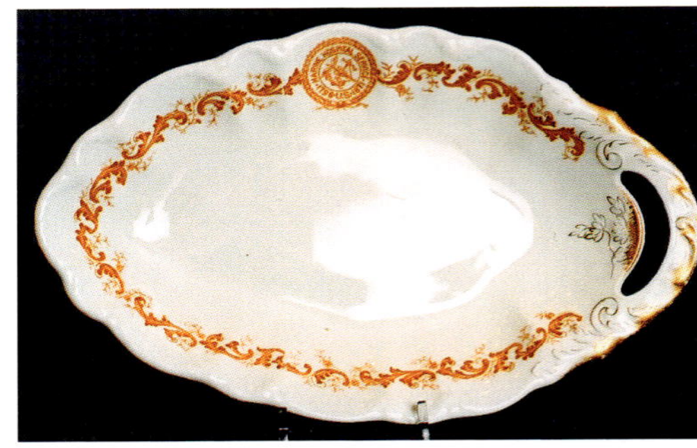

US Maritime Commission

USMC-1 **Maritime** ★★★

Beige china with a brown floral design in the centre and trimmed with a brown border stripe. Back stamped "USM Com."
Mfr: Wallace 1947.
Ref.: Conroy MCOM-1.

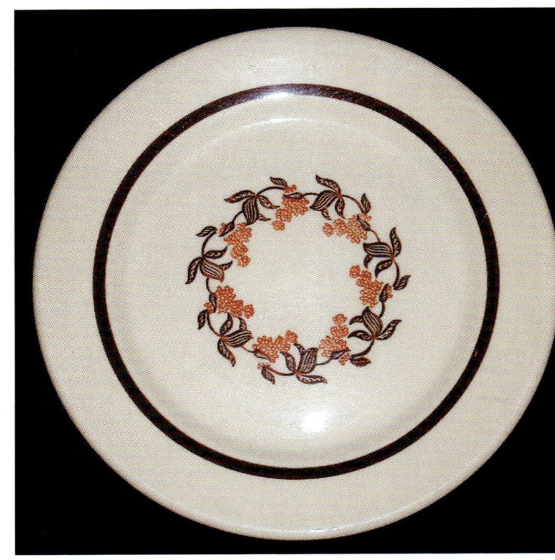

US National Oceanic and Atmospheric Administration

NOAA-1 **NOAA** ★★

White china top marked with the NOAA logo and trimmed with a gold border band. The logo is a circle with a white bird flying through the centre. The space below the bird is bright blue; the space above is dark blue, with "NOAA" in white.
Mfr: MYR 1973.

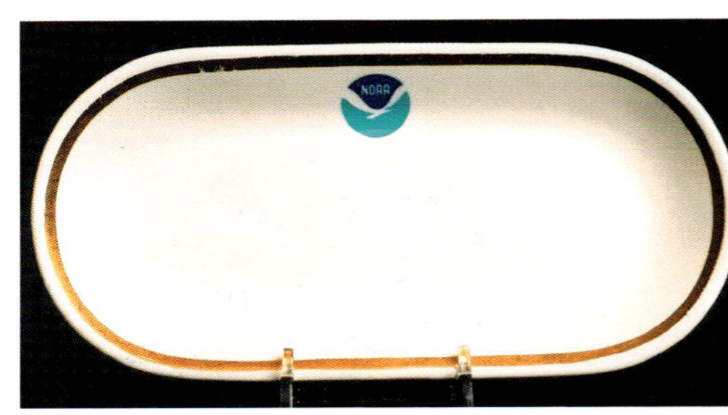

US Navy

USN-1 **Saranac** ★★★★★

White china top marked with green superimposed "USN". A platter bearing this badge was presented to the Kwakwaka'wakw people of Campbell River in appreciation for their help at the time of the USS *Saranac* sinking in Seymour Narrows on June 18, 1975.
Mfr: unknown 1875.
David Griffiths photo.

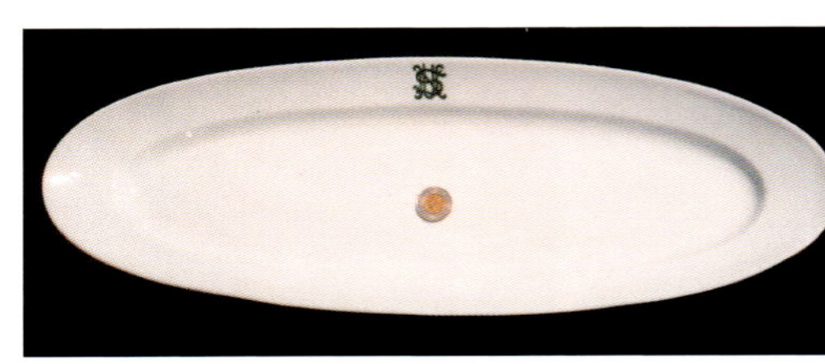

USN-1a **Superimposed Black** ★★★★★

White china top marked with black superimposed "USN".
Mfr: IMP ca 1891.

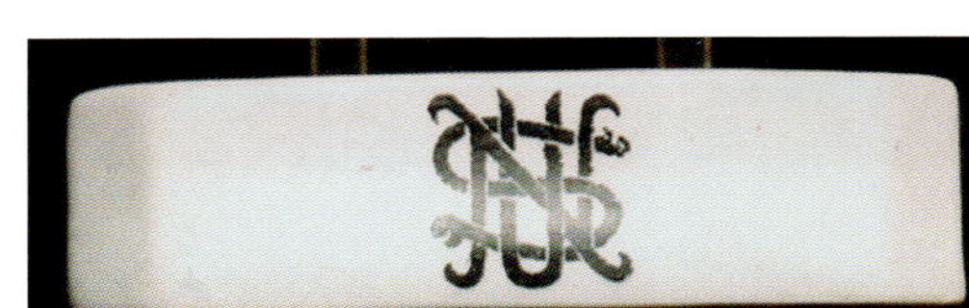

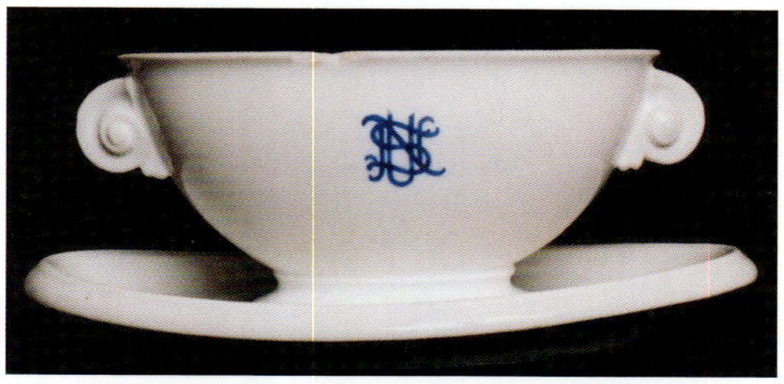

USN-1b Superimposed Blue ★★★★★
White china top marked with blue superimposed "USN".
Mfr: BUF ca 1915.

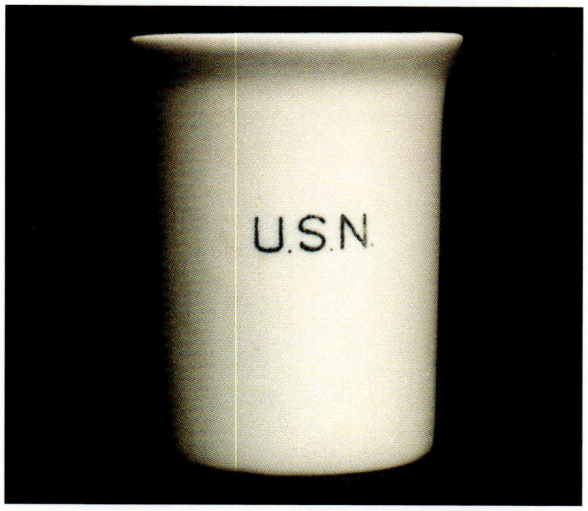

USN-2 Helvetica ★★★★
White china top marked with black Helvetica-style "U.S.N."
Provenance and use unknown.
Mfr: TRN ca 1894.

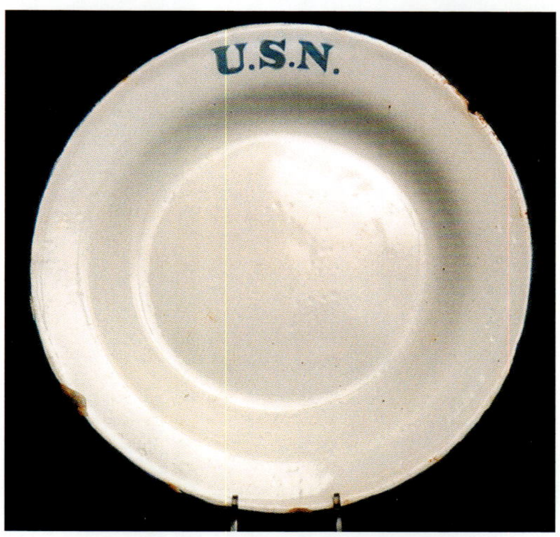

USN-3 Olympia ★★★
White enamelled metal ware top marked with "U.S.N." This pattern predates most other patterns and was used in the enlisted men's mess. A *Harper's Weekly* photograph in 1899 shows enlisted men eating off plates with this pattern on board the USS *Olympia*. The USS *Olympia* was a 5,870-ton cruiser built in 1894.
Mfr: unknown 1899 and earlier.

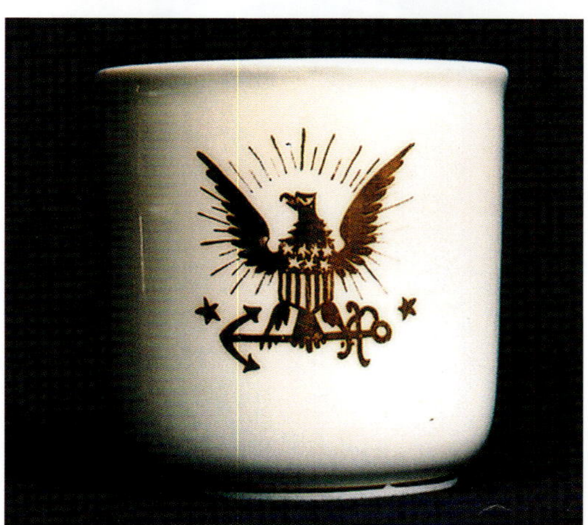

USN-4 Secretary of the Navy ★★★★★
White china top marked with a gold eagle clutching a fouled anchor. Highlights appear around the eagle with two stars, one on each side of the eagle. The stars are said to represent the Atlantic and Pacific oceans. This pattern was produced for the secretary of the USN.
Mfr: HCO ca 1894–98.

USN-5 Department of Navy (old) ★★★
White china top marked with a gold Department of the Navy seal and trimmed with a gold border stripe and inner blue pinstripe. The seal has an outer rope ring and an inner chain ring with "Department of the Navy" between them. In the centre is an eagle clutching an anchor, with highlights around it. This pattern has been recovered from the wreck of the USS *Maine*, which sank in 1898.
Mfr: SGO 1909, 1916, 1917; MYR 1917, 1918; GWD 1918; unknown 1917, 1918, 1928–29.

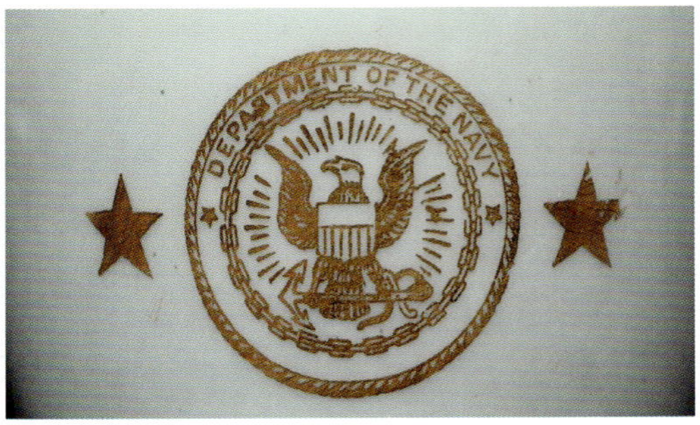

USN-5a Department of Navy (new) ★
White china top marked with a gold Department of the Navy seal and trimmed with a gold border stripe and inner blue pinstripe. This pattern was redesigned in the early 1940s. It differs from USN-5 in that there are no highlights at the sides of the eagle, the shield on the eagle is indistinct and the anchor is a modern one.
Mfr: numerous ca 1940–present.
Ref.: Conroy USN-8.

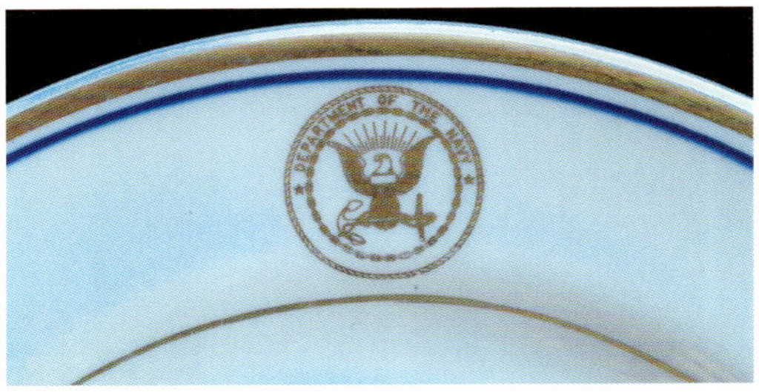

USN-6 Gold Anchor ★★
White china top marked with a gold anchor and trimmed with an outer gold border stripe flanked by an inner blue pinstripe. This pattern appears to have been produced for general wardroom use. It was used during World War I. China with this badge has been recovered from the cruiser USS *San Diego*, which was sunk by a German mine about 10 kilometres off New York in 1918.
Mfr: GWD 1908, 1910, 1918; BUF 1924, 1926, 1927,1928; CAR 1929, 1933–34; BWK 1929, 1937.

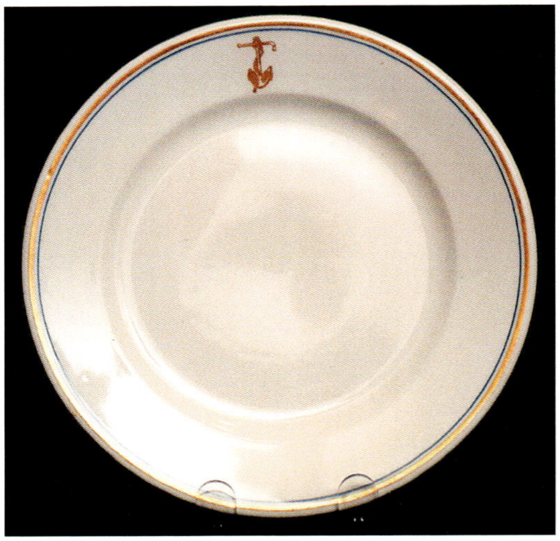

USN-7 DN Four Star ★★
White china top marked with a gold Department of the Navy seal and trimmed with a blue border stripe. Two gold stars on each side of the badge (four altogether) signifies this china's use by a fleet admiral. This pattern was likely the predecessor of USN-10.
Mfr: BUF 1928.

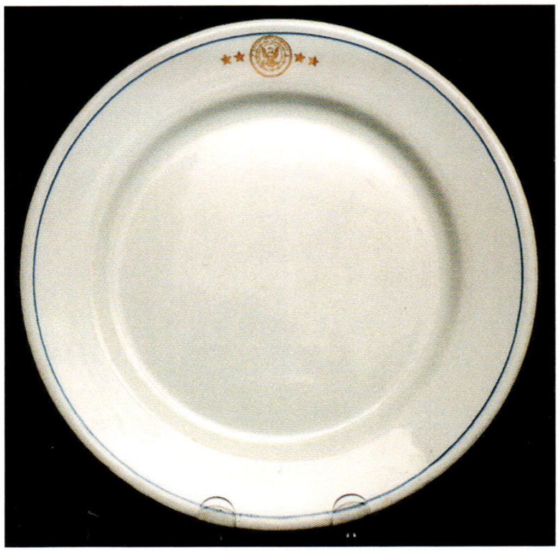

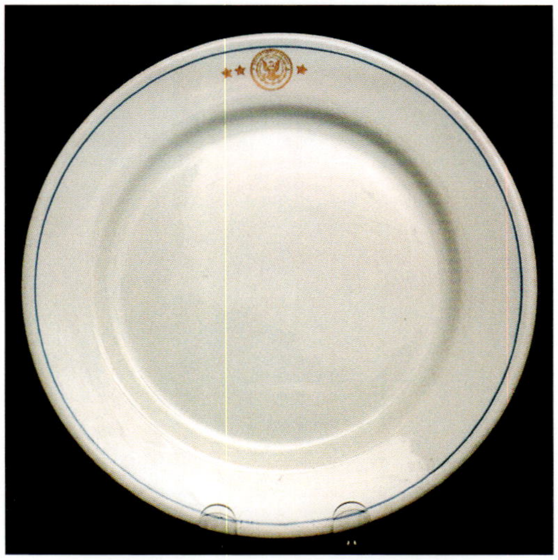

USN-8 **DN Three Star** ★★

White china top marked with a gold Department of the Navy seal and trimmed with a blue border stripe. Two gold stars on one side of the badge and one star on the other side (three altogether) signifies this china's use by a vice admiral. This pattern likely preceded USN-11.

Mfr: CAR 1930.

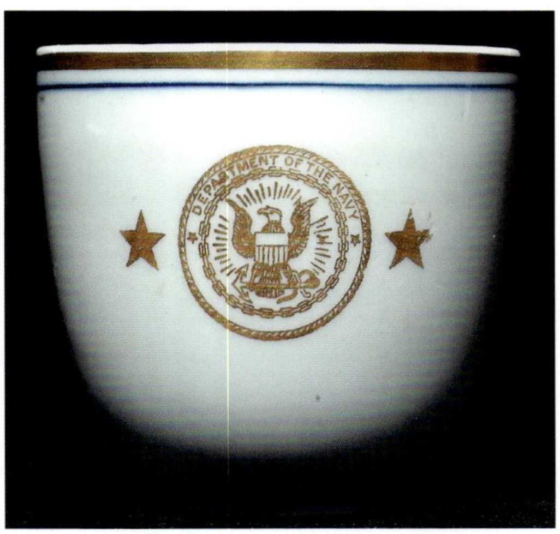

USN-9 **DN Two Star** ★★

White china top marked with a gold Department of the Navy seal and trimmed with a gold band and inner blue border stripe. A gold star on each side of the badge (two stars altogether) signifies this china's use by a rear admiral. This pattern likely preceded USN-12.

Mfr: SGO 1909, 1917, 1918.

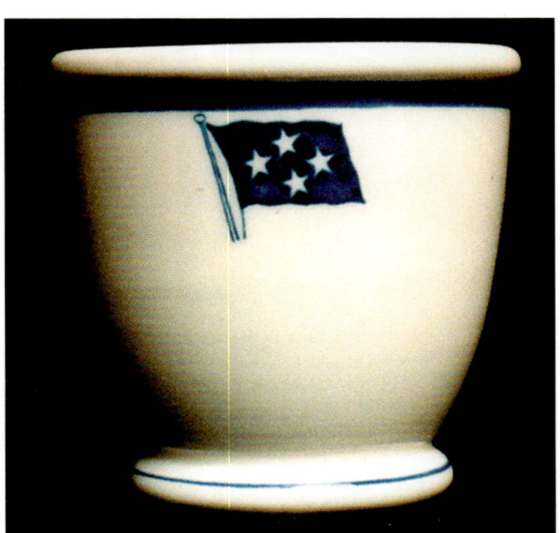

USN-10 **Fleet Admiral** ★

White china top marked with a blue flag adorned with four white stars and trimmed with a blue border stripe. A blue pinstripe trims the well of plates. This pattern was introduced in the early 1940s for the fleet admiral's mess.

Mfr: OPC 1941, SGO ca 1940s.

Ref.: Conroy USN-1.

USN-11 Vice Admiral ★
White china top marked with a blue flag adorned with three white stars arranged in a triangle and trimmed with a blue border stripe. A blue pinstripe trims the well of plates. This pattern was introduced in the early 1940s for the vice admiral's mess.
Mfr: TPC ca 1941.
Ref.: Conroy USN-2.

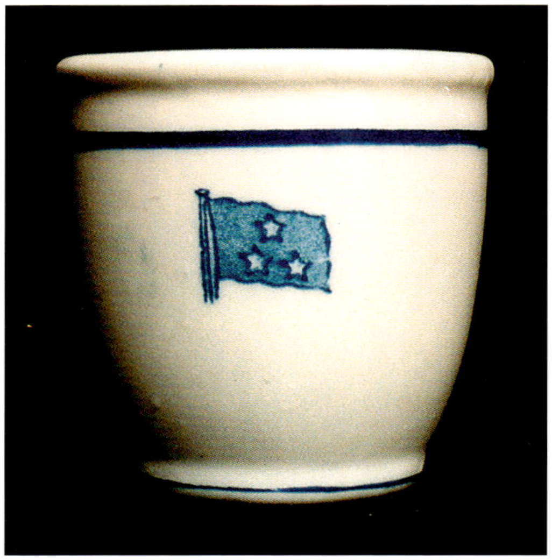

USN-12 Rear Admiral ★
White china top marked with a blue flag adorned with two white stars arranged one above the other and trimmed with a blue border stripe. A blue pinstripe trims the well of plates. This pattern was introduced in the early 1940s for the rear admiral's mess.
Mfr: TPC ca 1941, OPC 1942, SGO 1940s.
Ref.: Conroy USN-3.

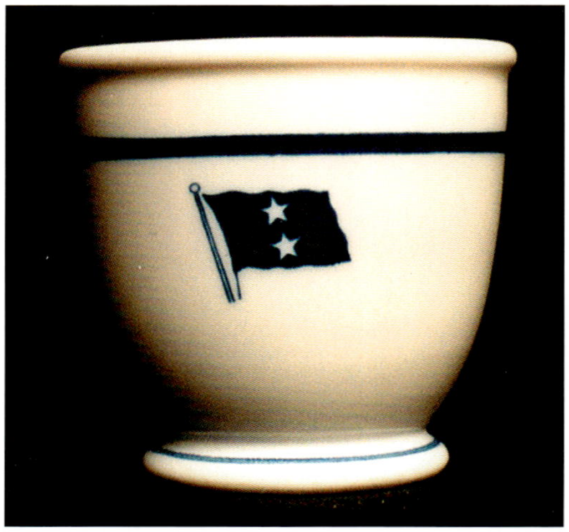

USN-13 Captain's Mess ★
White china top marked with a blue and white pennant adorned with four stars and trimmed with a blue border stripe. A blue pinstripe trims the well of plates. This pattern was introduced in the early 1940s and was used in the captain's mess.
Mfr: BUF ca 1940, SGO 1944, TPC ca 1940s, OPC 1942, HRL 1994.
Ref.: Conroy USN-4.

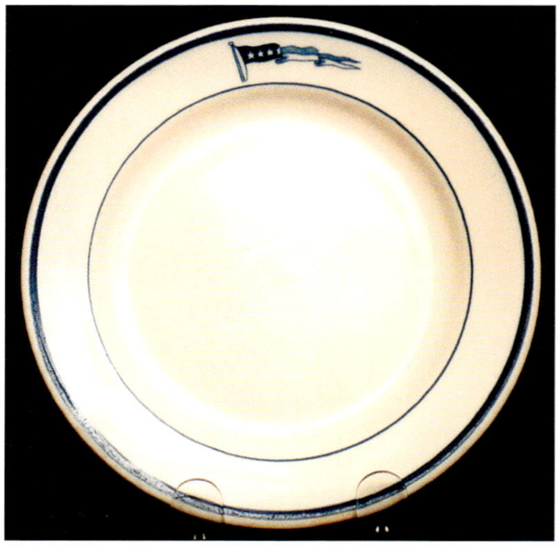

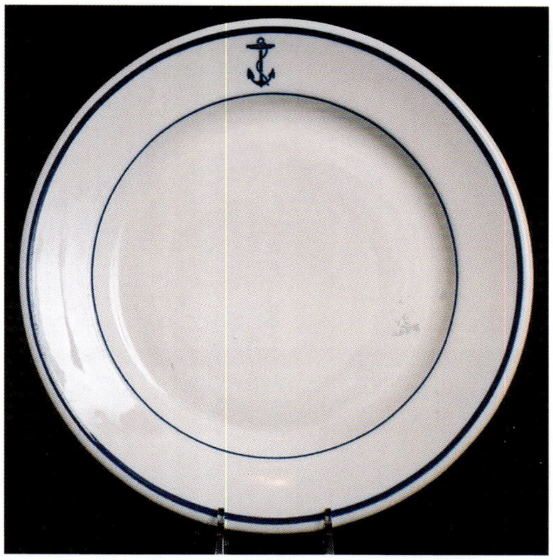

USN-14 Wardroom ★

White china top marked with a blue anchor and trimmed with a blue border stripe. A blue pinstripe trims the well of plates. This pattern was introduced in the early 1940s for use by officers in a ship's wardroom and most likely replaced the earlier Gold Anchor pattern (USN-6).
Mfr: numerous 1941–91.
Ref.: Conroy USN-5.

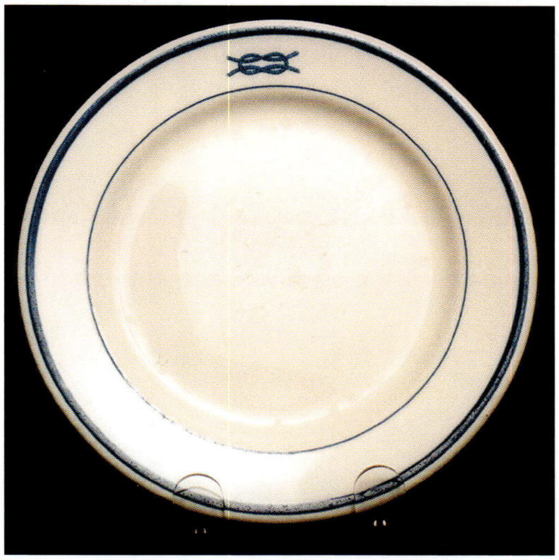

USN-15 Square Knot ★

White china top marked with a blue square knot and trimmed with a blue border stripe. A blue pinstripe trims the well of plates. This pattern was introduced in the early 1940s for use in the junior officers' mess.
Mfr: SGO ca 1940s, OPC 1941, MYR ca 1940–45.
Ref.: Conroy USN-6.

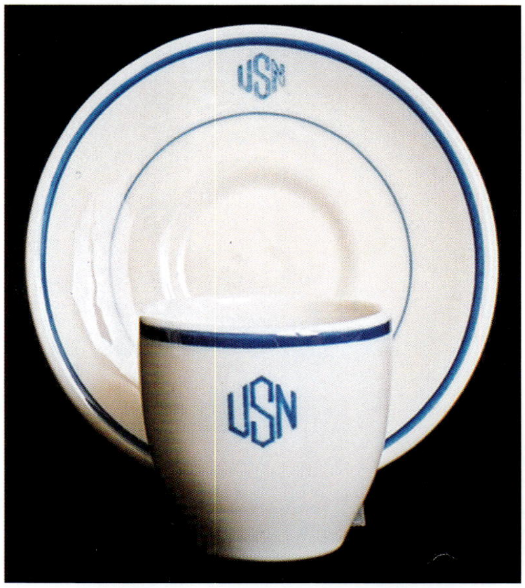

USN-16 Warrant Officer ★

White china top marked with "USN" and trimmed with a blue border stripe. A blue pinstripe trims the well of plates. This pattern was introduced in the early 1940s for use in the warrant officers' mess.
Mfr: TPC 1940s, SGO 1940s.
Ref.: Conroy USN-7.

USN-17 Chief Petty Officer ★★★

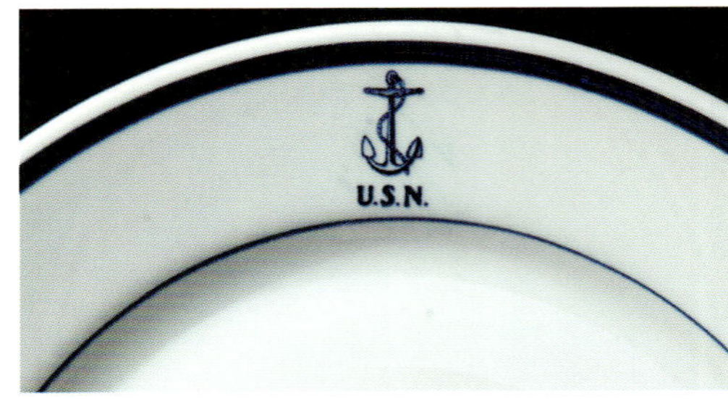

White china top marked with "USN" and a blue fouled anchor and trimmed with an outer blue border stripe. A blue pinstripe trims the well of plates. This pattern was introduced in the early 1940s and was used in the chief petty officer's mess.
Mfr: TPC 1940s.

USN-18 New Jersey ★★★

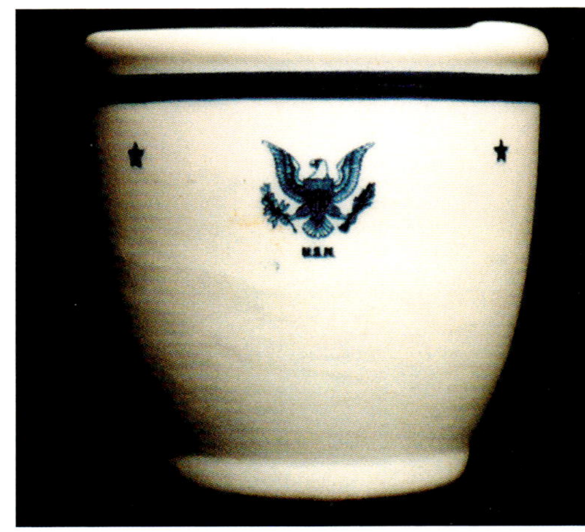

White china top marked with an eagle clutching an olive branch in one talon and a quiver of arrows in the other. "USA" appears beneath the eagle. The china is trimmed with a blue border stripe. This pattern is reported to have come from one of the World War II battleships, possibly the *New Jersey*, *Iowa*, *Ohio*, or another.
Mfr: TPC 1940s.

USN-19 Submariner ★★★

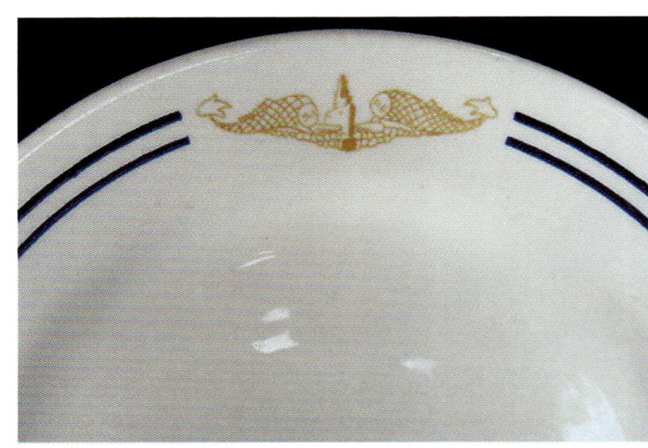

White china top marked with the bow of a submarine flanked on each side by a dolphin and trimmed with two blue side-by-side border stripes. This pattern is from the US Submarine Service.
Mfr: STR 1952.

USN-20 Enterprise ★

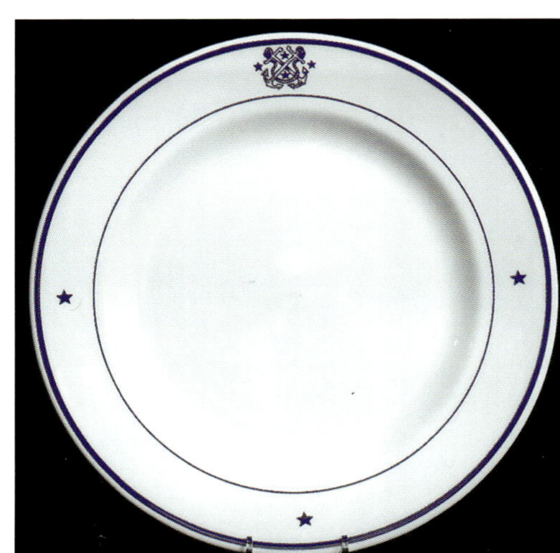

White china top marked with pair of crossed anchors connected by a rope and trimmed with a blue border stripe. Plates have a blue pinstripe around the well. Three blue stars decorate the border at positions south, east and west. This pattern is reported to have been produced for the aircraft carrier USS *Enterprise*, but this has not been verified.
Mfr: unknown.

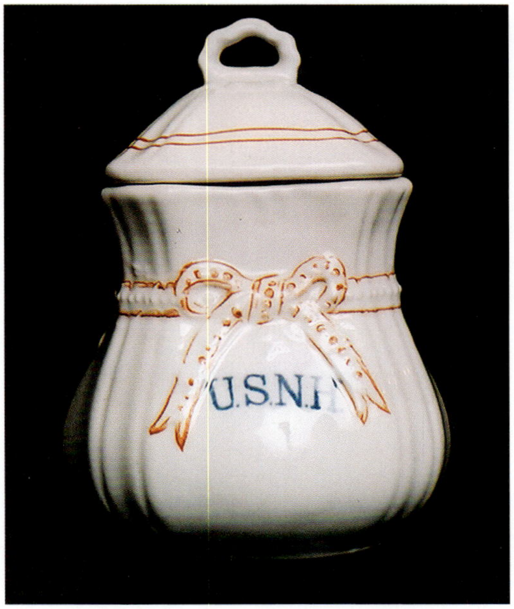

US Naval Hospital

Since 1898, the US Navy has operated hospital ships to evacuate sick and wounded personnel from combat areas and to provide humanitarian relief around the world. The US Navy Medical Department has also managed 22 naval hospitals worldwide since the 1860s.

USNH-1 Naval Hospital ★★★★★
White china top marked with blue "USNH" and trimmed with an orange bow and orange pinstripes. This is a very early piece and only one piece is known to exist.
Mfr: unknown.

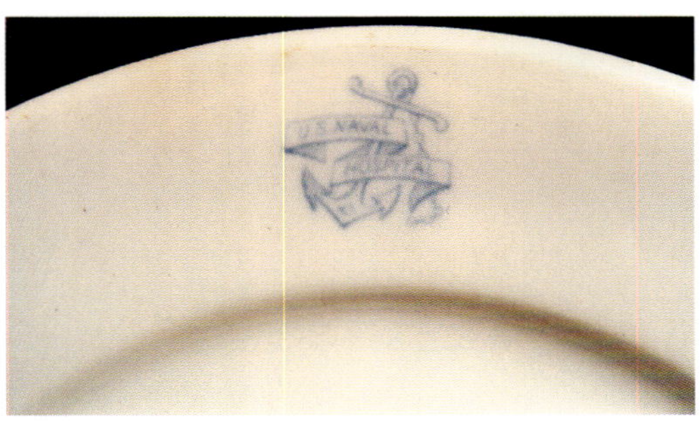

USNH-2 Fouled Anchor ★★★★★
White china top marked with a blue US Naval Hospital badge: a fouled anchor wrapped with a banner reading "US Naval Hospital".
Mfr: unknown.

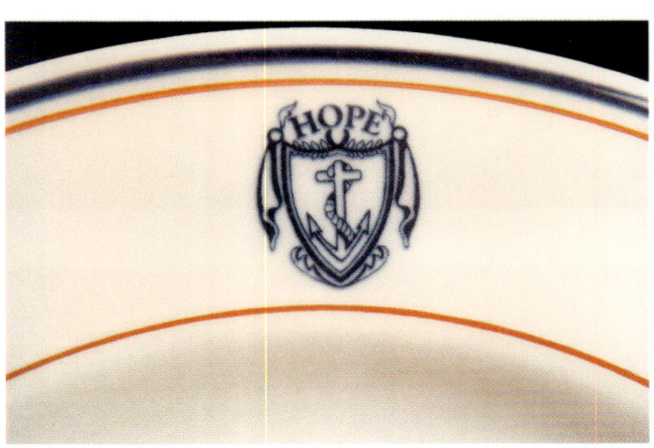

USNH-3 Hope ★★
White china top marked with a blue badge and trimmed with a blue border stripe flanked by an inner orange pinstripe. The badge consists of a blue shield containing a fouled anchor and the word "Hope" above it; white flags hang from each side of the shield. This china is from the US hospital ship *Hope,* which served in World War II. It also served as a civilian hospital ship from 1960 to 1973.
Mfr: SGO 1954–63.

US Revenue Cutter Service

USRC-1 Floral ★★★★

White china top marked with an orange USRCS logo and decorated with a wildflower transfer print. The logo consists of a double ring superimposed on two crossed anchors. Inside the ring is the badge of the US Treasury Department, and the ring text arched around it reads "US Revenue Cutter Service 1790".
Mfr: GWD 1894–1915.

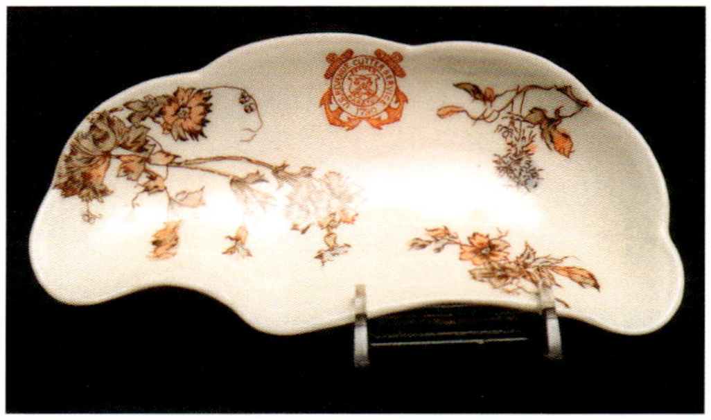

USRC-1a Floral ★★★★★

Same pattern as USRCS-1 except that the badge is green. This pattern was produced for the captain's mess.
Mfr: GWD 1894–1915.

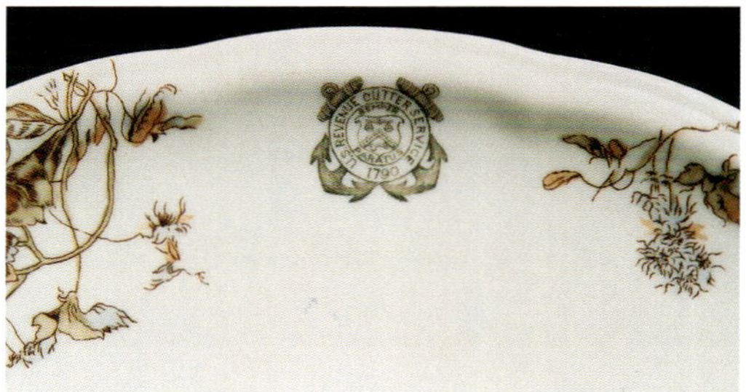

USRC-2 Orange Shield ★★★★

White china top marked with an orange chevron logo containing "USRCS" in script and trimmed with an orange border stripe flanked on each side by green pinstripes.
Mfr: LAM 1895–1911.

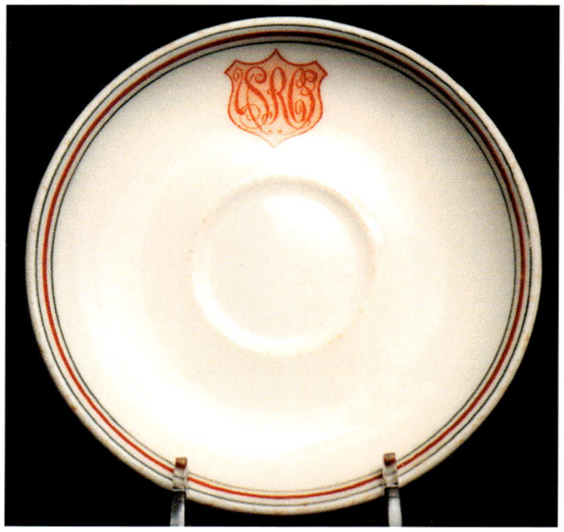

USRC-2a Green Shield ★★★★

Same pattern as USRCS-2 except that the badge is green and the colours of the trim stripes are reversed (green border stripe flanked by orange pinstripes). This pattern was produced for the captain's mess.
Mfr: LAM 1895–1911.

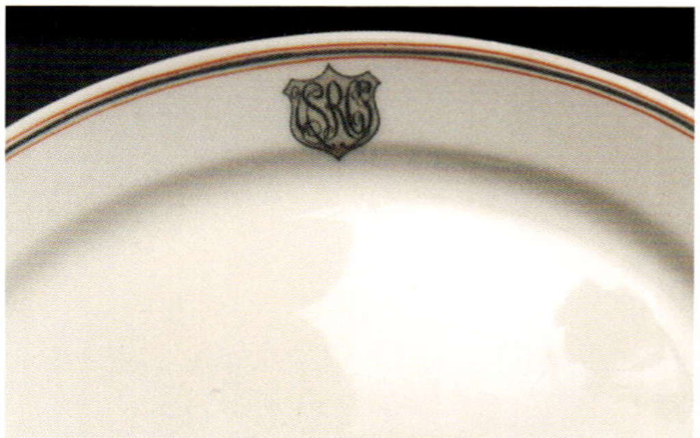

US Shipping Board

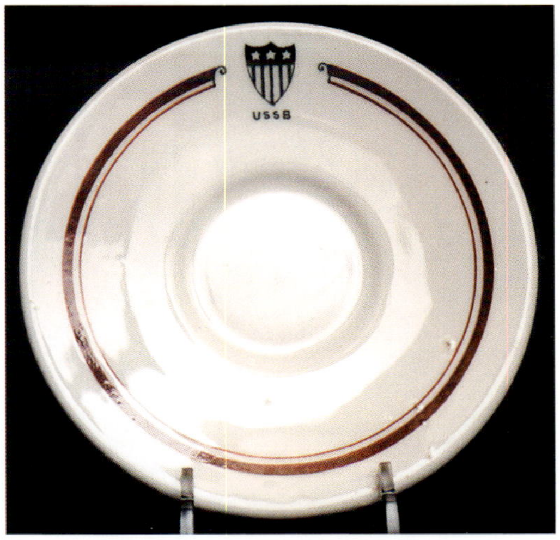

USSB-1 Centennial State ★★
White china top marked with a black USSB shield and trimmed with an orange border stripe and orange inner pinstripe.
Mfr: SGO 1921–23, BWK 1924.

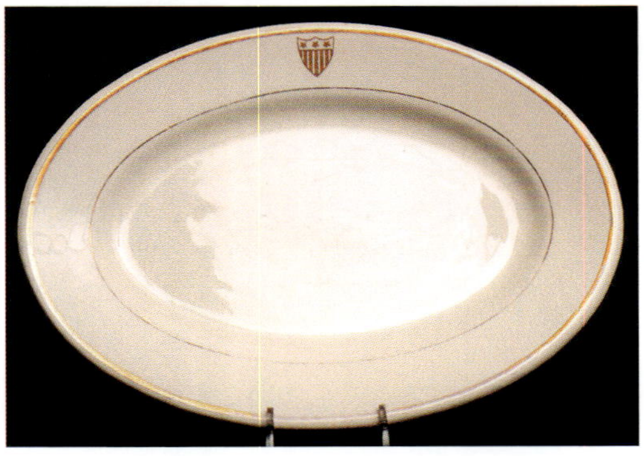

USSB-2 Granite State ★★
White china top marked with a gold USSB shield and trimmed with a gold edge stripe and a gold pinstripe around well of platters.
Mfr: BUF 1920–25.

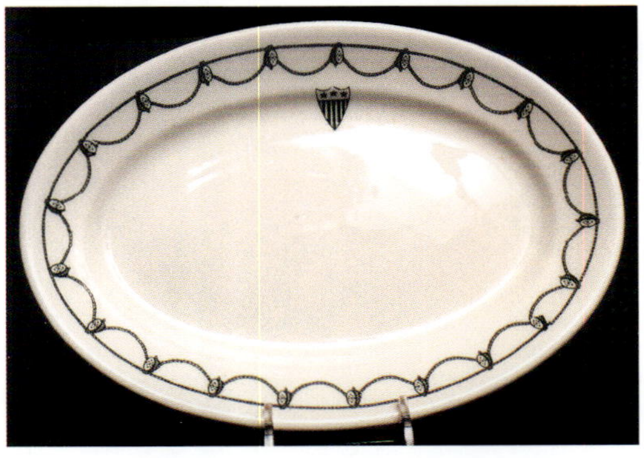

USSB-3 Hawkeye State ★★
White china top marked with a black USSB shield and trimmed with a black rope-and-pulley border trim.
Mfr: BUF 1924–25, JAC 1926–29.

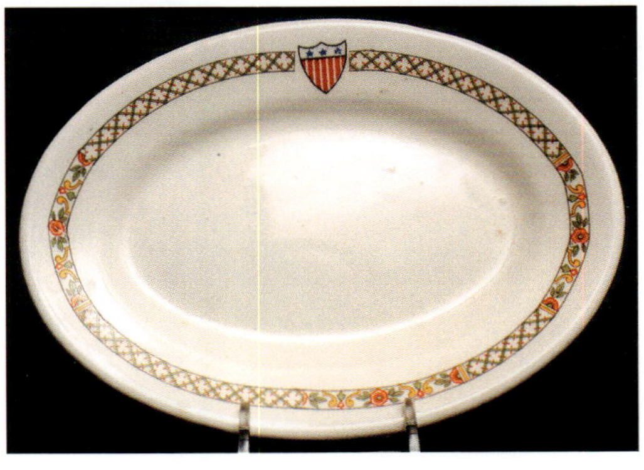

USSB-4 Red & Blue Shield ★★★★
White china top marked with a red-and-blue USSB shield and trimmed with a colourful floral and crisscross-design border band.
Mfr: BUF 1920.

USSB-5 **Black Shield** ★★★

White china top marked with a black Shipping Board shield and the "USSB" beneath it.
Mfr: BUF 1926–28, JAC 1928.

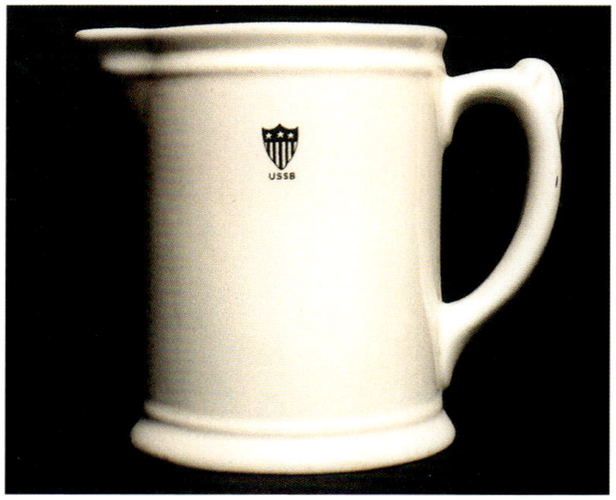

USSB-6 **Ring Logo** ★★★

White china top marked with a black USSB ring logo. The board's shield is inside the ring, and the ring text reads "United States Shipping Board".
Mfr: unknown.

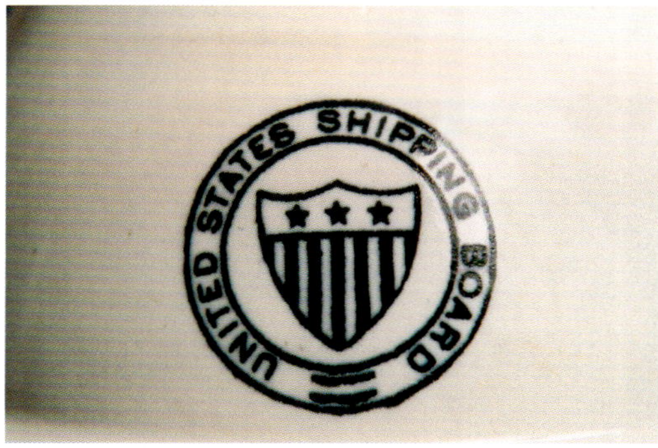

Pacific Coast Yacht Club Patterns

Corinthian Yacht Club

CYC-1 **Corinthian** ★★

White china top marked with a sailing ship, anchor and treasure chest. The Corinthian Yacht Club was established in 1886 at Tiburon, California.
Mfr: SYR 1979.
Ref.: Conroy vol. 2, p. 476.

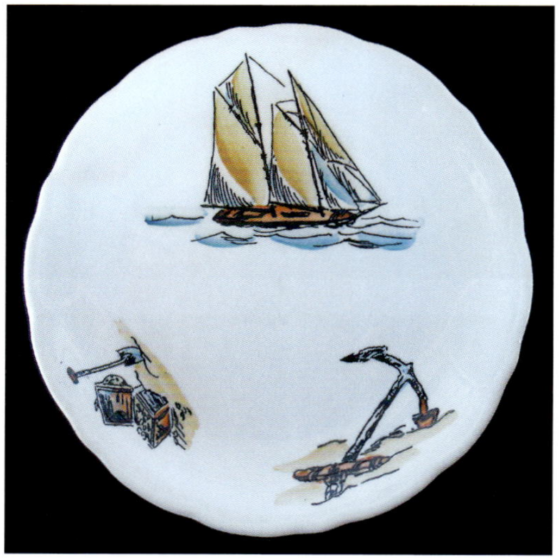

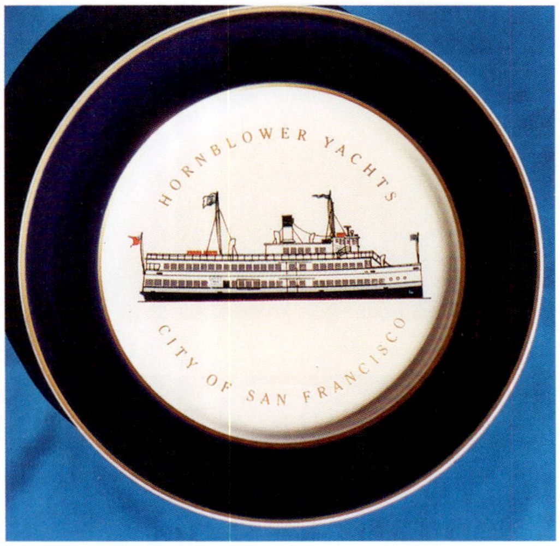

Hornblower Cruises and Events

HCE-1 **Hornblower** ★★

White china top marked with a drawing of a cruise ship in the centre and trimmed with a wide navy blue border and an outer gold pinstripe. Hornblower Cruises was established in 1980 by Terry MacRae and Michael Watson. They began operations with the single vessel, the *Admiral Hornblower*, a 75-person catamaran. The company currently operates 25 yachts located in six ports in California (Berkley, San Francisco, San Diego, Newport Beach, Marina del Rey and Los Angeles).
Mfr: Rego ca 1994.
Daniel Krummes photo.

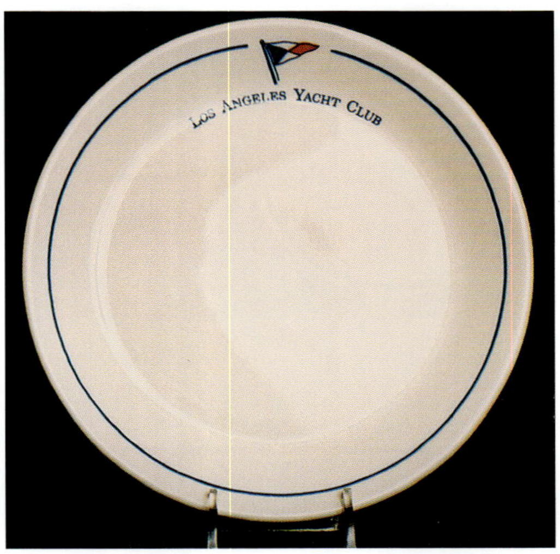

Los Angeles Yacht Club

LAYC-1 **Los Angeles** ★★

White china top marked with the club's burgee and trimmed with a blue border stripe. The Los Angeles Yacht Club was established in 1901 at San Pedro, California.
Mfr: unknown.

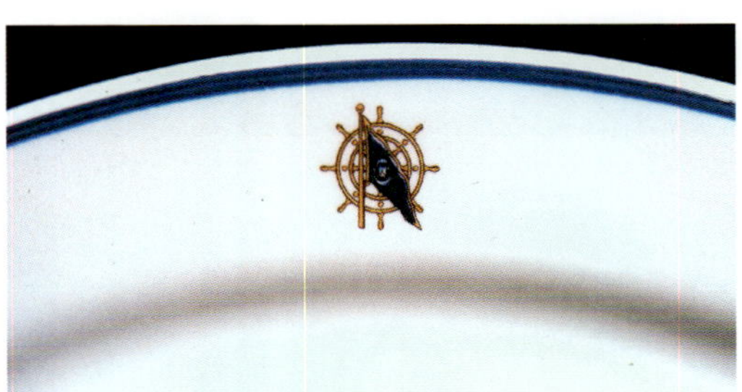

Newport Harbor Yacht Club

NHYC-1 **Newport** ★★

White china top marked with a brown ship's wheel superimposed by a dark blue flag with brown trim; the flag contains a white "N" in a circle. Plates have a fouled anchor on the bottom portion. The Newport Harbor Yacht Club was founded in 1916 at Balboa, California.
Mfr: SYR 1955–56, SGO 1979–82.
Ref: Krummes p. 110, Conroy vol. 2, p. 476.

Royal Vancouver Yacht Club

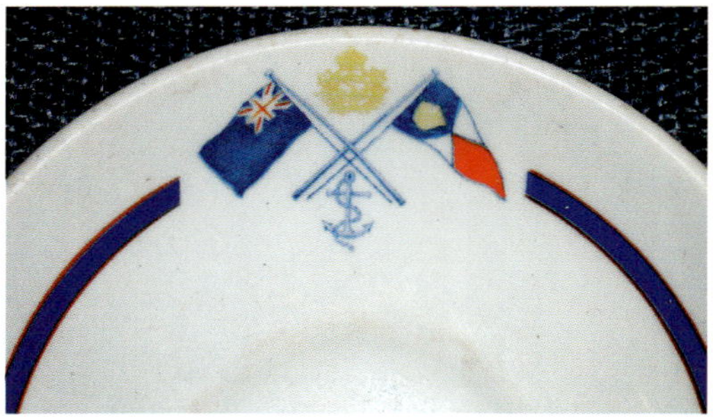

RVYC-1 Vancouver Old ★★
White china top marked with crossed flags, a blue ensign and the club's burgee. Above the flags is the RVYC crest and below is a blue fouled anchor. A blue border stripe trimmed with brown pinstripes rings the edge. The Royal Vancouver Yacht Club was founded in 1903 at Vancouver, BC.
Mfr: GLY 1950–69.

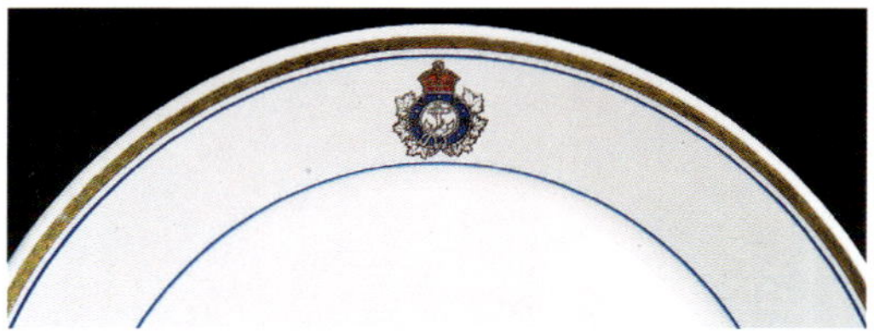

RVYC-2 Vancouver New ★★
White china top marked with a club crest in gold, blue and red, and trimmed with a gold border stripe flanked inside by two blue pinstripes.
Mfr: HTH 1978.

St Francis Yacht Club

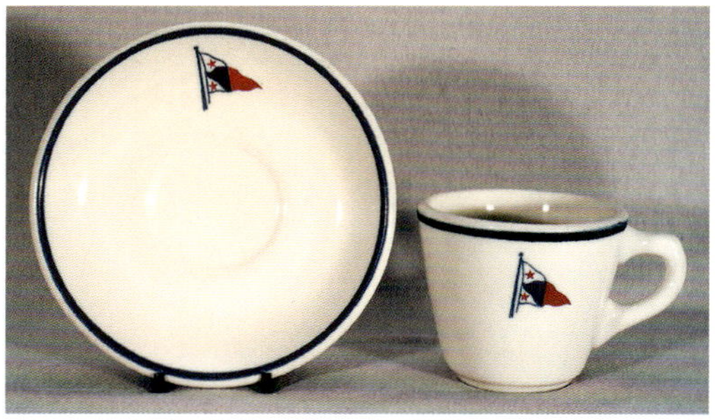

SFTY-1 St Francis ★★
White china top marked with the club's burgee and trimmed with a dark blue border stripe. The burgee is red and blue with white triangles in the upper corners. Within each triangle is a red five-point star. The St Francis Yacht Club was founded in 1927 and is located on Tinsley Island, on the Sacramento River delta, California.
Mfr: SYR 1968.
Ref.: Conroy vol. 2, p 476.
Barbara Conroy photo.

San Diego Yacht Club

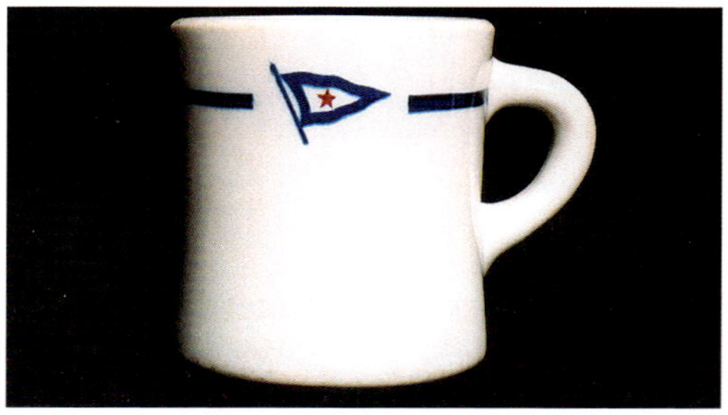

SDYC-1 San Diego ★★
White china top marked with the club's burgee and trimmed with a blue border stripe. The burgee is a blue pennant with a white centre that contains a red star. The San Diego Yacht Club was founded in 1886 in San Diego, California.
Mfr: JAC 1974, 1977.

San Francisco Yacht Club

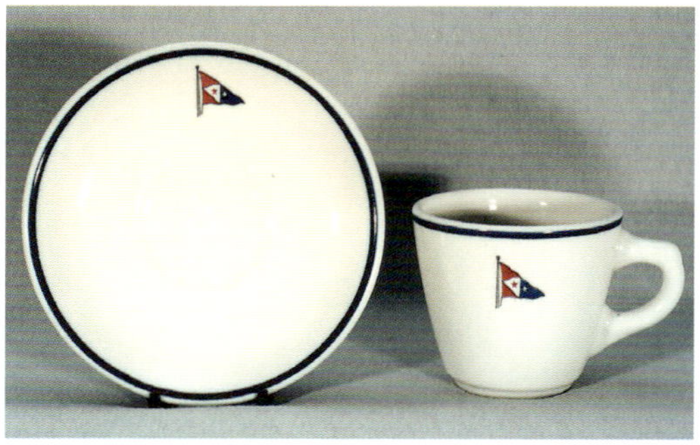

SFYC-1 San Francisco ★★
White china top marked with the club's burgee and trimmed with a blue pinstripe. The burgee consists of a red-and-blue pennant with a white triangle in the centre containing a red star. Founded in 1869, the San Francisco Yacht Club is the oldest yacht club on the west coast of North America. It is located in Belvedere, California.
Mfr: SYR 1963, 1977.
Ref.: Conroy vol. 2, p. 476.
Barbara Conroy photo.

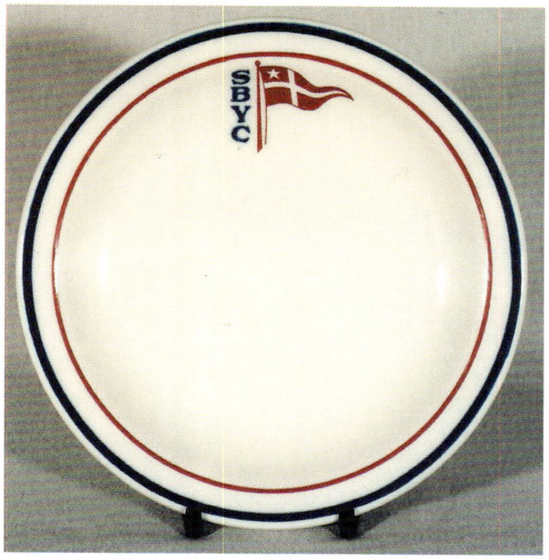

Santa Barbara Yacht Club

SBYC-1　　Santa Barbara　　★★

White china top marked with the club's burgee and trimmed with a dark blue border stripe and inner red stripe. The burgee is a red pennant with a white St George's cross and a white star in the upper right quadrant. The Santa Barbara Yacht Club was founded in 1872 in Santa Barbara, California.
Mfr: JAC 1979.
Ref.: Conroy vol 2, p. 214.
Barbara Conroy photo.

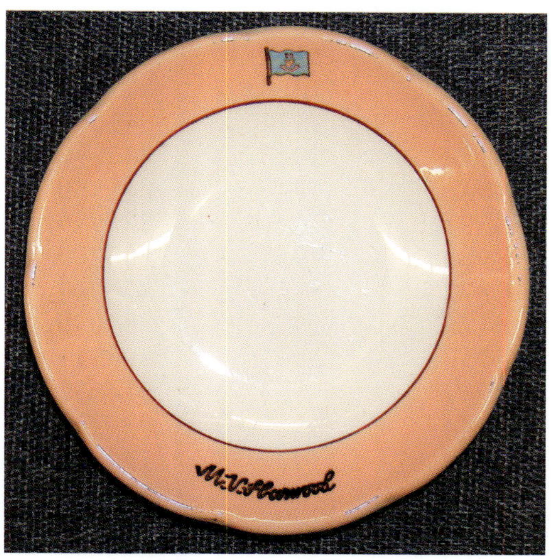

Other Yachts

MV *Harwood*　　★★★

White china top marked with a rose-coloured border. On the upper part of the border is a shipping company house flag and on the lower part "MV Harwood". The *Harwood* was built in Vancouver, BC, in 1943. She was a 107-foot (32.6-metre) wooden yacht registered to Albert L. McLennan of Vancouver, BC.
Mfr: MED 1953.

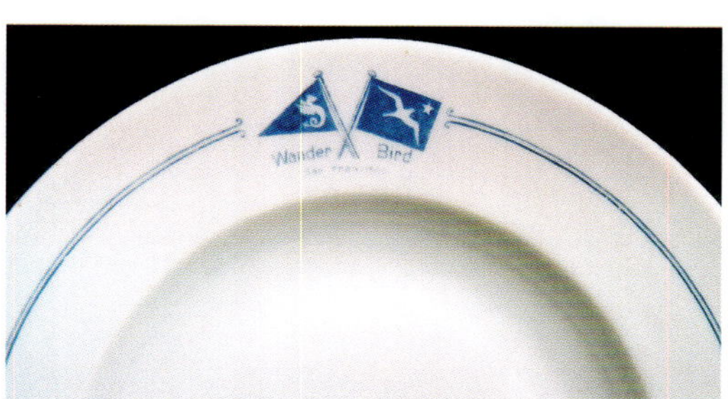

***Wander Bird*　　★★★**

White china top marked with two crossed flags, one the Sausalito Yacht club flag and the other most likely that of the *Wander Bird*. The *Wander Bird* was originally a German pilot schooner built in Hamburg in 1883. She was retired in the 1920s and purchased by journalist Warwick Thompkins, who moved her to the San Francisco Bay area in the 1930s, and operated her out of Sausalito. The vessel was laid up in the 1940s. Harold Sommer of Crowley Marine purchased *Wander Bird* in 1968 and completely rebuilt her. Today, the yacht is operated by a non-profit foundation in Seattle as a training vessel.
Mfr: J H Kleeberg ca 1930s.

Appendix 1. British Registration Marks

From 1842 to 1883 British ceramics manufacturers labelled their wares with a diamond-shaped registration mark. The purpose of the mark was to indicate that the design or shape had been registered with the London patent office and was protected for a period of three years against use by other manufacturers. The diamond could be printed or impressed on the china. Printed marks generally relate to the decoration and impressed marks to the shape of the object.

This registration system proved cumbersome and was replaced in 1884 with a simple numerical system. As most steamship china in this book post dates 1900, I have not included the interpretation of the diamond symbol. You can find information on the diamond symbology in Geoffrey Godden's book, *Encyclopedia of British Pottery and Porcelain Marks.*

From January 1, 1884, registered designs were numbered consecutively and these numbers appear on china preceded by "Rd." or "Rd. No.". The following table provides the time period covered by each registration number. The number opposite the year is the approximate first number issued for the new calendar year. A slight overlap may occur between the end of one year and the beginning of the next.

English Registration Marks

Year	Rd. No.	Year	Rd. No.
Jan 1884	**000001**	1891	**163767**
1885	**019754**	1892	**185713**
1886	**040480**	1893	**205240**
1887	**064520**	1894	**224720**
1888	**090483**	1895	**246975**
1889	**116648**	1896	**268392**
1890	**141273**	1897	**291241**

Year	Rd. No.	Year	Rd. No.
1898	**311658**	1929	**742725**
1899	**331707**	1930	**751160**
1900	**351202**	1931	**760583**
1901	**368154**	1932	**769670**
1902	**385500**	1933	**779292**
1903	**403200**	1934	**789019**
1904	**424400**	1935	**799097**
1905	**447800**	1936	**808794**
1906	**471860**	1937	**817293**
1907	**493900**	1938	**825231**
1908	**518640**	1939	**832610**
1909	**535170**	1940	**837520**
1910	**548920**	1941	**838590**
1911	**575817**	1942	**839240**
1912	**594195**	1943	**839980**
1913	**612431**	1944	**841040**
1914	**630190**	1945	**842670**
1915	**644935**	1946	**845550**
1916	**653521**	1949	**856999**
1917	**658988**	1952	**866280**
1918	**662872**	1955	**876067**
1919	**666128**	1958	**887079**
1920	**673750**	1961	**899914**
1921	**680147**	1964	**914536**
1922	**687144**	1967	**929335**
1923	**694999**	1970	**944932**
1924	**702671**	1973	**960708**
1925	**710165**	1976	**973838**
1926	**718057**	1979	**987910**
1927	**726330**	1982	**1005700**
1928	**734370**		

Examples of registration numbers used on Canadian Pacific Railway patterns: "Rd. No. 847730" (Foliage), "Rd. No. 878088" (Halifax) and "Rd. No. 774878" (Empress).

Appendix 2. China Manufacturers

Numerous manufacturers made china for west-coast shipping companies. The following list includes 94 manufacturers – some are still operating while others vanished decades ago. "Years of Operation" is the company's complete time span, including any predecessor that produced earthenware or bone china.

Many companies made subtle changes to their names, so the name in parentheses is an alternative name that a company may have used for marketing its china. The manufacturer's code appears in the pattern description guide.

Code	Manufacturer	Location	Years of Operation
ACP	Acropal	France	1990s
AWB	Ashworth Brothers	Hanley, England	1862–68
BAK	Bakewell Brothers	Hanley, England	1927–43
BRL	Burgess & Leigh Ltd	Burslem, England	1862–1999
BSH	Bauscher-Weiden	Weiden, Germany	1881–present
BST	Bishop & Stonier	Hanley, England	1891–1939
BTH	Booths Ltd	Turnstall, England	1891–1948
BUF	Buffalo	Buffalo, New York	1901–present
BWK	Bailey-Walker	Bedford Heights, Ohio	1923–41
CAF	Charles Ahrenfeldt	Limoges, France	1894–1969
CAR	Carr China Co.	Grafton, West Virginia	1916–52
CLD	Cauldon Potteries Ltd	Stoke-on-Trent, England	1905–20
CLD	Cauldon Potteries Ltd	Hanley, England	1920–62
COK	Cook China	Trenton, New Jersey	1894–1930
COP	Copeland-Spode	Stoke-on-Trent, England	1847–present
COR	Corning Glass Works	Corning, New York	1875–1989
CRB	Caribe China Co.	Vega Baja, Puerto Rico	1951–76
CRN	Crown Staffordshire	Fenton, England	1889–1985
CRD	Crown Ducal (AGR)	Cobridge, England	1925–60s
DNS	Deans (1910) Ltd	Burslem, England	1910–19
DUD	Dudson Brothers Ltd	Hanley, England	1898–present
DBN	Dunn Bennett & Co.	Burslem, England	1875–1983
FOL	Foley (E. Brain & Co)	Fenton, England	1903–63
FUR	Furnivals Ltd	Cobridge, England	1890–1968
GAM	Gamecho & Errandonea	Spain	1932–present
GDA	GDA Limoges	Limoges, France	1868–present
GCC	George Clews & Co.	Tunstall, England	1906–61
GIN	Richard Ginori	Milano, Italy	1896–present
GJS	George Jones & Sons	Stoke-on-Trent, England	1864–1957
GWS	W.S. George Pottery	East Palestine, Ohio	1909–60
GLD	Gladstone	Longton, England	1939–present
GLB	Globe Pottery Co. Ltd	Shelton, England	1914–57
GRM	Grimwades Ltd	Stoke-on-Trent, England	1900–1995
GWD	Greenwood Pottery	Trenton, New Jersey	1868–1933
GLY	Grindley Hotelware Co.	Tunstall, England	1908–79
HAL	Hall China Co.	East Liverpool, Ohio	1903–present
HAV	Theodore Haviland & Co.	Limoges, France	1892–1941
HCO	Haviland & Co	Limoges, France	1864–1930
HRL	Homer Laughlin	Newell, West Virginia	1873–present
HTH	J.E. Heath Ltd	Burslem, England	1950–80
IMP	Imperial China	Trenton, New Jersey	1891–1920s
INA	Inter American China	Long Beach, California	1977–92

Code	Manufacturer	Location	Years of Operation
IRQ	Iroquois (Warwick)	Syracuse, New York	1905–69
JAC	Jackson China	Falls Creek, Pennsylvania	1917–85
JOH	Johnson Brothers	Hanley, England	1883–present
KTK	Knowles, Taylor & Knowles	East Liverpool, Ohio	1870–1931
LAM	Lamberton (Maddock)	Trenton, New Jersey	1892–1924
LEX	Lenox, Inc	Lawrenceville, New Jersey	1906–present
LVT	Lovatt & Lovatt	Nottingham, England	1895–1931
McN	McNichol China Co.	Clarksburg, West Virginia	1892–1969
MAD	Maddock Pottery Co.	Trenton, New Jersey	1892–1924
MAL	C.T. Maling & Sons	Newcastle, England	1890–1963
MDK	John Maddock & Sons Ltd	Burslem, England	1842–1980
MAS	Maastricht China Co.	Maastricht, Holland	1834–present
MYR	Mayer China Co.	Beaver Falls, Pennsylvania	1881–1984
MEK	Alfred Meakin	Tunstall, England	1875–1976
MKN	J & G Meakin	Hanley, England	1851–1970
MED	Medalta	Medicine Hat, Alberta	1916–66
MIN	Mintons	Stoke-on-Trent, England	1793–present
NPT	Newport Pottery Co.	Burslem, England	1920–87
NSF	North Staffordshire Pottery	Cobridge, England	1940–52
OHO	Ohio Pottery Co.	Zanesville, Ohio	1900–1923
OPC	Onondaga Pottery Co.	Syracuse, New York	1871–1966
RFD	Radford (Samuel)	Fenton, England	1879–1957
RID	Ridgway's	Shelton, England	1879–1952
RLD	Royal Doulton	Burslem, England	1854–present
RGT	Royal Grafton (A.B. Jones)	Longton, England	1959–80s
RSW	Royal Schwarzburg	(Nathan Straus) Germany	1906–32
RLW	Royal Worcester	Worcester, England	1862–present
SCM	Scammell (Lamberton)	Trenton, New Jersey	1924–54
SGO	Shenango	New Castle, Pennsylvania	1901–90
SIM	Simpsons Potters Ltd	Cobridge, England	1944–2004
SND	Sneddon & Sons	Glasgow, Scotland	1911
SOV	Sovereign	Hamilton, Ontario	1933–73
SPD	Spode	Stoke-on-Trent, England	1770–present
STH	Sutherland (Hudson & Middleton)	Longton, England	1941–60s
SPL	Stanley Pottery Ltd	Longton, England	1928–31
STR	Sterling	East Liverpool, Ohio	1917–present
SYR	Syracuse China	Syracuse, New York	1966–present
SYC	Syracuse China	St Joliette, Quebec	1970–95
TPC	Tepco China Co.	El Cerrito, California	1918–68
TRN	Trenton Potteries Ltd	Trenton, New Jersey	1892–1969
TSK	Tuscan (Royal)	Longton, England	1933–49
UPW	Union Porcelain Works	Green Point, New York	1861–1920s
VAN	Vandesca-Syracuse Ltd	St Joliette, Quebec	1959–70
WAL	Wallace	Huntington Park, California	1931–64
WAR	Warwick	Wheeling, West Virginia	1887–1951
WEG	Wedgewood	Etruria, England	1769–1950
WEL	Wellsville China Co.	Wellsville, Ohio	1902–69
WGC	Guerin-Pouyat-Elite	Limoges, France	1841–1932
WHG	W.H. Grindley	Turnstall, England	1880–1987
WKR	Walker China	Bedford Heights, Ohio	1942–81
WLK	Wilkinson Ltd (Arthur J.)	Burslem, England	1885–present
WMG	William Guerin Co. (WG & Co.)	Limoges, France	1870–1932
WOD	Wood & Sons	Burslem, England	1865–1954

Important Ship China Manufacturers

John Maddock & Sons (Ltd), Burslem, England, 1855–1983

John Maddock and Sons began in 1830 as Maddock and Edwards in a premise on Newcastle Street, Burslem (now Stoke-on-Trent). From 1839 to 1842 John Maddock partnered with J. Seddon to make graniteware for the American market. In 1842, the firm became known as John Maddock, and evolved into John Maddock and Sons (Ltd) in 1855, when his sons joined the business. Over its history the company produced a variety of ironstone and vitrified earthenware products. John Maddock and Sons was the most prolific supplier of Pacific-coast steamship china at the turn of the 20th century. John Maddock and Sons were bought out in the late 1970s and ceased to exist in 1983.

Representative Back Stamps

Back stamps found on Pacific-coast ship china, beginning approximately in the years shown.

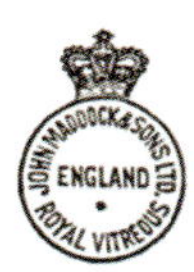

1896

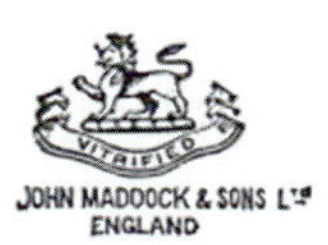

1896

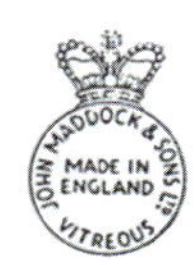

1945

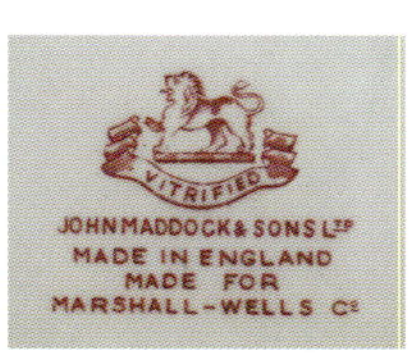

1945

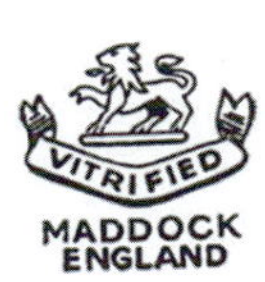

1955

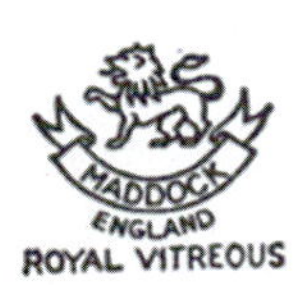

1961

Date Codes

Impressed in the body of Maddock china, date markings appear usually appear as Roman numerals, except in the late 1920s when actual date codes were inscribed. The first initials identify the king who reigned at the time the china was made. "EVIII" stands for Edward VIII (1901–10), "GV" for George V (1911–36) and "GVII" for George VII, 1937–52. The roman numerals indicate in which year of the monarch's reign the china was made.

Example: "GV – XX" means that the china was made in the 20th year of George V's reign, which was 1930.

In the late 1920s and again in the 1960s, Maddock china was occasionally marked with a month-year code; e.g. 7/30, meaning July 1930.

Known Shipping Company and Government Customers (19):

Alaska Commercial Co., Alaska Steamship, BC Salvage Co., California Navigation & Improvement Co., California Transportation Co., Canadian Coast Guard, Canadian National Steamships, Canadian Pacific, China Mail, Grand Trunk Pacific, Matson Navigation, Northwestern Steamship, Oceanic Steamship, Pacific Coast Steamship, Pacific Mail Steamship, Pacific Navigation, Red Collar Line, San Francisco & Portland Steamship Co., Terminal Steamship, Union Steamship of BC.

Grindley Hotel Ware Co. Ltd, Tunstall, 1908–79

W.H. Grindley established the Grindley Hotel Ware Company in 1908 when he built the Globe Works at Tunstall (now Stoke-on-Trent), specifically to manufacture hotelware. Grindley Hotel Ware produced vitrified earthenware china. In 1952, Dudson Brothers purchased the company. The Tunstall plant was enlarged in the 1970s and continued manufacturing under the Grindley name until 1979, when Dudson changed the name to Duraline Hotelware Co. to better reflect its product line. In 1992, Dudson further consolidated its holdings under the banner Dudson's Duraline Division. The Tunstall plant is still producing china today.

Representative Back Stamps

Back stamps found on Pacific-coast ship china in the approximate years shown.

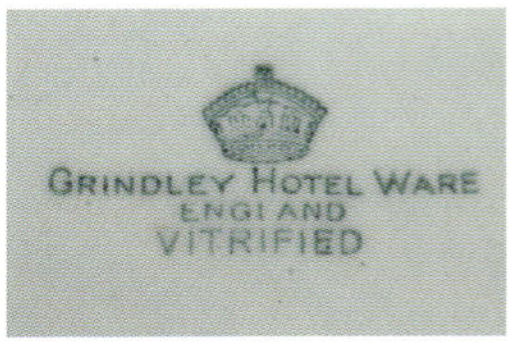

1908–20

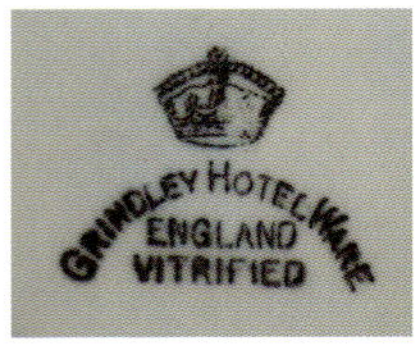

1922–36

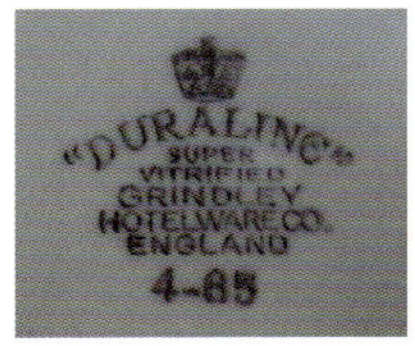

1936–78

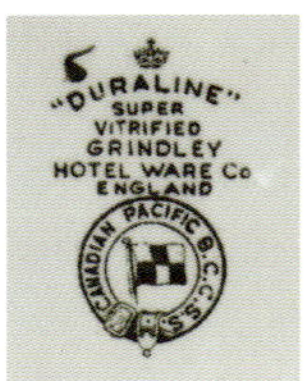

1948

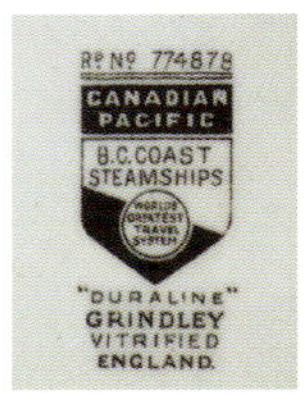

1948–57

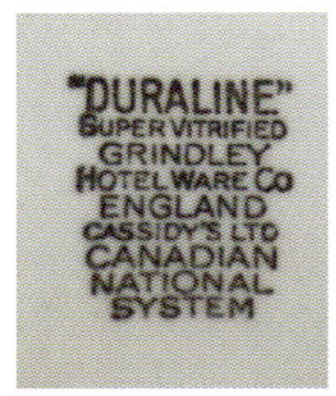

1938–47

Date Codes

Date markings on Grindley products appear both as impressed marks in the body of the china and as ink stamps. In both Grindley used a simple month-year code, which makes all pieces easy to date. Months are numbered 1 through 12, followed by a dash and the last two digits of the year. Example: "1–49" means January 1949.

Known Shipping Company and Government Customers (9):

Alaska Steamship, Canadian Coast Guard, Canadian National Steamships, Canadian Pacific, Department of Marine & Fisheries, Imperial Oil, Inter-Island Steamship, Pacific Steamship Company, Union Steamships.

Wood & Sons Ltd, Burslem, Stoke-on Trent, England 1865–2005

Thomas F. Wood and his father, Absalon Wood, founded Wood & Son in 1865 at Burslem (now Stoke-on-Trent). In 1907 it became Wood & Sons, and three years later "Ltd" was added. Wood & Sons started with the Trent Pottery, added the New Wharf Pottery and later the Stanley Potteries. This firm produced vitrified and non-vitrified china for the commercial market.

Representative Back Stamps

Back stamps found on Pacific-coast ship china in the approximate years shown.

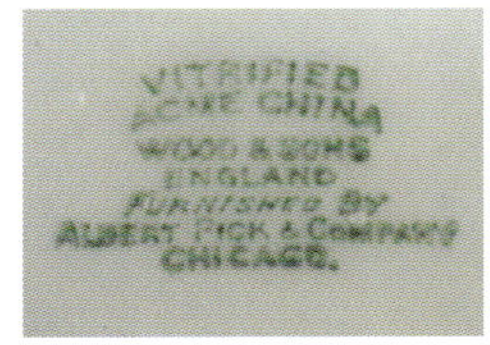

1907–11

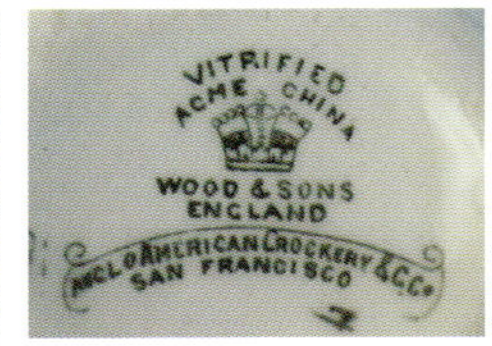

1907–11

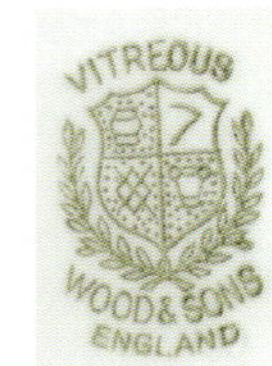

1910–20

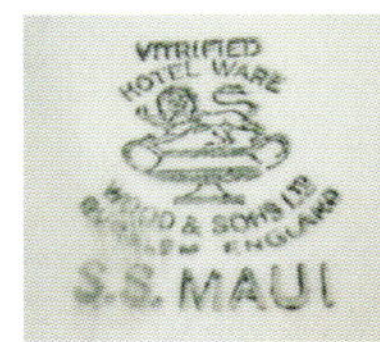

1920s

1942–45

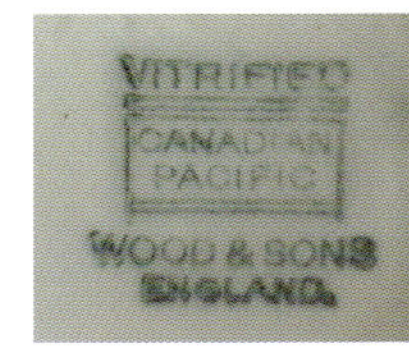

1942–45

Date Code

Wood & Sons china usually has an impressed three- or four-digit numerical date code. The first one or two digits represent the month and the last two the year, with usually a gap separating the month and year codes. Example: "11 43" means November 1943.

Known Shipping Company and Government Customers (8):

Canadian National Steamships, Canadian Pacific (BCCSS), California Transportation Company, Matson Navigation Co., Monticello Steamship Company, North Pacific Steamship Company, Pacific Coast Steamship, Oregon Railway & Navigation Company.

Ridgway's, Hanley, Stoke-on Trent, England, 1879–1983

Ridgway's began in 1835 when Edward Ridgway and Leonard Abington formed Ridgway & Abington to manufacture earthenware at the Church Works in Hanley (now Stoke-on-Trent). The company became Ridgways in 1879 and Ridgways (Bedford Works) Ltd in 1920. After absorbing several potteries in the 1950s, it changed the name to Ridgway Potteries Ltd in 1955. Ridgway merged with Royal Doulton in 1972 and became the hotelware division, Ridgway Steelite in 1975.

Representative Back Stamps

Back stamps found on Pacific-coast ship china beginning or in approximately in the years shown.

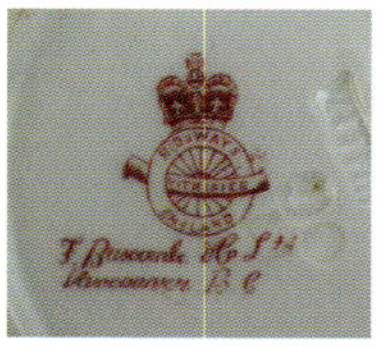

1905

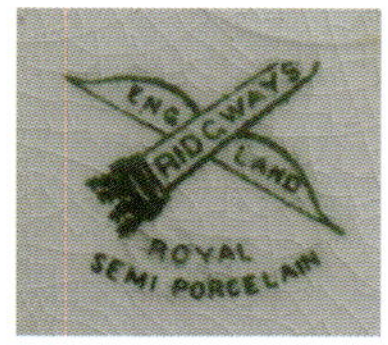

1912

1955–62

1955–?

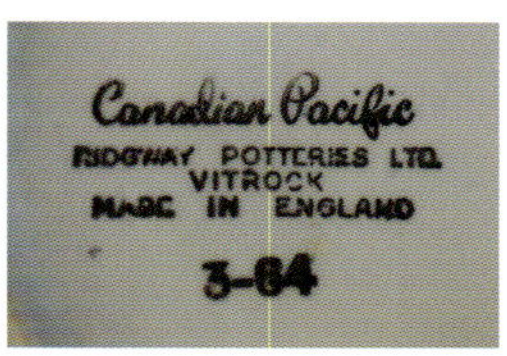

1960s–72

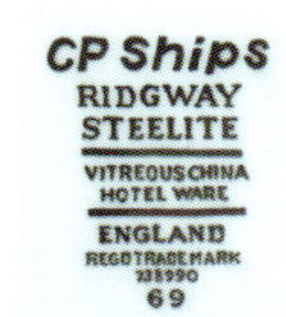

1968–72

The bow-and-quiver trademark was first registered in 1880: without any place mark ca 1880–85; with "Stoke-on-Trent" ca 1885–90; with "England" beginning ca 1891; and with "England" and "Royal Semi-Porcelain" ca 1912.

Date Codes

Some early Ridgway pieces (ca 1904–08) have an impressed date code consisting of four numbers. The first two appear to be the month and the last two the year. Example: "09 05" means September 1905.

During the 1960s and early 70s, Ridgway's trademark Vitrock china had an ink stamp. The first one or two digits represent the month and the second two identify the year. Example: "3-64" means February 1964.

Known Shipping Company and Government Customers (7):

Canadian National Steamships, Canadian Pacific (BCCSS), California Transportation Company, Department of Marine & Fisheries, North Pacific Steamship Company, Pacific Coast Steamship, Oregon Railway & Navigation Company.

Syracuse China Company, Syracuse, New York, 1871–present

The Syracuse China Company began in 1871 when a group of New York businessmen bought the struggling Empire Pottery on Fayette Street in Syracuse and renamed it the Onondaga Pottery Company (OPCo). Initially OPCo produced white graniteware, but by 1891 it had developed a full line of vitrified china. In 1924 the company built a new factory on Court Street specifically for the production of commercial china. From as early as 1879, OPCo incorporated the name Syracuse China in many of its back stamps. It was not until 1966 that the company name changed to Syracuse China Company to reflect the trade name it had been using for decades. Syracuse bought Vandesca Pottery of Canada in 1959, Mayer China in 1984 and Shenango China in 1987. Syracuse is still in business today producing commercial tableware.

Representative Back Stamps

Back stamps found on Pacific-coast ship china in the approximate years shown.

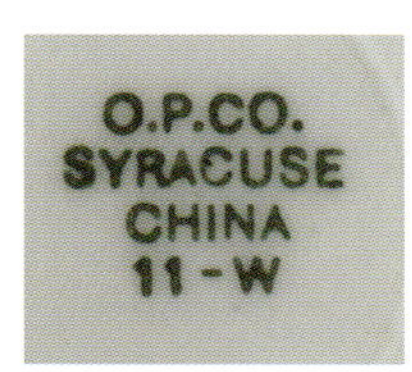

1897–1946

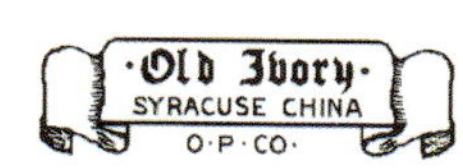

1827–50s

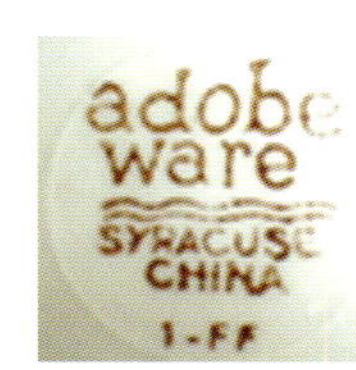

1932–72

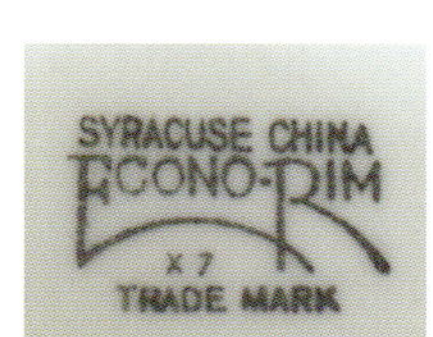

1933–67

1965–74

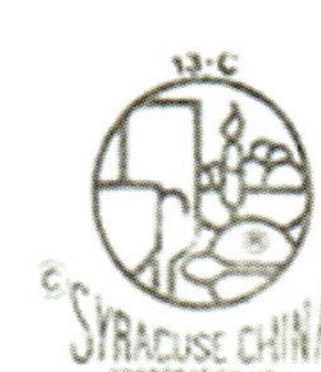

1982–90

Syracuse China Date Codes

Year	Code
1895	1-8
1896	9-20
1897	21-32
1898	33-44
1899	45-56
1900	57-68
1901	69-80
1902	81-92
1903	93-3★
1904	4-15★
1905	16-27★
1906	28-39★
1907	40-51★
1908	52-63★
1909	64-75★
1910	76-87★
1911	88-99★
1912	1-12#
1913	13-24#
1914	25-36#

Year	Code
1915	37-48#
1916	49-60#
1917	60-72#
1918	73-84#
1919	85-90#
1920	A
1921	B
1922	C
1923	D
1924	E
1925	F
1926	G
1927	H
1928	I
1929	J
1930	K
1931	L
1932	M
1933	N
1934	O

Year	Code
1935	P
1936	Q
1937	R
1938	S
1939	T
1940	U
1941	V
1942	W
1943	X
1944	Y
1945	Z
1946	AA
1947	BB
1948	CC
1949	DD
1950	EE
1951	FF
1952	GG
1953	HH
1954	II

Year	Code
1955	JJ
1956	KK
1957	LL
1958	MM
1959	NN
1960	OO
1961	90
1962	91
1963	92
1964	93
1965	94
1966	95
1967	96
1968	97
1969	98
1970	99

Year	Code
1971	100
1972	101
1973	2
1974	3
1975	4
1976	5
1977	6
1978	7
1979	8
1980	9
1981	10
1982	11
1983	12
1984	13
1985	14
1986	15

Year	Code
1987	16
1988	17
1989	18
1990	19
1991	20
1992	21
1993	22
1994	23
1995	24
1996	25
1997	26
1998	27
1999	28
2000	29
2001	30
2002	31

★ Number inside an impressed circle.
Number inside an impressed diamond.

Date Codes

1903–11	Impressed circle with number inside.
1912–19	Impressed diamond with number inside.
1920–45	Ink stamp containing letter: A (1920) to Z (1945).
1946–60	Ink stamp containing double letters: AA (1946) to OO (1960).
1961-1973	Ink stamp of two or three digits to identify the year: 90 (1961) to 101 (1972).
1973–82	Ink stamp of a number for the year, followed by a letter for the month: 2 (1973) to 11 (1982); A (January) to L (December).
1983–	Ink stamp of a number for the year (continuing on from the previous sequence) followed by a dash and a letter representing the quarters of the year separated. Example: "12–B" means the second quarter of 1983.

Known Pacific-Coast Customers (19):
Alaska Steamship, Dollar Line, Great Northern Steamship, Great Northern Pacific Steamship, Inland Navigation, Matson Navigation, Monticello Steamship, Oregon Railway and Navigation, Pacific Coast Steamship, Pacific Far East Line, Pacific Mail Steamship, Panama Mail Steamship, Puget Sound Navigation Company, Regulator Line, Union Oil of CA, Western Steam Navigation Co., Royal Canadian Navy, United States Army Transport, United States Navy.

Shenango China Company, New Castle, Pennsylvania, 1901–90

The Shenango China Company was incorporated in New Castle, Pennsylvania, in 1901.

It began making semi-porcelain dinner sets and toilet sets, and by 1909 was producing durable hotelware products. In 1912, Shenango purchased the New Castle Pottery Company and moved its operation to the new plant. From 1936 to 1958, it made china for the Theodore Haviland Company of France. Shenango bought controlling interest in Castleton China in 1951 and bought Wallace China of California in 1959. In June 1979, the Shenango China Company became the property of the Anchor Hocking Corporation. In 1987 it was purchased by Syracuse China's parent company, CPI.

Representative Back Stamps

Back stamps found on Pacific-coast ship china in the years shown.

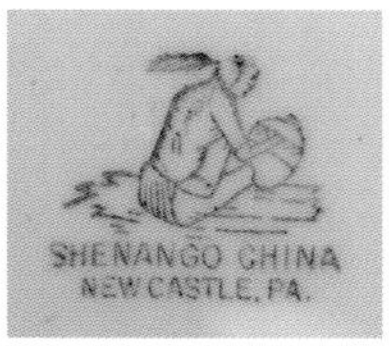

1912–20s

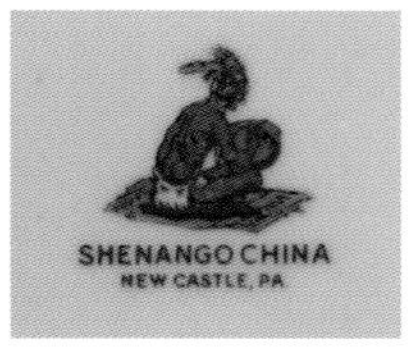

1920s–50s

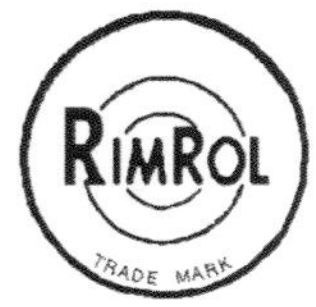

1939–50s

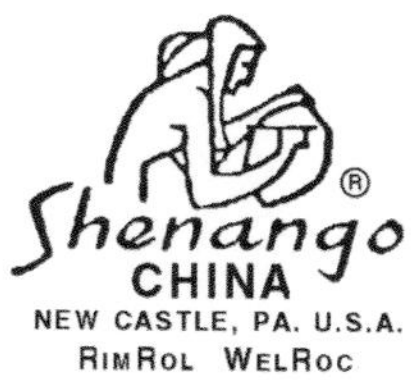

ca 1954–63

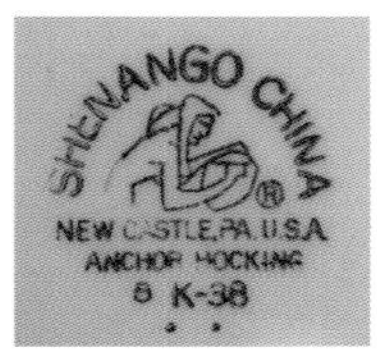

1979–87

Date Code

Early Shenango china rarely has a date code. During the 1920s the company occasionally marked china with a month-and-year date code: the first two digits represent the month of production and the second two the year.

Interpreting Shenango Date Codes After 1950

Year	Jan-Jun	Jul-Dec
1950	M-8	A-8
1951	N-9	B-9
1952	O-10	C-10
1953	P-11	D-11
1954	Q-12	E-12
1955	R-13	F-13
1956	T-14	G-14
1957	U-15	H-15
1958	V-16	I-16
1959	W-17	J-17
1960	X-18	K-18
1961	Y-19	L-19
1962	Z-20	A-20
1963	M-21	B-21
1964	N-22	C-22
1965	O-23	D-23
1966	P-24	E-24
1967	Q-25	F-25
1968	R-26	G-26
1969	T-27	H-27
1970	M-28	A-28

Year	Jan-Jun	Jul-Dec
1971	N-29	B-29
1972	O-30	C-30
1973	P-31	D-31
1974	Q-32	E-32
1975	R-33	F-33
1976	T-34	G-34
1977	U-35	H-35
1978	V-36	I-36
1979	W-37	J-37
1980	X-38	K-38
1981	Y-39	L-39
1982	Z-40	A-40
1983	M-41	B-41
1984	N-42	C-42
1985	O-43	D-43
1986	P-44	E-44
1987	Q-45	F-45
1988	R-46	G-46
1989	T-47	H-47
1990	U-48	I-48
1991	V-49	J-49

Known Pacific-Coast Customers (6):

Alaska Steamship, American President Lines, States Line, US Shipping Board, US Navy, US Coast Guard.

Buffalo China Inc., Buffalo, New York, 1901–83

In 1901, William J. Rea formed the Buffalo Pottery Company to supply the needs of the Larkin Soap Company. It fired its first kiln in 1903 and initially produced semi-porcelain dinner sets and toilet sets to be distributed with Larkin products. In 1916 Buffalo switched to producing fully vitrified commercial china. The company's name changed to Buffalo China Inc. in 1956, and it purchased Allegheny China of Warren, Pennsylvania, in 1962. Buffalo China became a subsidiary of Oneida Ltd of Oneida, New York, on November 29, 1983.

Representative Back Stamps

Back stamps found on Pacific-coast ship china in the approximate years shown.

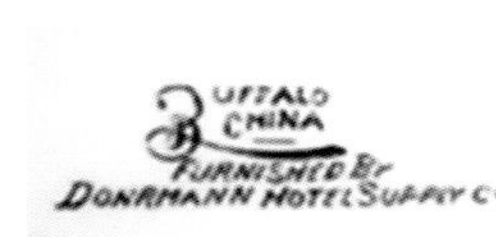

1905

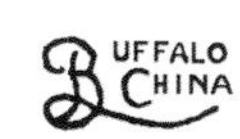

1915–40s

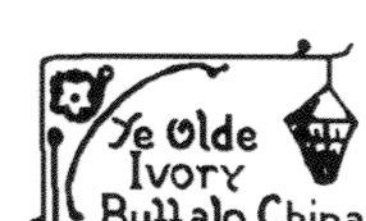

1929–40s

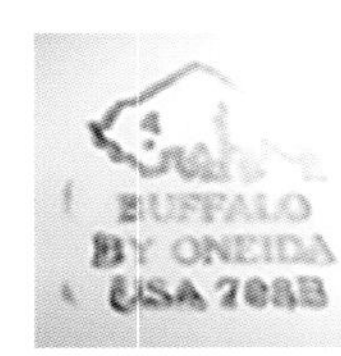

1980–?

Date Code

Many of the early Buffalo back stamps (1907–29) incorporated the year of manufacture. In the late 1920s and through the 1930s Buffalo back stamped pieces with a three-letter ink stamp (e.g., DGR), but no information exists to decipher this code. From 1940 until 1962 Buffalo used a letter-number combination, presumably "A" meaning 1940 and "W" 1962, but this has not been confirmed. From 1962 to 1966 it used a simple month-year code, making all pieces from this period easy to date: the months are numbered 1 to 12, followed by a dash and two numbers representing the year (e.g. "12–66" = December 1966).

Known Pacific-Coast Customers (12):
Alaska Steamship, American Mail Line, California Transportation Co., Chevron Shipping, Monticello Steamship, Pacific Coast Steamship, Pacific Steamship, Union Oil of California, US Bureau of Fisheries, US Coast & Geodetic Survey, US Navy, United States Shipping Board.

Mayer China Company, Beaver Falls, Pennsylvania, 1881–1990.

Brothers Joseph and Ernest Mayer formed the Mayer Pottery Company in 1881. The company was incorporated in 1923 as the Mayer China Company. The company initially produced white ironstone china. By 1900 it was producing white granite, semi-porcelain dinner sets and toilet sets. In 1915 it began producing hotelware and discontinuing its household tableware. Shenango China purchased Mayer in 1964, and four years later both companies were sold to Interpace. After several more transactions, Syracuse China bought Mayer on December 14, 1984. Syracuse operated the Mayer plant until 1990, when it closed.

Representative Back Stamps
Back stamps found on Pacific-coast ship china in the approximate years shown.

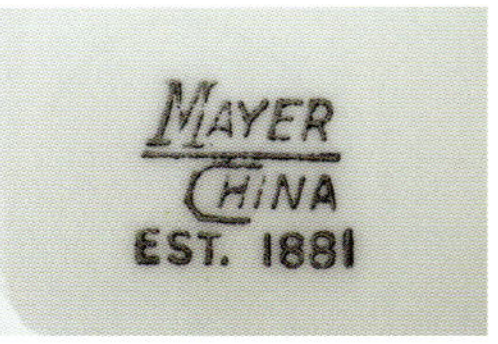

1930s–40s

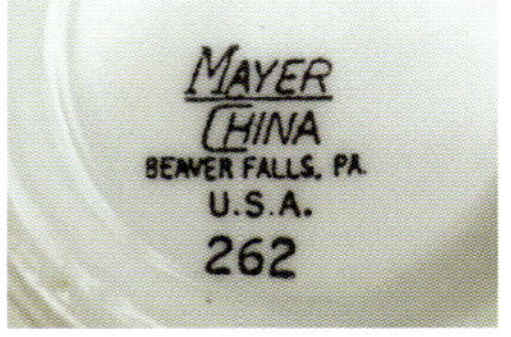

1940s–60s

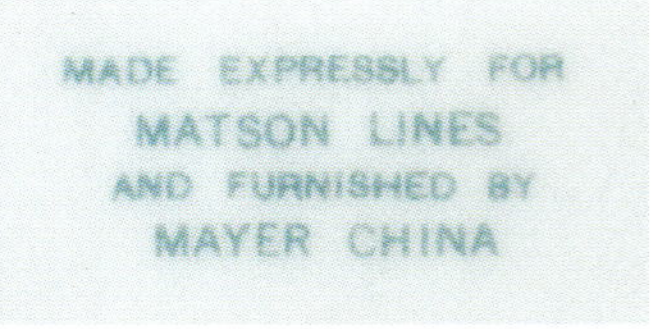

1952–63

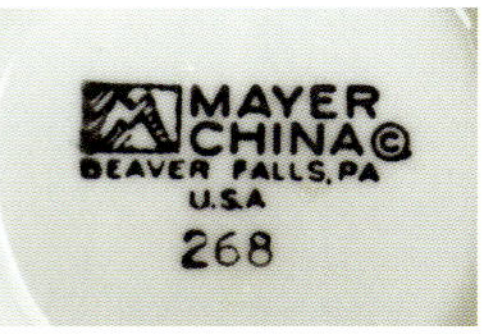

1968–78

Date Codes
During the 1890s and early 1900s, Mayer used an indecipherable date code that consisted of one number and one letter. From 1950 on, it used a three-digit numerical code: the first number represents the quarter of the year and the last two numbers the year of production. Example: "154" means the first quarter of 1954.

Known Pacific-Coast Customers (8):
Alaska Steamship, American Mail Line, California Transportation Company, Matson, US Coast & Geodetic Survey, NOAA, US Coast Guard, US Navy.

Appendix 3. China Distributors

It was not uncommon for steamship companies to order their china through a distributor. Accordingly, many pieces of nautical china have the name of the distributor on the back stamp. Correlating the manufacturer's back-stamp information with the years that a distributor operated can help establish the time period that a piece of china was used.

Some of the companies in this list may have operated beyond the years shown. The distributor's code may appear in the pattern description guide.

Code	Distributor	Location	Years of Operation
AAM	Anglo-American Crockery	San Francisco, CA	1893–1911
ALC	A.L. Cahn & Sons	New York, NY	1944
BRC	Bell Rinfret Co.	Montreal, Quebec	1939–82
BTH	L. Barth & Sons Inc.	New York, NY	1868–1926
BRN	J. Bornstein & Sons	Seattle, WA	1905–24+
BRD	Broduer Co.	Montreal, Quebec	ca 1910
BUS	Buscombe & Company	Vancouver, BC	1932–56
CAS	Cassidy's Ltd	Vancouver, BC	1928–2000
CBS	Cassidy-Buscombe	Vancouver, BC	1956–64
DOR	Dohrmann Hotel Supply Co.	San Francisco, CA	1904–50s
FBC	Frederick Buscombe Co.	Vancouver, BC	1901–32
GIM	Gimbel Brothers Inc.	New York, NY	1922–87
GRN	Greene-Winkler	Seattle, WA	1926–50
HBC	H. Barron Co.	New York, NY	1940s
KDC	Kniffin & Demurest Co.	New York, NY	1903–61
MAR	Marshall-Wells BC Ltd	Victoria / Vancouver, BC	1899–1957
MHC	Maxwell Howe Co.	Oakland, CA	1930s
NDR	Nathan-Dohrmann Co.	San Francisco, CA	1886–1905
NID	Nicol & Davidson Co.	New York, NY	1870–1900
NST	Nathan Straus & Sons	New York, NY	1923–36
OHT	Ottenheimer Brothers	San Francisco, CA	1906–10?
PCK	Albert Pick & Co.	Chicago, IL	1895–1965
RCL	Renwick & Cunliffe Ltd	Vancouver, BC	1911–35
ROB	Robinson & Co. Ltd	Winnipeg, Manitoba	ca 1916–17
SHW	James M. Shaw Co.	New York, NY	1828–1936
STN	Sternheim & Co.	San Francisco, CA	1882–92
STN	Wangenheim, Sternheim	San Francisco, CA	1893–1903
STN	S. Sternheim & Sons	San Francisco, CA	1904–05
SEL	M. Seller & Company	Portland / Seattle / Spokane	1885–1916
SEL	M. Seller & Company	Portland / Seattle / Tacoma	1917–61
WAN	John Wanamaker	New York / Philadelphia	1896–1978
WIL	A.T. Wiley & Sons	Montreal, Quebec	1880–1936

Appendix 4. Pacific Coast Steamship Line and Government Service House Flags

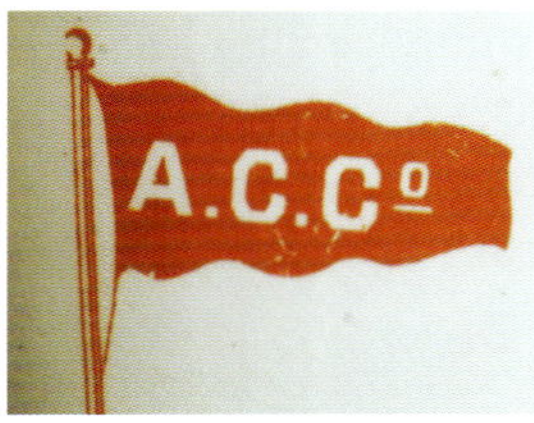

Alaska Commercial, 1868–1901

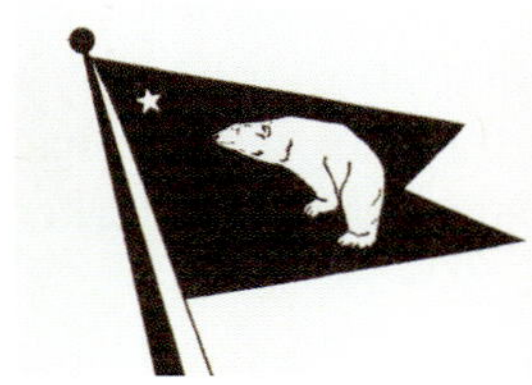
Alaska Cruise Lines, 1957–71

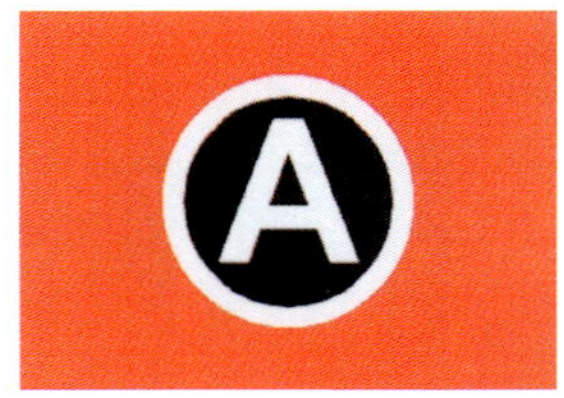

Alaska Steamship, 1912–71

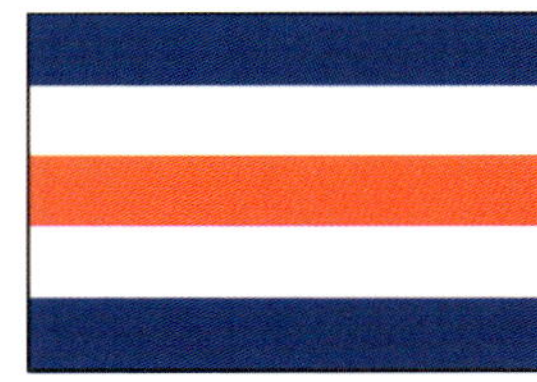
American Mail, 1926–71

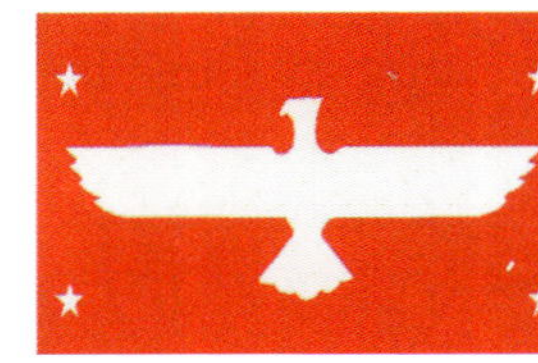
American President Lines, 1938–present

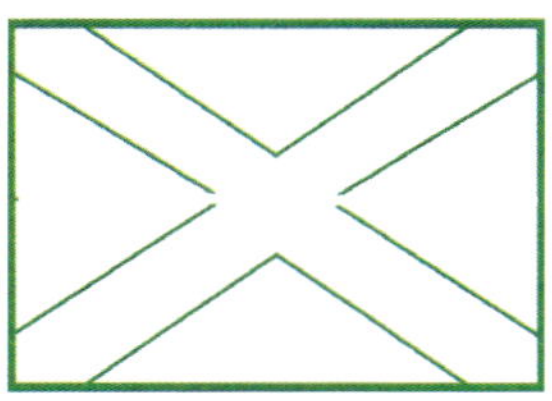
American Yukon Navigation, 1913–43

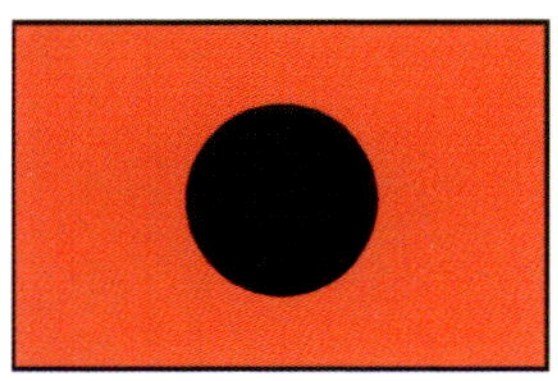
Black Ball Lines of Canada, 1951–61

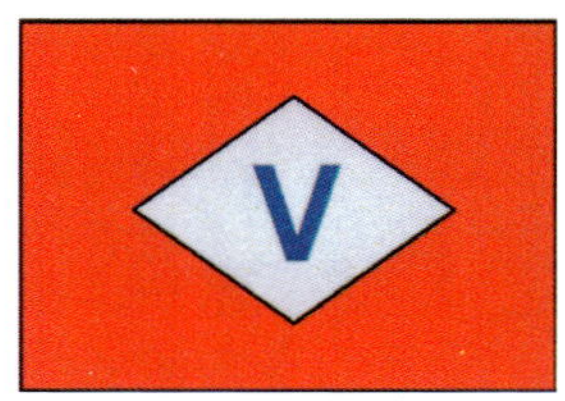

Boscowitz Steamship, 1899–1911

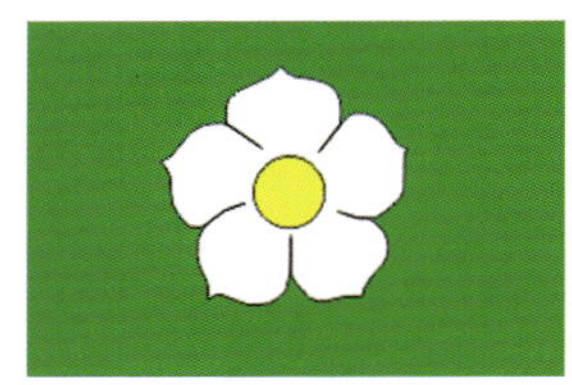
BC Ferries, 1960–77

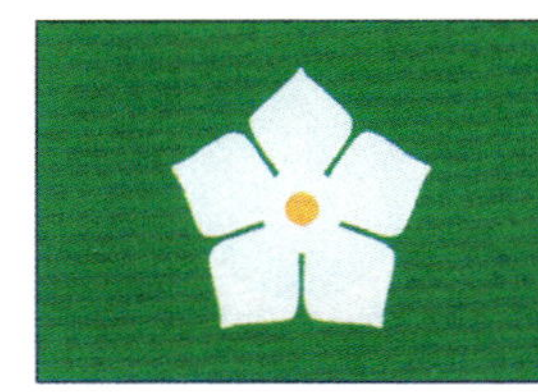
BC Ferries, 1978–2002

BC Ferries, 2003–present

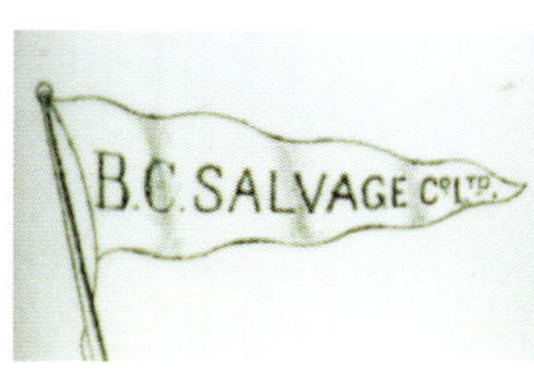

BC Salvage, 1901–16

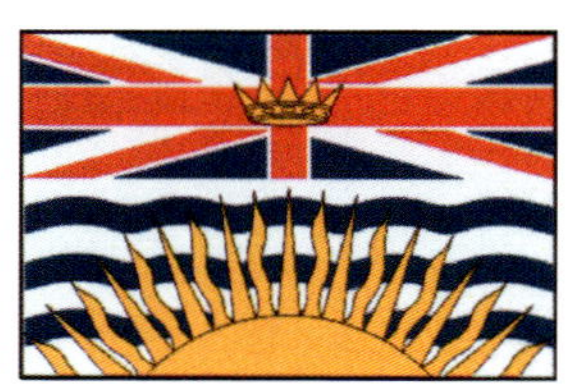
BC Steamship, 1975–88

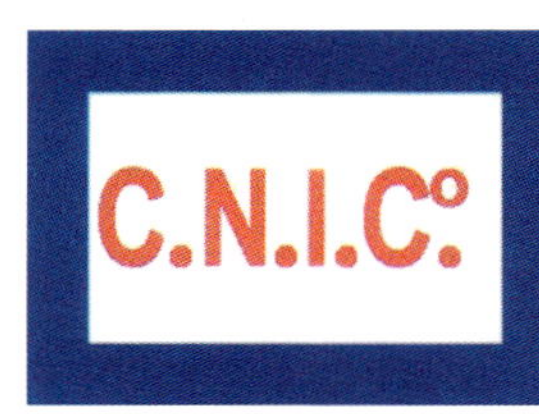

California Navigation & Improvement, 1889–1927

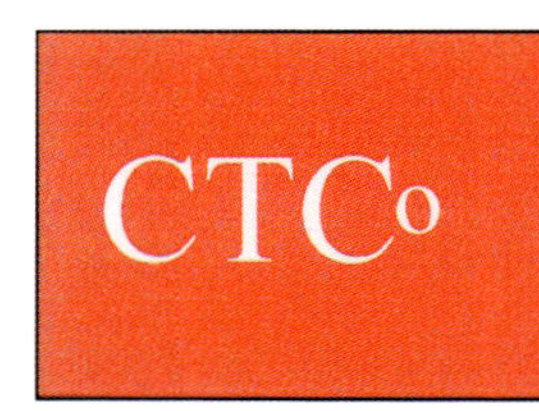

California Transportation, 1875–1947

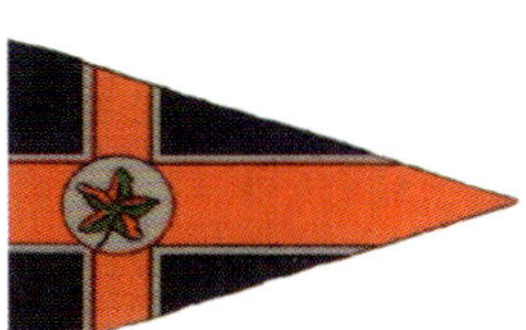
Canadian National Steamships, 1925–75

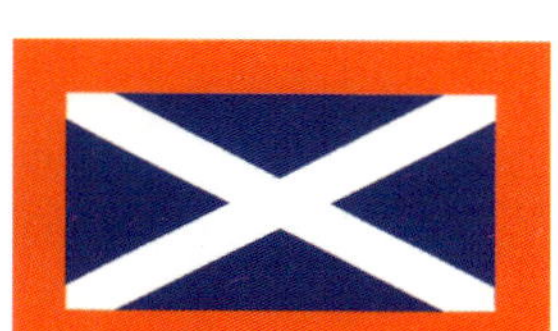
Canadian Pacific Navigation, 1883–1901

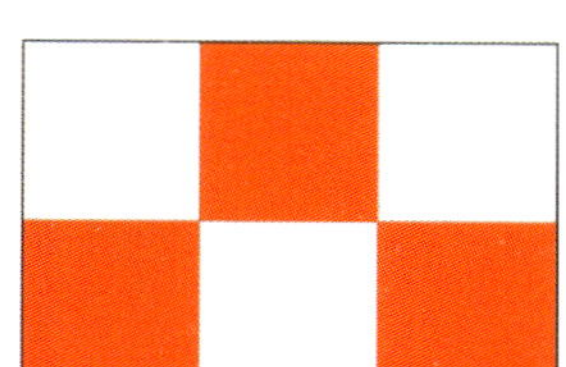
Canadian Pacific Railway, 1891–1968

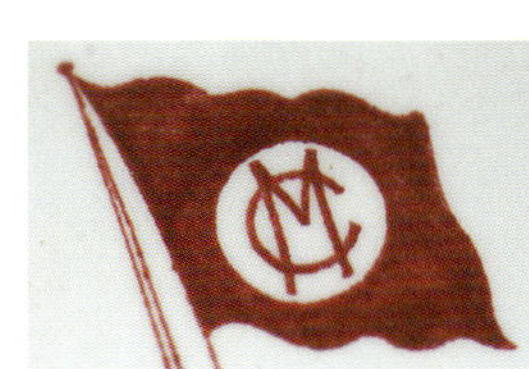
China Mail Steamship, 1915–23

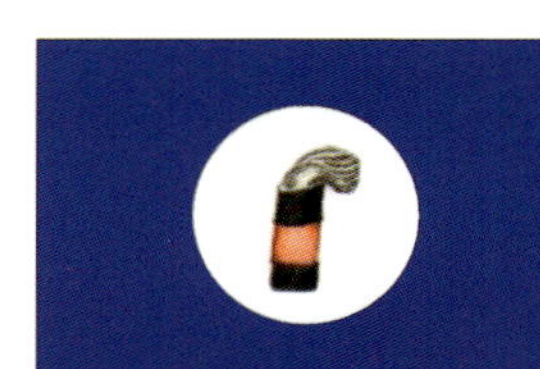
Couer d'Alene & St Joe Transportation, 1903–22

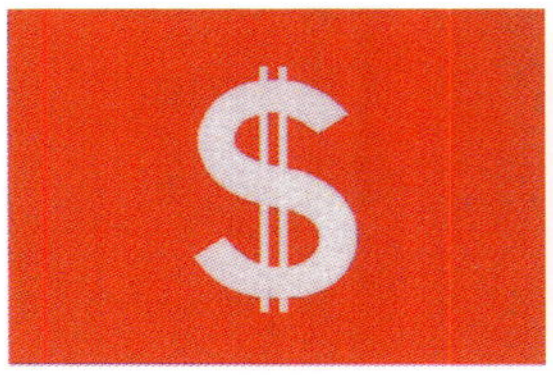

Dollar Steamship, 1900–1938

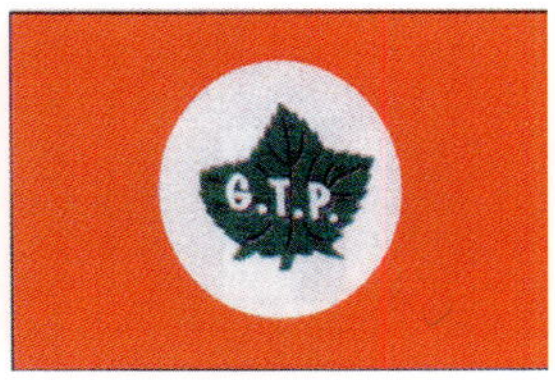

Grand Trunk Pacific Coast Steamship, 1910–25

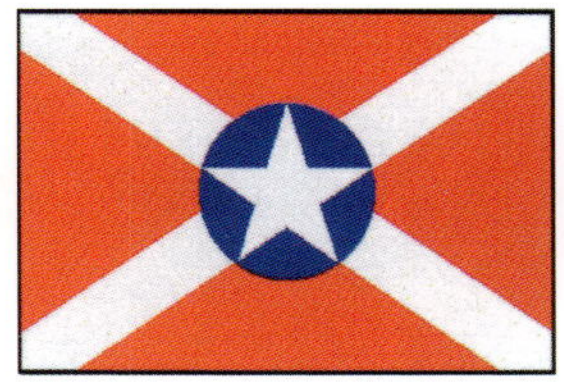

Great Northern Pacific Steamship, 1914–17

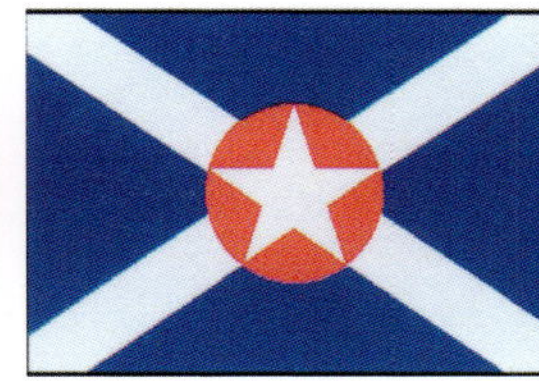

Great Northern Steamship, 1904–15

Imperial Oil, 1880–1978

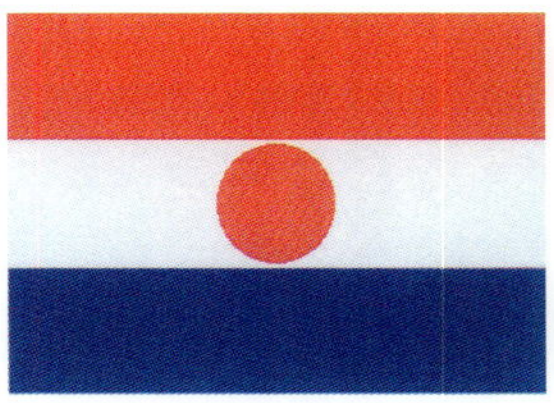

Inter-Island Steam Navigation, 1883–1950

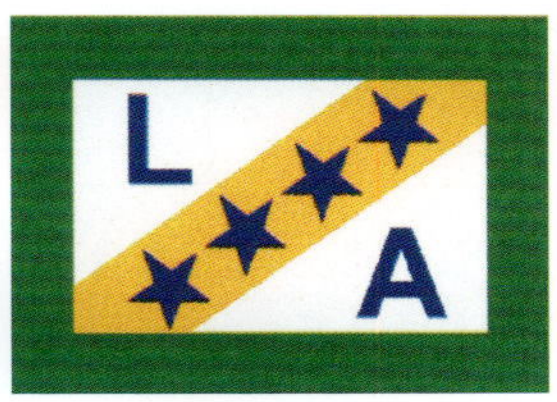

Los Angeles Steamship, 1920–37

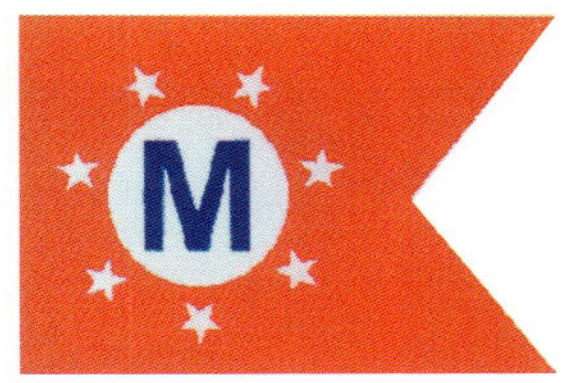

Matson Navigation, 1901–present

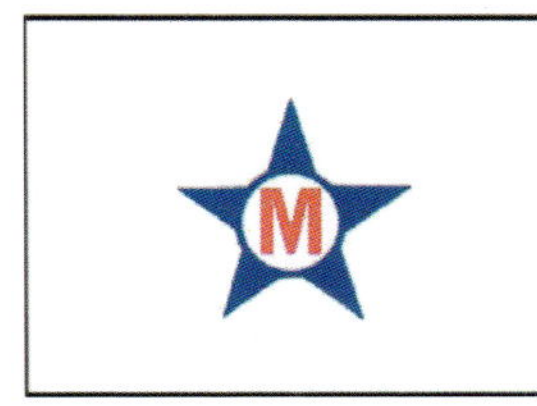

McCormick Steamship, 1923–40

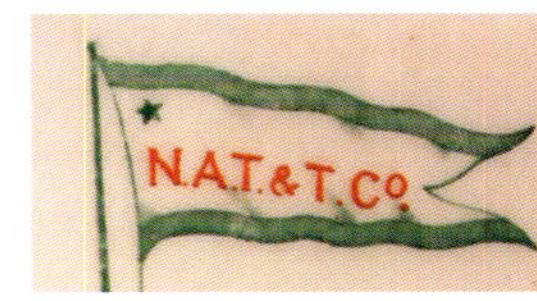

North American Transportation and Trading, 1892–1906

Northland Navigation, 1951–80

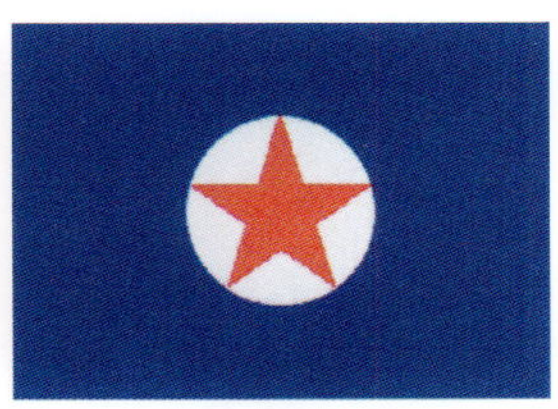

North Pacific Steamship, 1905–17

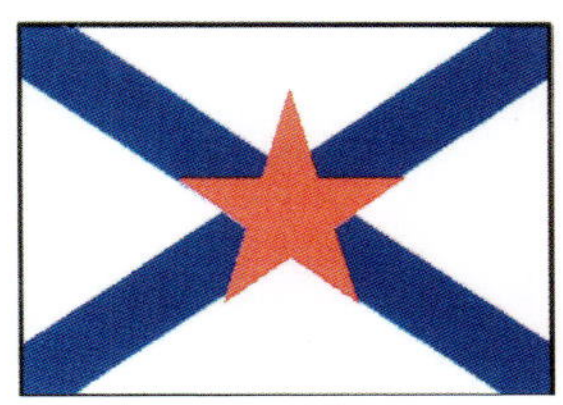

Northwestern Steamship, 1904–08

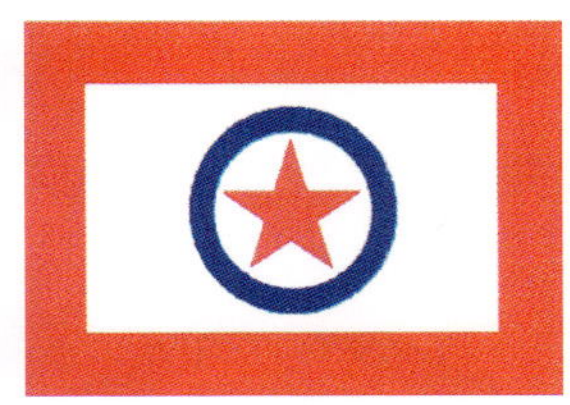

Oceanic Steamship, 1880–1926

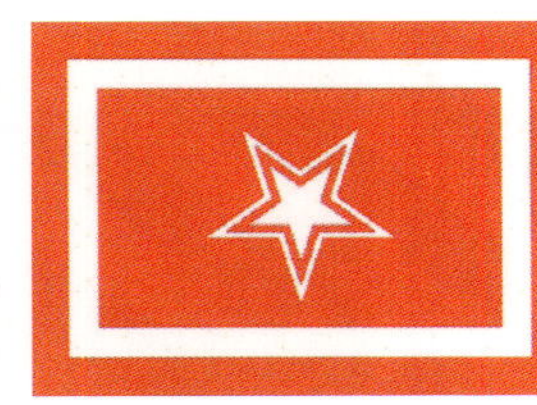

Oregon Railway & Navigation, 1879–1910

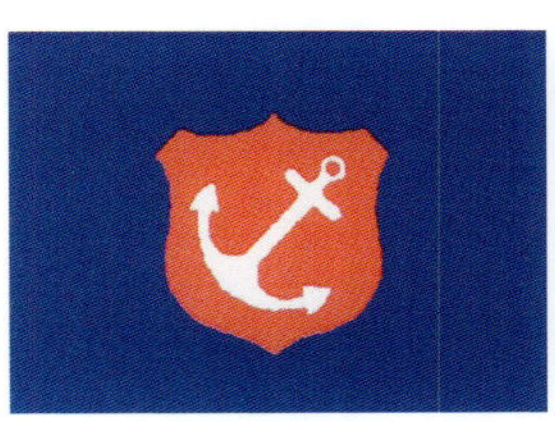

Pacific Alaska Navigation, 1912–16

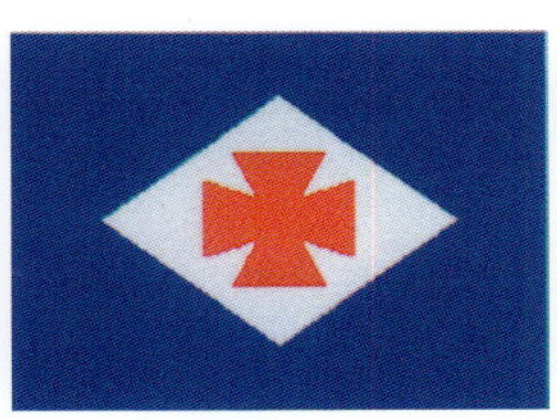

Pacific Coast Steamship, 1877–1916

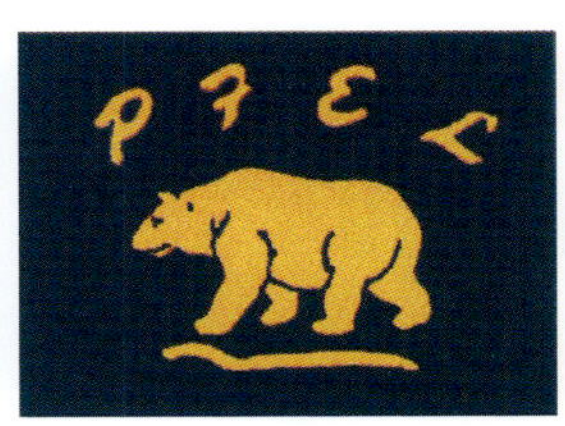

Pacific Far East Line, 1946–78

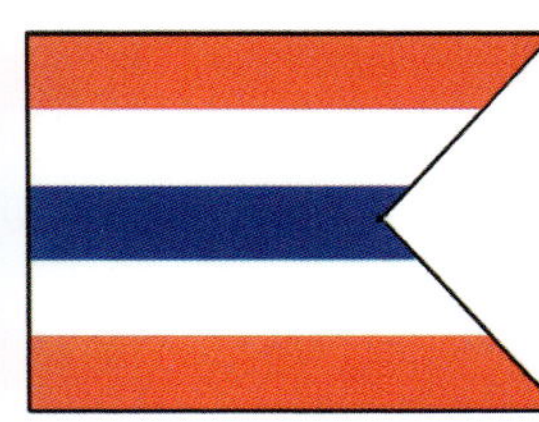

Pacific Mail Steamship, 1848–1925

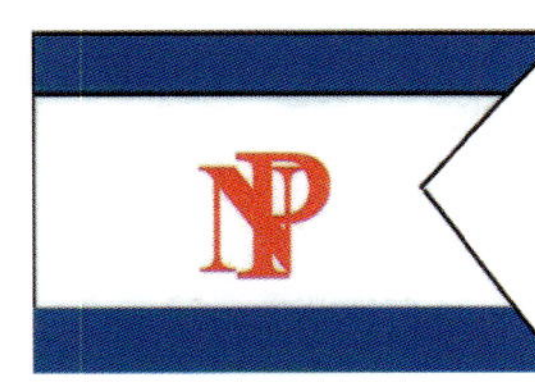

Pacific Navigation, 1910–16

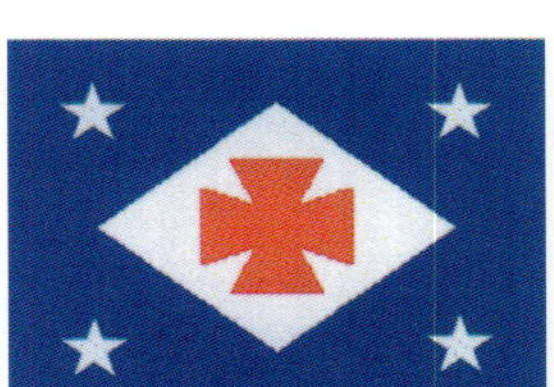

Pacific Steamship, 1916–35

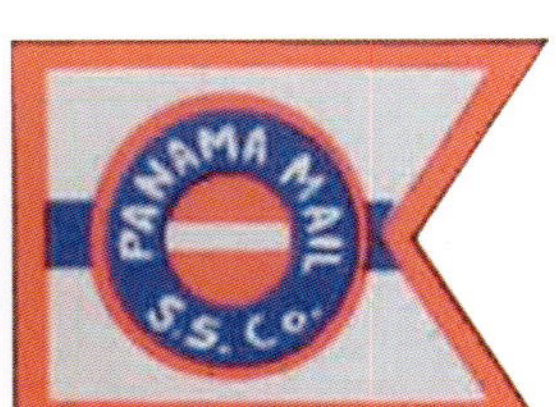

Panama Mail Steamship, 1925–31

Pope & Talbot, 1947–61

Puget Sound Navigation, 1898–1926.

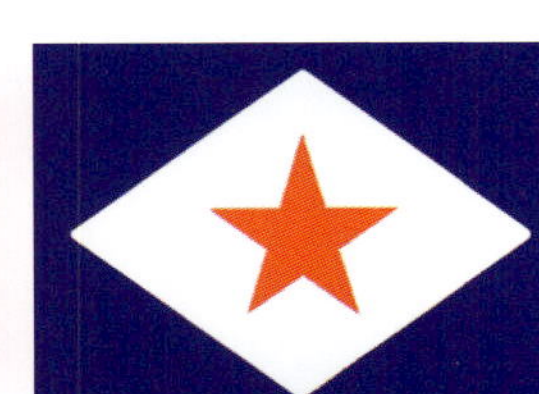

Puget Sound Navigation, 1927–51

San Francisco & Portland Steamship, 1904–24

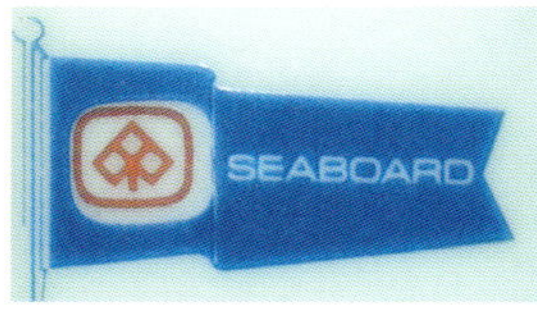

Seaboard Shipping, 1936–present

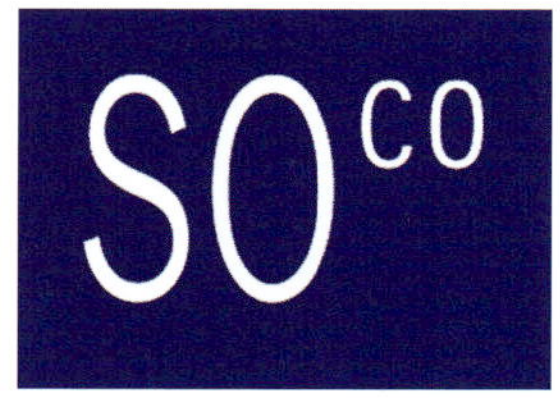

Standard Oil of California

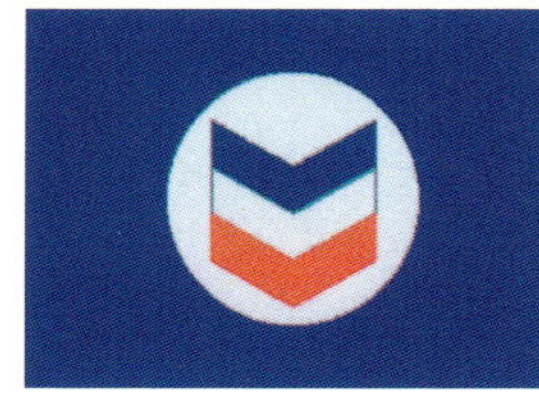

Standard Oil, 1931–83

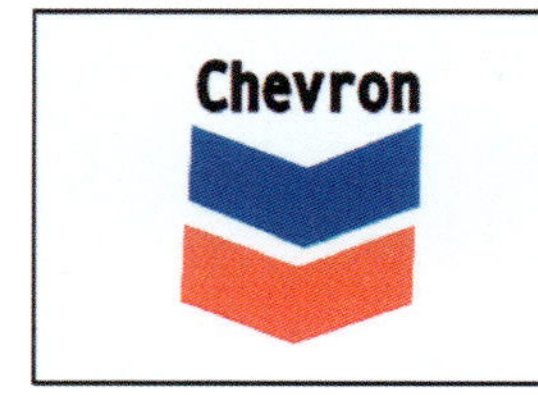

Standard Oil of California, 1984-present

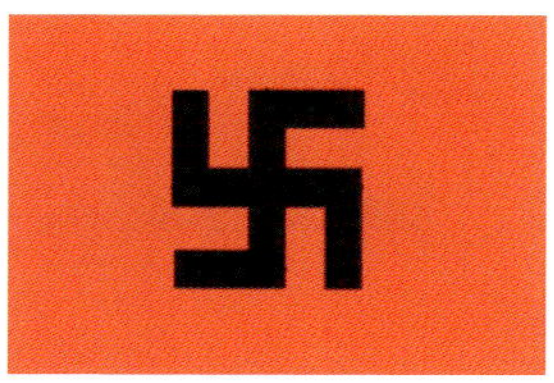

States Steamship, 1928–40

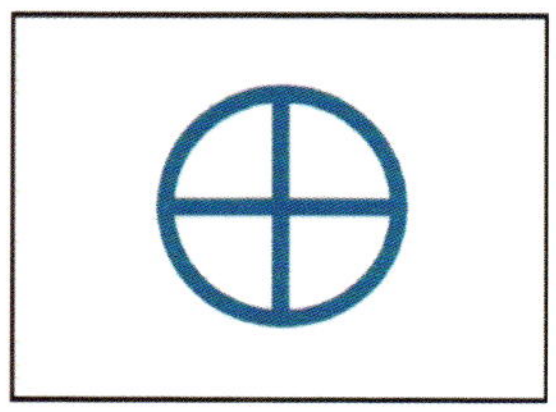

States Steamship, 1941–53

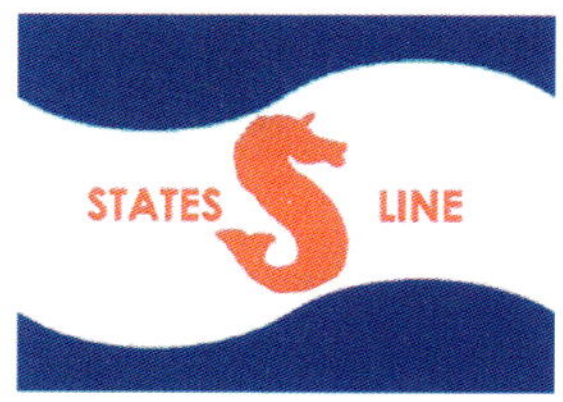

States Steamship, 1954–73

Terminal Steamship, 1902–20

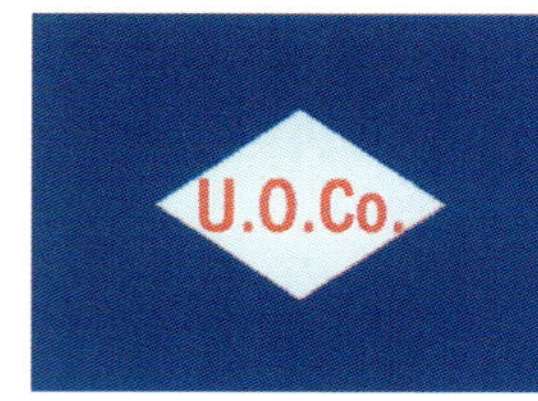

Union Oil of California, 1890–1946

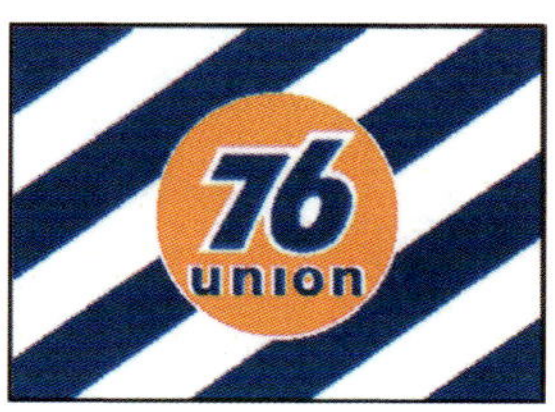

Union Oil of California, 1947–present

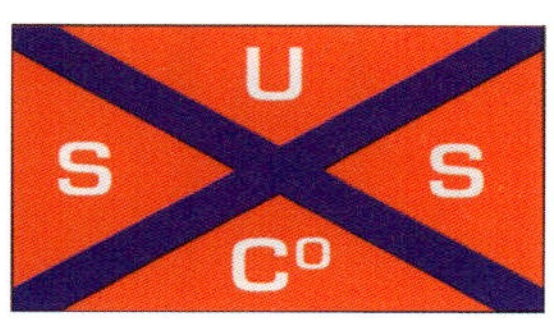

Union Steamship of BC, 1889–1958

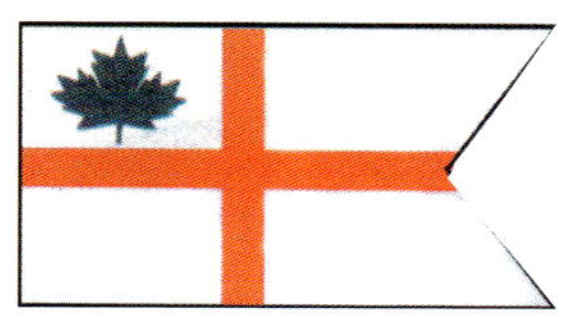

Western Canada Steamship, 1946–65

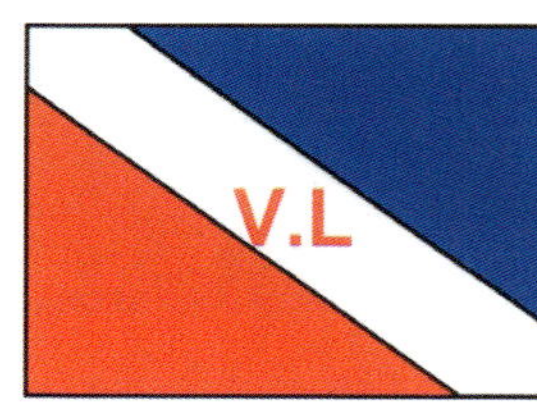

Western Steam Navigation, 1900–1907

Canadian Government Flags

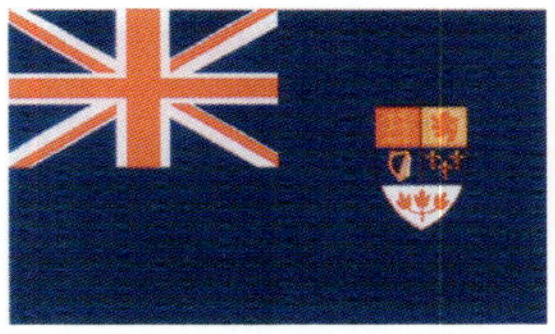

Canadian Government ship, 1892–1965

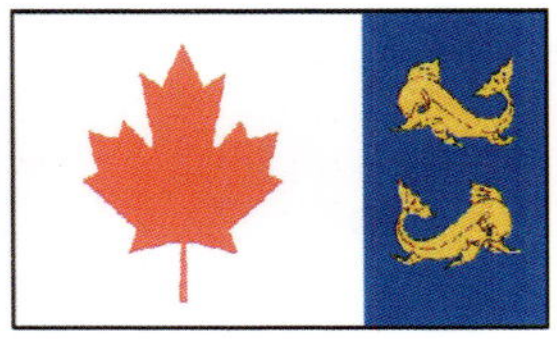

Canadian Coast Guard, 1962–present

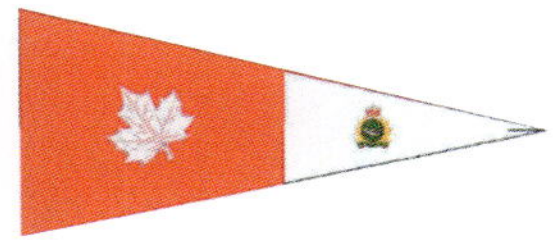

Fisheries Canada

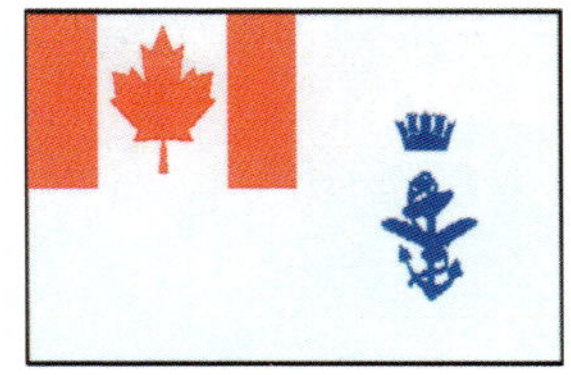

Royal Canadian Navy, 1965–present

US Government Flags

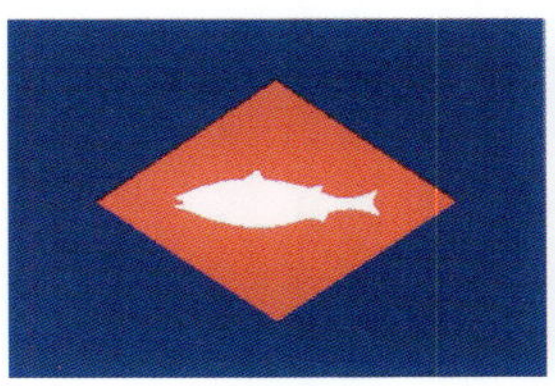

US Bureau of Fisheries, 1903–36

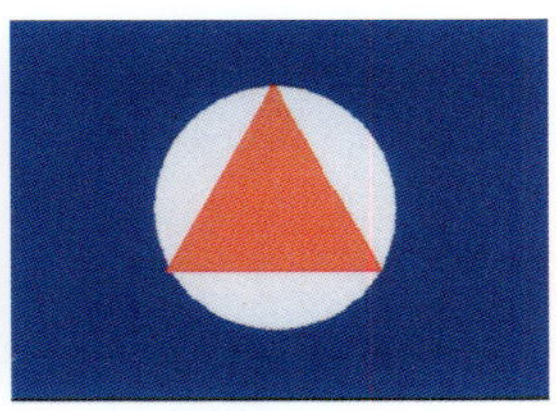

US Coast & Geodetic Survey, 1898–1970

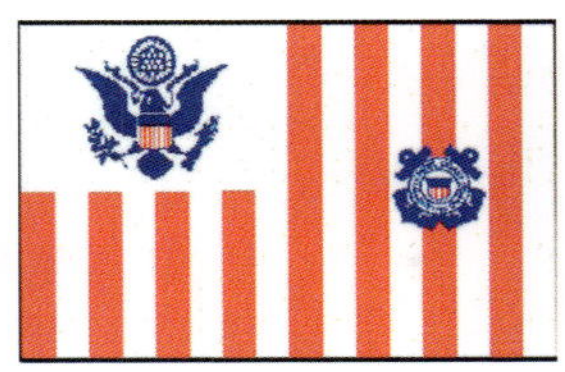

US Coast Guard 1915–present

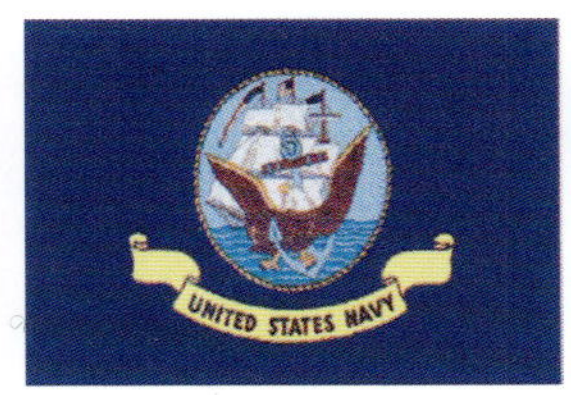

US Navy

US Navy Commissioning Pennant

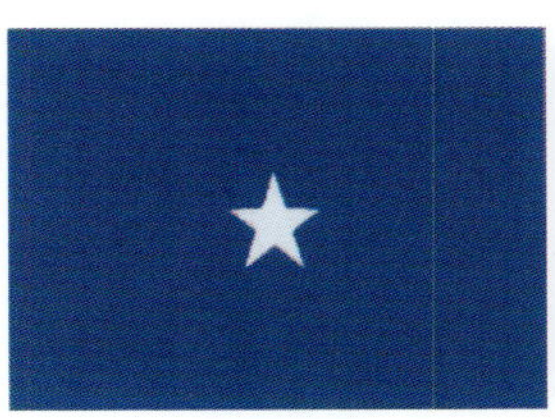

US Navy Rear Admiral (lower half)

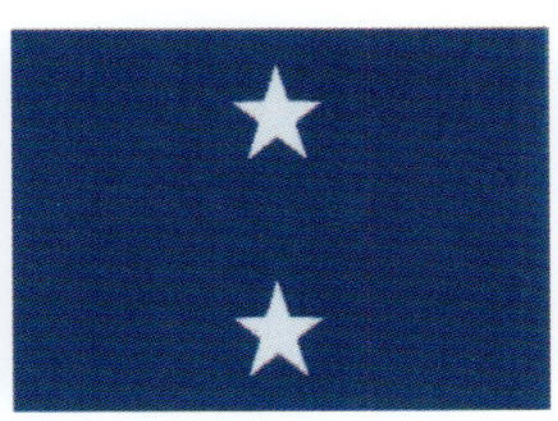

US Navy Rear Admiral (upper half)

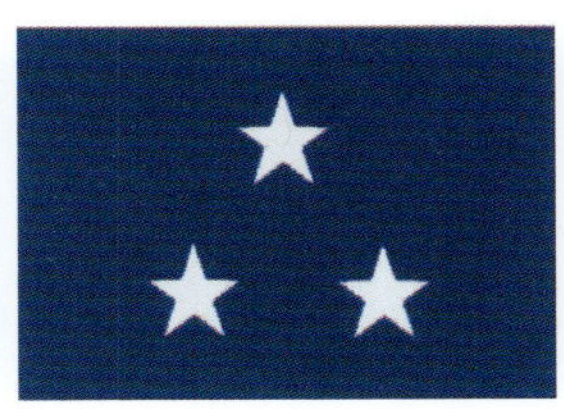

US Navy Vice Admiral

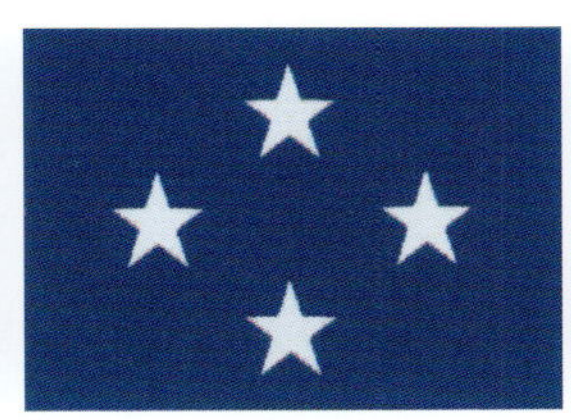

US Navy Admiral

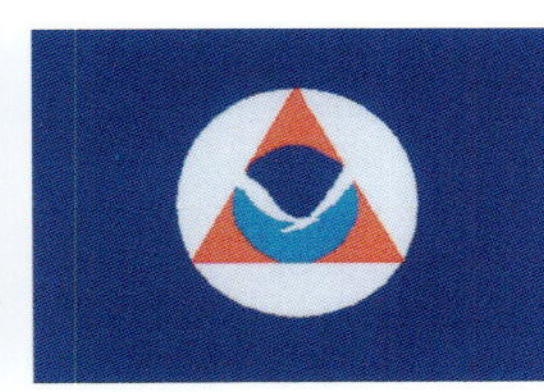

US National Oceanic & Atmospheric Administration, 1970–present

Appendix 5. Conserving Dive-Recovered China

Many ship-china patterns are recovered by divers from old dock sites and shipwrecks. When recovered from salt water these finds must be treated immediately or irreparable damage will occur. Submerged china has been subjected to increased pressure that pushes salt water into it. After recovery, as water evaporates out of the china, the salt remains and forms crystals. These crystals exert extreme pressure on the china body and glaze. If the china is not treated, it will eventually crack and the glaze will exfoliate (flake off). In extreme cases, the china may break into several pieces.

Here I offer some general guidance on how to treat dive-recovered china. But to ensure success, consult professional advice from a conservator at a museum, college or university. The Conservation Research Lab at Texas A&M University has some good general information on its website: http://nautarch.tamu.edu/class/anth605/File4.htm.

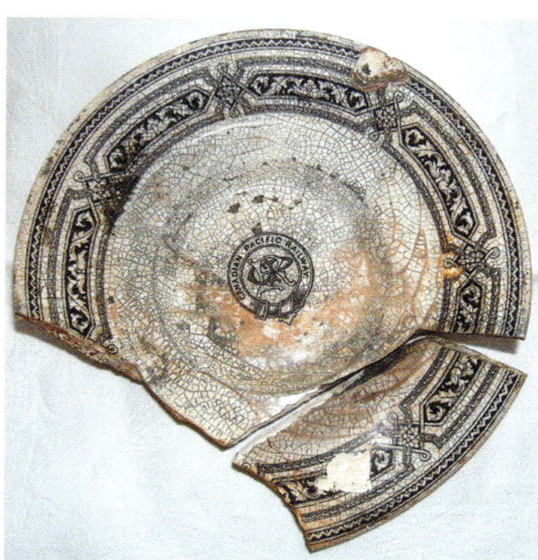

Untreated china from the wreck of the *Empress of Ireland*: a shard has broken off and the glaze is exfoliating.

Removing Salt

To prevent saltwater damage, place dive-recovered china into fresh water as soon as possible. The fresh water must be not contain any chlorine or fluorine. Distilled water is best, but filtered tap water will do. Chlorinated water simply replaces the ocean salt with chloride salts, which have the same devastating effect as salt. Chlorine also has bleaching properties that can damage the transfer prints on some china. The CPR Empress pattern and the CNS Jasper patterns are extremely vulnerable to chlorine bleaching.

It can take up to a year to extract the salts from china. Generally, the longer an object has been submerged in salt water and the softer and more porous the material, the longer it will take to leach out the salts. In a professional lab, a conservator uses a conductivity meter to measure the progress of desalination. For those of us who don't have this equipment, the best procedure is to change the fresh water routinely, once a week at first, and then, after a couple of months, every two weeks. Using de-ionized water in the final baths should remove the last of the salts.

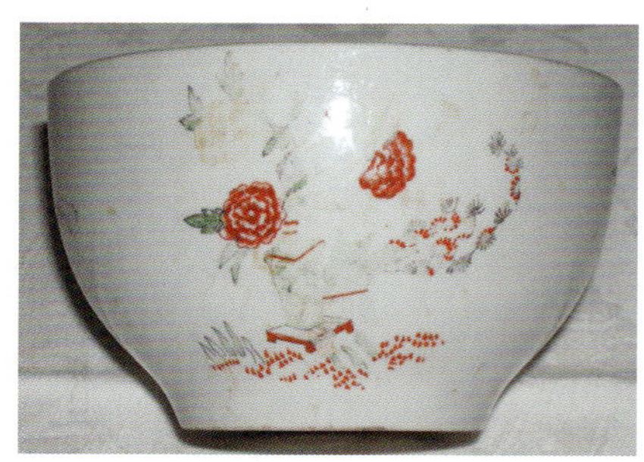

CPR Empress china soaked in chlorinated water to remove stains, resulting in a washed-out transfer print.

Removing Encrustations

Dive-recovered china can be covered in marine growth or stained by iron oxide (rust). Calcareous encrustations and rust can be removed by bathing the china in diluted muriatic acid (a product used by bricklayers to remove excess mortar from brickwork). Wash only the affected area with the acid. Once the encrustations are gone, immediately place the china in a neutralizing solution. A little baking soda dissolved in water works well to clean dive-recovered china; the household cleaner CLR (Calcium, Lime, Rust) also works. When using either of these products wear protective eyewear and rubber gloves and avoid breathing the fumes.

Removing Stains

Do not soak dive-recovered china in bleach. Bleach is a chloride. When it dries, it forms destructive crystals, just like salt, that can damage the china. The safest product for cleaning china is hydrogen peroxide, a mild disinfectant readily available from most pharmacies. Professional curators use hydrogen peroxide because it is non-intrusive and does not introduce chlorides into the china.

After you have removed the salt from the china, take the piece from its freshwater bath and place it in a bath of hydrogen peroxide. Leave it in until the stains have diminished. Peroxide does not act quickly, so don't expect results in hours. It may take days or even weeks to remove persistent stains and it may take several rotations from fresh water into the peroxide to remove staining.

Glossary

Arched upward In logos, lettering that begins at the centreline and arches above an object.

Arched downward In logos, lettering that begins at the centreline and arches beneath an object.

Back stamp The information that is transfer printed or stamped on the bottom or back of china. It may relate to the manufacturer, distributor or the shipping company.

Band In patterns, a line wider than four millimetres.

Burgee A broad, tapering pennant, with or without a swallowtail.

Coastwise Coastal shipping – travelling to ports along the same coast.

Custom pattern China produced and decorated for a specific customer. The pattern may include the company's flag or other custom decorations ordered by the customer.

Distributor A wholesaler or dealer who represents specific china manufacturers and would order the china products required by a retailer or customer.

Flag Rectangular flag.

Garter In logos, a ring with a belt buckle finish to it.

Glaze A glass-like transparent finish that seals the china and makes it watertight.

Helvetica A style of lettering in which letters have straight sides and squared angles.

Incised back stamp Manufacturer information or date coding is impressed into the body of the china. See *back stamp*.

Intercoastal Travelling between coasts on either side of a continent, as in the east and west coasts of North America.

Interscroll Intertwined or overlapping letters.

Ironstone A medium- to heavy-gauge semi-vitrified china manufactured at the turn of the 20th century and known for its durability and longevity.

Life ring In logos, a ring with four quadrants, usually defined with a rope border.

Ornate Decorative (flowery), double-stamped lettering with a lot of serifs.

Pennant Triangular flag.

Pinstripe In patterns, a line one millimetre or less in width.

Ring In logos, two circles forming a ring, much like the top view of a donut.

Saint Andrew's cross A cross from corner to corner of a flag, forming an "X" (also called a saltire).

Saint George's cross A vertical/horizontal cross, forming a "+", its arms reaching the edges of the flag.

Script A style of lettering that flows like handwriting.

Stock pattern An off-the-shelf china pattern produced by manufacturer for restaurants, railroads and shipping companies without any custom decorations or logos.

Stripe In patterns, a line with a width of two to four millimetres.

Superimposed initials In patterns, one letter placed on top of or in front of another

Swallowtail A flag with a triangular notch cut into the end.

Transfer printing The process used to transfer a design or pattern to the china body. An image is first engraved onto a copper plate. Colour pigments are then poured into the engraving. The image is transferred to a piece of tissue-like paper and then the paper applied to the surface of the china.

Trans-Pacific Travelling from west coast of North American across the Pacific Ocean to the Orient, Australia or New Zealand.

Under-glaze transfer A transfer print applied to china after its initial firing but before glazing. Glazing makes the image permanent. See *glaze* and *transfer printing*.

Vitrified china China fired at an extremely high temperature to fuse the ceramic body together like glass.

Wardroom The officers' mess aboard a warship. It is a place of recreation as well as a dining room, and usually has an adjoining galley (kitchen).

Selected References

Books

American Association of Masters, Mates and Pilots, The. *The Revival of Steamboat Days on the Sacramento River. The Master, Mate and Pilot*, vol. 7, no. 9, February 1915.

Appleton, Thomas E. *Usque ad Mare: A history of the Canadian Coast Guard and Marine Services*. Ottawa: Department of Transport, 1968.

Betz, Clinton H. The North Pacific Steamship Co. of San Francisco. *The Sea Chest, Journal of the Puget Sound Maritime Historical Society*, March 1986.

Coman, Edwin T. and Gibbs, Helen M. *Time, Tide and Timber: A Century of Pope & Talbot*. Stanford, California: Stanford University Press, 1950.

Conroy, Barbara J. *Restaurant China: Identification and Value Guide for Restaurant, Airline, Ship and Railroad Dinnerware*, volume 1. Paducah, Kentucky: Collector Books, Schroeder Publishing Co., 1998.

Conroy, Barbara J. *Restaurant China: Identification and Value Guide for Restaurant, Airline, Ship and Railroad Dinnerware*, volume 2. Paducah, Kentucky: Collector Books, Schroeder Publishing Co., 1999.

Cushion, J.P. *Handbook of Pottery and Porcelain Marks*. Boston: Faber & Faber, 1980.

Emmons, Frederick E. *Pacific Liners 1927–1972*. Newton Abbot, UK: David & Charles, 1973.

Emmons, Frederick E. *American Passenger Ships: The Ocean Lines and Liners, 1873–1983*. Newark: University of Delaware Press, 1985.

Godden, Geoffery. *Encyclopedia of British Pottery and Porcelain Marks*. New York: Crown Publishers, 1964.

Harlan, George, and Clement Fisher Jr. *Of Walking Beam and Paddle Wheel: A Chronicle of San Francisco Bay Ferryboats*. San Francisco: Bay Books.

Harwood M/V. *Department of Transport List of Steam and Motor Vessels on Registry Books in Canada*, December 31, 1947.

Henderson's Directory of British Columbia. Incorporated companies up to February 17, 1910.

Hittell, John S. *The Commerce and Industries of the Pacific Coast of North America*. San Francisco: A.L. Bancroft & Co, 1882.

Kovel, Terry, and Ralph Kovel. *Dictionary of Pottery and Porcelain Marks*. New York: Crown Publishers, 1953.

Krummes, Daniel. *Dining on Inland Seas*. Traverse City, Michigan: Nautical Works Press, 1998.

Laister, Peter. *Mariners' Memorabilia: A guide to British Shipping Company China of the 19th and 20th Centuries*. Longfield, UK: Peter and Pam Laister, 2006.

Lawson, Will. *Pacific Steamers*. Glasgow: Nautical Press, 1927.

Lehner, Lois. *Complete Book of American Kitchen and Dinner Wares*. Des Moines, Iowa: Wallace-Homestead Book Company, 1980.

Lehner, Lois. *Lehner's Encyclopedia of U.S. Marks on Pottery, Porcelain and Clay*. Paducah, Kentucky: Schroeder Publishing, 1988.

Luckin, Richard. *Dining on Rails*. Golden, Colorado, 1994.

McIntyre, Douglas. *The Official Guide to Rail Road Dining Car China*. Williamsville, New York: Golden Spike Enterprises, 1990.

Newell, Gordon. *Pacific Steamboats*. Seattle: Superior Publishing, 1958.

Newell, Gordon. *Ocean Liners of the 20th Century*. Seattle: Superior Publishing, 1963.

Newell, Gordon. *The H.W. McCurdy Marine History of the Pacific Northwest*. Seattle: Superior Publishing, 1966.

Newell, Gordon. *The H.W. McCurdy Marine History of the Pacific Northwest (1966–1976)*. Seattle: Superior Publishing, 1977.

Outwater, Myra Y. *Ocean Liner Collectibles*. Atglen, Pennsylvania: Schiffer Publishing, 1998.

Pugh, Gordon. *Naval Ceramics*. Newport, UK: Ceramic Book Company, 1971.

Sandknop, Steven S. *Nothing Could be Finer: A Compendium of Rail Road Dining Car China*. Edina, Montana: Sandknop Publications, 1977.

Spence, Karl D. *How to Identify and Price Oceanliner Collectibles*. Clovis, New Mexico: Spence Publications, 1991.

Spence, Karl D. *Oceanliner Collectibles*, vol. 2. Clovis, New Mexico: Spence Publications, 1996.

Spence, W.F. *Report of the Royal Commission on Coasting Trade*. Ottawa: Queens Printer, 1957.

Tate, E. Mowbray. *Transpacific Steam*. Cranbury, New Jersey: Cornwall Books, 1986.

Thoma, Mifflin. *Schooner from Windward: Two Centuries of Hawaiian Interisland Shipping*. Honolulu: University of Hawaii Press, 1983.

Turner, Robert D. *The Pacific Princesses*. Victoria, BC: Sono Nis Press, 1977.

Turner, Robert D. *The Pacific Empresses*. Victoria, BC: Sono Nis Press, 1988.

Turner, Robert D. *Those Beautiful Coastal Liners*. Victoria, BC: Sono Nis Press, 2001.

Twigg, Art. *Union Steamships Remembered 1920–1958*. Campbell River, BC: A.M. Twigg, 1997

Williams' British Columbia Directory 1892.

Wright, E.W. *Lewis & Dryden's Marine History of the Pacific Northwest*. New York: Antiquarian Press, 1961.

Periodical Articles

Doyle, Jim. "This Bird has Flown: Rejuvenated Schooner sets Sail for Seattle, Ending an Era in Sausalito". *San Francisco Chronicle*, Oct. 21, 1998.

Hemion, Austen D. "History of Pacific Alaska Navigation Co., or The Rise of H.F. Alexander". *The Sea Chest*, Dec. 1982, pp. 59-71.

Krummes, Daniel C. "Dating Ships China". Steamboat Bill, no 199, pp. 182-85.

McDonald, Lucile. "The Revenue Cutters: Cutter Service Created in 1790, Northwest Service begins in 1854". *The Sea Chest*, Jun. 2001, pp. 171-75.

Seeley White, James. "Ship's Chinaware of the Pacific Northwest". Paper presented at 38th Northwest Anthropological Conference, Central Washington University, April 18-20, 1985.

Seeley White, James. "The Lore of Ships Ware". *Diver* 11:6 (Sep. 1985), p. 12.

Stadum, Lloyd M. "Passenger Steamers *President* and *Governor*". *The Sea Chest*, Sep. 1979, pp. 13-26.

Stadum, Lloyd M. "Steamer Mainlander, A Pioneer Design". *The Sea Chest*, Dec. 1979, pp. 50-52.

Stadum, Lloyd M. "The Disappointing *Haleakala* of 1923". *The Sea Chest*, Dec. 1986, pp. 61-65.

Stadum, Lloyd M. "Monticello Steamship Company's Fleet", part 5. *The Sea Chest*, Dec. 1990, pp. 65-85.

Stritzel, Bill. "Dispatching the Product of the Soil: The Fruit and Grain Trade on the Sacramento River". *Journal of the Sacramento County Historical Society* 6:1-4 (2006), pp. 205-24.

Stritzel, Bill. "Steam Navigation on the Interior Bays and Rivers of Northern California". Unpublished manuscript, January 2008.

Yost, Allan. "Sailing Day in the Thirties". *The Sea Chest*, Dec. 1972, pp. 72-80.

Internet Sites

A-Z of Stoke-on-Trent Potters. www.thepotteries.org/allpotters/index.htm

Blanchard, David M. "U.S. Shipping Board Records in the National Archives". *Nautical Research Journal* 45:1 (1997). p. 49. www.naut-res-guild.org/Journals/0300add/0300add.htm

Central Pacific Railroad Photographic History Museum. http://cprr.org/Museum/Sacramento_Steamers.html

Flags of the World. www.crwflags.com/fotw/flags/

Hodder, Edwin. "Life of Sir George Burns (1890)". The Ships List – Cunard Line. www.theshipslist.com

Maritime Administration (MARAD). www.nvr.navy.mil/marad.htm

Mercogliano, Salvatore. "One Hundred Years in the Making: Birth of Military Sea Transportation Service (MSTS)". www.USMM.org

Niero, Mike. "History of Naval Tableware", "US Navy China and Dinnerware", "Worldwide Naval Tableware". The Pirate's Lair: www.thepirateslair.com

Oregon Historic Photograph Collections. www.photos.salemhistory.org

Peterson, Ronald G. "Collectibles of the SP&S". http://sites.onlinenw.com/ronaldgp/default.htm

Sacramento History Online. www.sacramentohistory.org/index.html

Spokane, Portland and Seattle Railway. "Collectibles of the Spokane, Portland and Seattle Railway". www.home.att.net/~spsrailway/northbank.html

Superior Court Records, King County. Gronvold vs Whaley 1951. www.mrsc.org/mc/courts/supreme/039wn2d0710.htm

Wikipedia. "Ferries of San Francisco Bay". http://en.wikipedia.org/wiki/Ferries_of_San_Francisco_Bay

Wikipedia. "Southern Pacific Railroad". http://en.wikipedia.org/wiki/Southern_Pacific_Railroad

Acknowledgements

No publication of this sort can be produced in isolation. It requires the collective knowledge of many people to be complete. I have been fortunate to make the acquaintance of many generous people who have supported me by offering encouragement, feeding me information, providing photographs, sending me china and allowing me to take photographs of their collections. (Even with all this help I am sure that as this book is published someone will produce a pattern never previously seen.) The following people deserve special thanks for great efforts in helping me with this book.

I can't give enough thanks to my diving and antiquing buddy, Brian Fardoe. He travelled with me to countless shows recording date codes, taking backup digital pictures and often buying pieces when I wasn't around to ensure I had a sample to photograph. My friends, Alice and Gary Cannon, welcomed me into their home and provided many rare US Navy and Coast Guard patterns for me to photograph. They also paved the way for me to meet, visit and photograph several other collections that would have been otherwise inaccessible.

Thanks to Bruce Lanham for sending me samples of many china patterns from shipwrecks and for sharing his knowledge of shipwreck patterns. I learned a lot about marked and unmarked patterns used on the *State of California, Princess Kathleen, Princess Sophia,* SS *Columbia,* SS *Aleutian* and others. And to Daniel Krummes, who without hesitation sent me photos and write ups on a number of Pacific-coast patterns.

I thank Barbara Conroy for use of her ship china shapes drawings; Mike Clement, a friend and avid Union Steamship memorabilia collector, for opening doors to people with collections and ferrying me to remote dock sites in search of china; Brian Gisborne for lending me countless books from his library to help with steamship company write-ups; Michael Rice and Lorrie Belveal for lending me ship postcards to scan; Mike Paris for helping me photograph many pieces in the early stages of this project; Greg Cochkanoff, Richard Luckin, Alberto Bisgano, Ed Mannings, David Smith, Lester Silva, John Wood and Bart Malone for sending photos of hard-to-get patterns; Dr Wallace Chung, Karey Litton, Phil Beaudry, Will Smith and Lorne Campbell, all passionate Canadian Pacific collectors, for sharing their collections and knowledge of CPR patterns; Kathleen Lathom, one of the region's most knowledgeable ship-china collectors, for opening her home and allowing me to photograph many hard-to-find patterns; Karl Spence, who never hesitated to provide information on request; Charles Dragonette for answering my many questions during the early stages of my research; Annette Smith for making sure that I had all the Alaska patterns she could find; Merle and Diane Thompsen for opening their home and van-load of china to help me improve the information in my database; and Bill Stritzel for educating me with respect to San Francisco Bay operations.

Thanks also to those who shared their collections with me: Lance and Judy Floer, Gary Bateman, Dave Dickerson, Karl Gagnon, George Meszaros, Ross McLeod, Barry O'Sullivan, Paul Smood, Richard Reisinger, John and Linda Sendy, Lorrie Belveal, Duncan MacKenzie, Roger Porter, Roger Pinfield, Ken Sutherland, Gerry Millar, Lorne Carroll, Paul Wright, Derrick Prince, Kent Barnard, Bob Briggs, Rick McIntosh and Adrian R. Cook.

Libraries, maritime museums and historical organizations played an important role in my research. I am indebted to the following institutions for answering my requests for information: Dawson City Museum, Maritime Museum of British Columbia, Nelson Museum of Art and History, Museum of Northern Idaho, Puget Sound Maritime Historical Society,

Royal British Columbia Museum, J. Porter Shaw Library, Vancouver Maritime Museum, and Yukon Archives.

Thanks also to the china manufacturing companies that responded to my requests for help: Syracuse China, Homer Laughlin, and Dudson.

Last but not least, I am indebted to the Flags of the World website and Joseph McMillan for allowing me to use many of the shipping company house flag images in this book.

Index of Shipping Companies and Government Services

Index of China Manufacturers